70+

OPTIMIZATION IN STATISTICS

STUDIES IN THE MANAGEMENT SCIENCES

Editor in Chief
ROBERT E. MACHOL

Volume 19

NORTH-HOLLAND PUBLISHING COMPANY—AMSTERDAM·NEW YORK·OXFORD

OPTIMIZATION IN STATISTICS

With a view towards applications
in Management Science and Operations Research

Edited by

S.H. ZANAKIS
J.S. RUSTAGI

under the Departmental Editorship of
Arie Y. Lewin for *Management Science*

1982

NORTH-HOLLAND PUBLISHING COMPANY—AMSTERDAM·NEW YORK·OXFORD

© *North-Holland Publishing Company - 1982*

All right reserved. No part of this publication may be reproduced, stored in a retrieval system, or transmitted, in any form or by any means, electronic, mechanical, photocopying, recording or otherwise, without the prior permission of the copyright owner.

This North-Holland/TIMS series is a continuation of the Professional Series in the Management Sciences, edited by Robert E. Machol.

ISBN: 0444 86544 6

Published by:

NORTH-HOLLAND PUBLISHING COMPANY
AMSTERDAM, NEW YORK, OXFORD

Sole distributors for the U.S.A. and Canada:

ELSEVIER SCIENCE PUBLISHING COMPANY, INC.
52 VANDERBILT AVENUE
NEW YORK, NY 10017

Printed in The Netherlands

TABLE OF CONTENTS

Preface ix

A Prelude to Optimization in Statistics
S.H. Zanakis and J.S. Rustagi 1

I. REGRESSION AND CORRELATION

Introduction to Contributions in Regression and Correlation
S.H. Zanakis and J.S. Rustagi 7

Optimization Techniques in Linear Regression: A Review
S.C. Narula 11

LAV (Least Absolute Value) Estimation in Linear Regression: A Review
T. Dielman and R. Pfaffenberger 31

Generalized Network Approaches for Solving Least Absolute Value and Tchebycheff Regression Problems
D. Klingman and J. Mote 53

An Algorithm to Select the Best Subset for a Least Absolute Value Regression Problem
R.D. Armstrong and M.T. Kung 67

An Absolute Deviations Curve-Fitting Algorithm for Nonlinear Models
A. Tishler and I. Zang 81

Some Algorithms for Concave and Isotonic Regression
C.-F. Wu 105

Concordant and Discordant Monotone Correlations and Their Evaluation by Nonlinear Optimization
G. Kimeldorf, J.H. May and A.R. Sampson 117

II. MULTIVARIATE DATA ANALYSIS AND DESIGN OF EXPERIMENTS

Introduction to Contributions in Multivariate Data Analysis and Design of Experiments
J.S. Rustagi and S.H. Zanakis — 133

Constrained Multivariate Analysis
R.E. Hausman, Jr. — 137

Derivation of a Maximum Rank Sum Statistic and Application to Discriminant Analysis
C. Wang — 153

An Algorithm Using Lagrangian Relaxation and Column Generation for One-Dimensional Clustering Problems
L.E. Stanfel — 165

Algorithmic Improvements for Obtaining the Upper Multinomial Bound
R. Plante and P. Sinha — 187

Error Localization for Erroneous Data: A Survey
G.E. Liepins, R.S. Garfinkel and A.S. Kunnathur — 205

Selection of Cost-Optimal Fractional Factorials, $2^{m-r}\ 3^{n-s}$ Series
C.A. Mount-Campbell and J.B. Neuhardt — 221

III. PARAMETER ESTIMATION, RELIABILITY AND QUALITY CONTROL

Introduction to Contributions in Parameter Estimation, Reliability and Quality Control
J.S. Rustagi and S.H. Zanakis — 235

Parameter Estimation Under Progressive Censoring Conditions for a Finite Mixture of Weibull Distributions
J. Mandelbaum and C.M. Harris — 239

Optimal Outlier Tests for a Weibull Model – To Identify Process Changes or to Predict Failure Times
N.R. Mann — 261

A Bayesian Scheme for Estimating Reliability Growth Under Exponential Failure Times
N.D. Singpurwalla   281

Extreme Points of the Class of Discrete Decreasing Failure Rate Average Life Distributions
N.A. Langberg, R.V. León, J. Lynch and F. Proschan   297

A Bicriterion Model for Acceptance Sampling
H. Moskowitz, A. Ravindran, G. Klein and P.K. Eswaran   305

Notes about Authors   323

Authors' Addresses   331

PREFACE

Need for optimization arises often in statistics implicitly or explicitly. Although significant developments have been made in both fields, their interface has not received the attention it deserves. Statisticians on one hand and management scientists/operations researchers on the other have made many important contributions to this interdisciplinary subject through their own *separate* publications and conferences. There has been very little participation in each other's activities – even those devoted to the common subject. The first two symposia dedicated to "Optimizing Methods in Statistics" (Rustagi) were attended mostly by statisticians, while participants in "Optimization in Statistics" sessions at recent ORSA/TIMS National Meetings (Zanakis) were primarily operations researchers/management scientists.

Encouraged by the responses of participants in these activities, we felt that a *TIMS Studies in the Management Sciences* volume devoted to this subject would bring statisticians and management scientists closer together and stimulate further interest in this important area. A call for papers was announced in late 1979/early 1980 through national and international journals of statistics, operations research, management science, and industrial engineering. We sought original and expository papers in the area of optimization as utilized in statistical problems arising in the above disciplines. Development of new and intelligent application or comparison of existing optimization algorithms to important statistical problems were encouraged.

This volume contains the papers which survived the usual *Management Science* refereeing and editing processes. We believe that the volume will appeal to management scientists, operations researchers, statisticians, and industrial engineers. It can be used as a reference or textbook in colleges and universities.

The papers in this volume are grouped in three sections according to the statistical applications area: (a) regression and correlation; (b) multivariate data analysis and design of experiments; and (c) statistical estimation, reliability and quality control. An introduction to the articles of each section is provided, as well as an introduction to the overall subject. The reader is assumed to be familiar with elements of statistics, especially in the categories listed above, and with elements of optimization theory, especially linear programming. No great mathematical sophistication is required for reading this book – a knowledge of algebra and elementary calculus will suffice for most of the chapters.

As editors of this volume, we would like to express our sincere appreciation to the authors, whose work and interest made this volume possible; to about fifty referees, whose detailed and critical reviews were indispensable to improve the quality of so many different papers; to Professor Robert E. Machol, Editor-in-Chief of the TSMS publication series, for reviewing carefully the draft of this volume; and to our secretaries at the West Virginia College of Graduate Studies, Florida International University and The Ohio State University for efficiently handling the large volume of correspondence during all phases of this effort.

Miami, Florida
Columbus, Ohio

Stelios H. Zanakis
Jagdish S. Rustagi *

September 1981

Editors

* The work by J.S. Rustagi was partially supported by contract no. N00-14-78-C-0543 (NR 042-403) of the Office of Naval Research.

A PRELUDE TO OPTIMIZATION IN STATISTICS

S.H. ZANAKIS and J.S. RUSTAGI

Lack of certainty creates a need for statistics in any field of human endeavor. Statistical techniques are widely used in many areas of study, including sociology, medicine, business, and engineering – to mention only a few. All statistical problems contain elements of optimization. In general, this necessitates estimation of some unknown parameters (variables) that will optimize some measure of performance while often satisfying certain restrictions (constraints) on the parameters and input/output data.

The following examples occur frequently in statistics and are examined in more detail later in this volume.

- Linear and nonlinear curve fitting by regression analysis for predictions that minimize some measure of error (absolute, squared, maximum, etc.).
- Design of experiments and sample surveys that minimize costs while they restrict errors, or vice versa.
- Clustering, classification or discrimination of multivariate data according to different attributes in order to minimize improper handling of data and maximize estimation accuracy.
- Estimation of parameters of statistical distributions using criteria such as maximum likelihood, minimum chi-square, minimum variance, etc. (Bard [2] 1979, Zanakis [13,14]). The last criterion is used also in optimal response surface designs as well as in canonical correlation and principal component analysis of multivariate data (linear transformations of variables to maximize or minimize their correlations).
- Test of statistical hypotheses that minimize the probability of type II error while holding the probability of type I error at a certain level (Neyman–Pearson [6] problem).
- Quality control design problems arise in both acceptance sampling (e.g. minimize inspector errors) as well as in process control (e.g. detect outliers, and design minimum cost control charts).
- Reliability design and maintenance policies that minimize costs and restrict probabilities of breakdown or vice versa.

In contrast to classical statistics, Bayesian statistics is concerned with statistical procedures when some prior information is available on the unknown parameter. Bayesian statistics uses this prior information with the sample to derive the posterior distribution on the parameters. The procedures are based on this posterior distribution. The evaluation of these procedures often requires the assumption of a loss function. Such procedures require the application of optimization techniques.

Optimization methods to solve all previous problems may be classified into one of the following categories: classical (using calculus), numerical, variational (including dynamic programming) and other mathematical programming (constrained linear, nonlinear and integer optimization). The availability of high-speed computers has accelerated the growth of management science/operations research and produced many algorithmic and software developments in optimization. The use of such new optimization techniques in statistics, however, has not been extensive, primarily due to four reasons.

First, the topics of optimization and statistics have not been sufficiently blended in graduate education. Secondly, no books integrating the two subjects have been available until recently (Rustagi [9] and Arthanari and Dodge [1] have provided expositions of variational and mathematical programming techniques in statistics). Thirdly, most of the optimization techniques have been developed by nonstatisticians and it is true that there always exists some time lag between the development of some techniques (e.g. mathematical programming) and their use in another area (e.g. statistics). Finally, discourses on the optimization–statistics interface have been scattered in the literature, and until recently professional society meetings on this interface were infrequent.

The organization of two statistical symposia on "Optimizing Methods in Statistics" (Rustagi [8,11]) and recognition of the importance of this interdisciplinary area for management scientists and operations researchers (e.g. see Zanakis [13,15]) have increased the awareness of the field. We were very pleased to see our first "Optimization in Statistics" session at the Joint National Meeting of the Operations Research Society of America and The Institute of Management Sciences on 2 April 1976 followed by similar sessions in each subsequent semiannual joint national conferences of these two societies, and more recently of the American Institute for Decision Sciences. A special issue of *Communications in Statistics* was devoted to optimization in statistics (Rustagi [10]) and an international conference on "Optimizing Methods in Statistics" was sponsored in 1977 by the Bernoulli Society for Statistics in the Physical Sciences.

The present volume is an outgrowth of all these efforts. Contributors come from a variety of disciplines, including statistics, operations research/management science, business administration, industrial engineering, etc. We hope that this effort to cross the boundaries of single disciplines on this common theme will be soon followed by others.

In order to restrict this volume to a manageable size, it was decided not to include subjects of the opposite interface, namely use of statistics in optimization. Interest in this subject has been renewed, as is evident from the inclusion of such sessions in several recent ORSA/TIMS national meetings (Golden and

Zanakis). Considerable attention has been given recently to the development and application of statistical inference procedures for point and interval estimation of the true (global) optimum when this result cannot be guaranteed by the optimization technique used. (See Sielken and Monroe [12] for a comprehensive exposition). This is a vital need when solving nonconvex mathematical programming problems, or when employing heuristics, for example to solve large, otherwise unsolvable, integer programming problems (Zanakis and Evans [15]). Other efforts have included search procedures and stopping rules guided by Bayesian learning (Hill [3]), sequential hypothesis testing (Robbins [7]), or regression curve-fitting for response surface analysis (Myers [4]). Use of statistics to handle uncertainties in linear programming (stochastic and chance-constrained programming) have been well known for some time. The growing interest in the subject of statistics in optimization will, hopefully, result soon in a book on this topic.

References

[1] T.S. Arthanari and Y. Dodge, *Mathematical Programming in Statistics* (Wiley, New York, 1981).
[2] Y. Bard, *Nonlinear Parameter Estimation* (Academic Press, New York, 1974).
[3] J.D. Hill, "A Search Technique for Multimodal Surfaces", *IEEE Trans. on Syst. Sc. & Cybernetics* 5 (1969) 2–8.
[4] R.H. Myers, *Response Surface Methodology* (Allyn and Bacon, Boston, 1971).
[5] M.F. Neuts (Ed.), *Algorithmic Methods in Probability*, TIMS Studies in the Management Sciences, vol. 7 (North-Holland, Amsterdam, 1977).
[6] J. Neyman and E.S. Pearson, "Contributions to the Theory of Testing Statistical Hypothesis", Stat. Res. Mem. I (1936) 1–37 and II (1938) 25–57.
[7] T.C. Robbins, "Go–Stop: Global Optima by a Sequential Testing Optimization Procedure", Paper presented at 39th National ORSA Meeting, Dallas, 5–7 May 1971.
[8] J.S. Rustagi (Ed.), *Optimizing Methods in Statistics* (Academic Press, New York, 1971).
[9] J.S. Rustagi, *Variational Methods in Statistics* (Academic Press, New York, 1976).
[10] J.S. Rustagi (Ed.), "Optimization in Statistics", *Communications in Statistics*, Special Issue (1978).
[11] J.S. Rustagi (Ed.), *Optimizing Methods in Statistics* (Academic Press, New York, 1979).
[12] R.L. Sielken, Jr. and H. Monroe, "How Near is a Near-Optimal Solution: Confidence Limits for the Global Optimum", Technical Reporet no. 62, Institute of Statistics, Texas A&M University (May 1980).
[13] S.H. Zanakis, "Optimization and Statistics: A Synergism", Southeast TIMS Meeting Proceedings (1973) 427–434.
[14] S.H. Zanakis, "Computational Experience with Some Nonlinear Optimization Algorithms in Deriving Maximum Likelihood Estimates for the Three-Parameter Weibull Distribution", in: M.F. Neuts, ed., *Algorithmic Methods in Probability*, TIMS Studies in The Management Sciences, vol. 7 (North-Holland, Amsterdam, 1977) 63–77.
[15] S.H. Zanakis and J.R. Evans, "Heuristic Optimization: Why, When and How to Use It", *Interfaces* 11 (October 1981) 84–91.

PART I

REGRESSION AND CORRELATION

INTRODUCTION TO CONTRIBUTIONS IN REGRESSION AND CORRELATION

S.H. ZANAKIS and J.S. RUSTAGI

Analysts, experimenters, and decision-makers use regression analysis extensively in many areas of business, science, and technology. Surveys of Fortune 500 firms have shown that regression analysis is the most often used technique in operations research/management science, particularly in the areas of accounting (Jagetia [7]), quality control and maintenance (Ledbetter and Cox [8]).

Correlation analysis provides a measure of linear relationship (association) among a set of variables *without* regard to cause–effect direction. Regression analysis, however, requires such a direction for prediction purposes: i.e. the response (dependent variable y) is assumed to be related to one or more independent variables (x) through a functional form or model $y = f(p, x)$. The parameters, p, of the model are estimated by minimizing, over a set of n observations $y = (y_1, y_2, \ldots, y_n)$, some function of deviations of the hypothesized model response $\hat{y}$ from the observed value y, e.g.

$$\text{minimize} \sum_{i=1}^{n} |y_i - \hat{y}_i|^K.$$

The power constant K is usually set at $K = 1$ (the least-absolute deviation criterion or L_1 norm, also called the Tchebycheff criterion), $K = 2$ (the least-squares deviation criterion or L_2 norm), or $K = \infty$ (the least maximum deviation criterion or L_∞ norm, which is equivalent to minimize $\max|y_i - \hat{y}_i|$). There is extensive literature on various methods as applied to regression problems using different optimization criteria.

The classical procedure of least squares has been used extensively because of its analytical tractability and its highly developed theory and widespread literature. For a more complete account the reader is referred to Rao [9], Draper and Smith [4], and Chatterjee and Price [3]; to Wesolowsky [11] for a computer-oriented treatment; and to Harter [6] for a historical exposition.

The criterion of minimizing the sum of absolute deviations is preferable to that of least squares in the presence of large disturbances (outliers) or when the classical assumption of a normal distribution of error, $e_i = y_i - \hat{y}_i$, is violated due to "contamination" or "heavy tails". Nevertheless, the use of the least absolute criterion has been restricted because of (i) the lack, until recently, of good statistical inference procedures, and (ii) the need for specialized computer algorithms, to which statisticians and many practitioners often do not have access. Charnes, Cooper and Ferguson [2] were the first to show how to estimate parameters in least absolute linear regression models using linear programming (LP), the dual formulation of which provides the same estimates much more quickly than the primal formulation (Wagner [10]). An added advantage of the LP approach over the classical approach is that it can easily accomodate restrictions on the model parameters. Recently, high-speed computers, development of special simplex-based algorithms, and new statistical inference procedures have greatly increased the use of the least absolute criterion.

The first five papers in this section deal with the least absolute value (LAV) approach.

Narula, in his paper "Optimization techniques in linear regression: A review", provides a survey of formulations and optimization procedures for solving the linear regression problem under the three criteria (L_1, L_2, and L_∞). He also reviews algorithms for finding the minimum set of independent variables that best describes the functional relationship under these criteria. The comparison review paper by Dielman and Pfaffenberger "LAV (least absolute value) estimation in linear regression: A review" summarizes recent developments, examines properties of these estimators under different error distributions, and provides various statistical estimation procedures. These confidence-interval and hypothesis-testing results void a gap in L_1 estimation, and make it more competitive to the traditional least-squares method where inferential procedures have been well known for a long time.

An innovative approach for solving the linear least-absolute and maximum-absolute regression problems (L_1 and L_∞) via generalized networks is presented by Klingman and Mote in their paper on "Generalized network approaches for solving least absolute value and Tchebycheff regression problems". This method holds considerable promise for computational efficiencies in large problems as compared to present LP-based algorithms, due to powerful new software developments in network theory (Glover et al. [5]).

In "An algorithm to select the best subset for a least absolute value regression problem" by Armstrong and Kung, a numerical procedure is provided for this problem. It is based on a primal/dual LP code coupled with a branch-and-bound scheme.

While the previous papers concentrated on the least absolute *linear* regression problem, the paper by Tishler and Zang, "An absolute deviations curve-fitting algorithm for nonlinear models", presents a new method applicable to either linear or nonlinear least-absolute regression, with the latter promising very competitive and efficient results. As is often done in fitting nonlinear models, it uses a smooth approximation (convex to ensure a global optimum) to replace the absolute function, thus avoiding discontinuous differentiability complications. Properties, sensitivity, computational experience and guidelines for implementing the algorithm are also provided.

When the specific form of the functional relationship cannot be assumed, general assumptions of concavity or convexity of the model can be used. Isotonic regression models have resulted from many such general assumptions. A comprehensive account of statistical inference under order restrictions, including isotonic regression, is given by Barlow, Bartholomew, Bremner and Brunk [1]. Applications of concave and isotonic regression abound in the context of operations research and management science, and in many other fields such as in enzyme-kinetics. In his paper "Some algorithms for concave and isotonic regression", Wu provides numerical procedures to solve these problems exactly or approximately. These procedures are compared to those based on other mathematical programming methods.

The relationship between two variables traditionally has been statistically measured by the correlation coefficient. The correlation coefficient measures the linear association and is not sensitive to other kinds of associations that are also useful to statisticians. Kimeldorf, May and Sampson, in their paper "Concordant and discordant monotone correlations and their evaluation by nonlinear optimization", give four new measures of monotone relationships that are particularly useful in developing ordinal scales. These measures require the solution of optimization problems, for which the authors propose a nonlinear programming algorithm.

References

[1] R.E. Barlow, D.J. Bartholomew, J.M. Bremner and H.D. Brunk, *Statistical Inference Under Order Restrictions* (John Wiley and Sons, New York, 1972).
[2] A. Charnes, W.W. Cooper and R.O. Ferguson, "Optimal Estimation of Executive Compensation by Linear Programming", *Management Science* 1 (1955) 138–151.
[3] S. Chatterjee and B. Price, *Regression Analysis by Example* (John Wiley and Sons, New York, 1977).
[4] N.R. Draper and H. Smith, *Applied Regression Analysis*, 2nd edn. (John Wiley and Sons, New York, 1981).
[5] F. Glover et al., "Improved Computer-Based Planning Techniques: Parts I & II", *Interfaces* 8 (1978) 16–25 and 9 (1979) 12–20.
[6] H.L. Harter, "The Method of Least Squares and Some Alternatives – Parts I & II", *Intl. Stat. Review* 42 (1974) 174–177 and 235–264.
[7] L.C. Jagetia, "A Survey of Management Science Techniques in Accounting", Paper presented at ORSA/TIMS national meeting, Miami, November 1976.
[8] W.N. Ledbetter and J.F. Cox, "Are OR Techniques Being Used?", *Industrial Engineering* (February 1977) 19–21.
[9] C.R. Rao, *Linear Statistical Inference and its Applications* (John Wiley and Sons, New York, 1973).
[10] H.W. Wagner, "Linear Programming Techniques for Regression Analysis", *J. Amer. Statist. Assoc.* 54 (1959) 206–212.
[11] G.O. Wesolowsky, *Multiple Regression and Analysis of Variance: An Introduction for Computer Users in Management and Economics* (John Wiley and Sons, New York, 1976).

OPTIMIZATION TECHNIQUES IN LINEAR REGRESSION: A REVIEW

Subhash C. NARULA
Rensselaer Polytechnic Institute

    This paper briefly reviews the optimization techniques that play a significant role in linear regression analysis. These techniques are studied in relation to the three main criteria used in the estimation of parameters of a linear model, namely, minimization of the sum of squared errors (the "least squares" criterion), minimization of the sum of absolute errors, and minimization of the maximum absolute error. For each of the criteria, the problem is formulated as an optimization problem and the development of efficient algorithms to estimate the parameters for the full model as well as the selection of variables are discussed. These algorithms exploit the special structure of the problem.

1. Introduction

    Optimization techniques play an important role in the estimation of the unknown parameters in a multiple linear regression model,

$$y = X\beta + \epsilon, \tag{1}$$

where y is an $n \times 1$ vector of response variables, X an $n \times k$ matrix of predictor variables, β a $k \times 1$ vector of unknown parameters, and ϵ an $n \times 1$ vector of random errors. It is generally assumed that (1) contains all relevant variables and their appropriate functions plus, possibly, some extraneous variables and their functions. Often there are many practical and economic reasons for including fewer than k variables in the model. In terms of the statistical properties of the estimators also (see Narula and Ramberg [87] and Walls and Weeks [125]), it may be desirable to delete variables from (1). This problem is referred to as the "variable selection problem".
    Whatever the model, we would like the predicted value of y to be close to the observed value of y. This "closeness" can be stated, in general, in terms of a L_p-norm. Here the L_p-norm problem can be stated as: find a vector β that minimizes

$$\sum_{i=1}^{n} |y_i - x_i'\beta|^p, \tag{2}$$

Received August 8, 1980; revised March 27, 1981.

for $p \geq 1$, where y_i is the ith element of the vector y and x_i is the ith row of the matrix X. The values of p most widely studied in the literature are $p = 1, 2$, and ∞. Clearly, for $p = 1$ the criterion minimizes the sum of absolute errors (MSAE); for $p = 2$, it minimizes the sum of squared errors (MSSE); and for $p = \infty$, (2) is equivalent to the minimization of the maximum absolute error (MMAE). In the literature, these criteria are also well known by other names: the MSSE criterion as the least squares criterion and L_2-norm; the MSAE criterion as the least absolute deviation (LAD), least sum of absolute errors (LSAE), least absolute value (LAV), minisum criterion and L_1-norm; and the MMAE criterion as the uniform solution, Tchebycheff criterion, minimax criterion and L_∞-norm.

The objective of this paper is to review the optimization techniques discussed in the literature for the estimation of β in (1) for the MSSE, the MSAE, and the MMAE criteria. The reader may refer to Ekblom [51], Forsythe [52], and Ramsay [100] for the general L_p-norm; and to Tishler and Zang [122] for the MSAE estimation for nonlinear regression models. In section 2 the mathematical programming formulations are given for the different criteria. The optimization techniques are discussed for the MSSE, the MSAE and the MMAE criterion in sections 3, 4, and 5, respectively. The paper concludes with a few remarks in section 6.

2. Problem formulation

Although (2) provides a general formulation of the problem for $p \geq 1$, the problem can be restated, for special values for p, in a form that is easier to recognize and more convenient to solve.

2.1. The MSSE criterion

For $p = 2$, formulation (2) reduces to: find a vector β that minimizes

$$Q(\beta) = y'y + \beta'X'X\beta - 2\beta'X'y. \tag{3}$$

Clearly (3) is an unconstrained quadratic programming problem. For the problem, Sposito [113] gives a linear programming (LP) formulation. Before the formulation is given here, note that β in (1) is unrestricted in sign. We write $\beta = \beta^+ - \beta^-$, where β^+ and β^- are $k \times 1$ vectors and $\beta^+, \beta^- \geq 0$. The LP formulation is:

minimize $0'(\beta^+ - \beta^-)$
subject to
$X'X(\beta^+ - \beta^-) = X'y,$ $\quad (4)$
$\beta^+, \beta^- \geq 0,$

where $\boldsymbol{0}$ is a $k \times 1$ vector of zeros. This formulation is given here for the sake of completeness only since linear programming would not generally be used to solve (4).

2.2. The MSAE criterion

For $p=1$, formulation (2) reduces to: find a vector $\boldsymbol{\beta}$ that minimizes,

$$\sum_{i=1}^{n} |y_i - x_i'\boldsymbol{\beta}|. \tag{5}$$

This can be viewed as a geometric programming problem. Recently, Pfaffenberger and Dinkel [99] gave the dual formulation of (5) as:

maximize $y'\boldsymbol{\delta}$
subject to
$X'\boldsymbol{\delta} = \boldsymbol{0}$, (6)
$-\boldsymbol{1} \leq \boldsymbol{\delta} \leq \boldsymbol{1}$,

where $\boldsymbol{\delta}$ is an $n \times 1$ vector of dual variables, and $\boldsymbol{1}$ is an $n \times 1$ vector of ones.
By making a change of variable, $\boldsymbol{\delta} = \boldsymbol{\Delta} - \boldsymbol{1}$, we can rewrite (6) as:

maximize $y'(\boldsymbol{\Delta} - \boldsymbol{1})$
subject to
$X'\boldsymbol{\Delta} = X'\boldsymbol{1}$, (7)
$\boldsymbol{0} \leq \boldsymbol{\Delta} \leq 2\boldsymbol{1}$,

which is a LP problem. This is also the dual LP formulation given by Wagner [124]. Before writing the primal LP formulation given by Wagner [124], we note that e, an $n \times 1$ vector of differences between the predicted and the observed value of the response variable, is unrestricted in sign. We write, $e = e^+ - e^-$, where e^+ and e^- are $n \times 1$ vectors and $e^+, e^- \geq \boldsymbol{0}$. Note that e^+ and e^- denote the over- and under-prediction of y. Now the primal LP problem is

minimize $\boldsymbol{1}'(e^+ + e^-)$
subject to
$X(\boldsymbol{\beta}^+ - \boldsymbol{\beta}^-) + e^+ - e^- = y$, (8)
$e^+, e^- \geq \boldsymbol{0}$,
$\boldsymbol{\beta}^+, \boldsymbol{\beta}^- \geq \boldsymbol{0}$.

This formulation is also known as a goal programming problem, Charnes and Cooper [39, 40], and Ignizio [73]. It may be noted that (7) is the dual of (8).

Other extensions of the criterion are known in the literature. Narula and Wellington [91] proposed the minimum sum of weighted absolute errors (MSWAE) criterion that subsumes the MSAE and the minimum sum of (absolute) relative errors (MSRE criterion, Narula and Wellington [90]) as special cases. For the MSWAE formulation, only the objective function of the problem in (8) is changed to

minimize $w'(e^+ + e^-)$,

where w is an $n \times 1$ vector of weights, $w > 0$.

More recently, Koenker and Bassett [80] have given yet another generalization of the MSAE criterion and called it regression quantiles. For regression quantiles, we need to change the objective function in (8) to

minimize $\theta 1'e^+ + (1-\theta)1'e^-$,

where $0 < \theta < 1$. For $\theta = 1/2$, the regression quantile is the same as the MSAE criterion.

2.3. The MMAE criterion

For $p = \infty$, an equivalent formulation for (2) is: find a vector β that minimizes

$$\underset{i=1,\ldots,n}{\text{maximum}} |y_i - x_i'\beta|. \tag{9}$$

For (9), Wagner [124] gives equivalent primal and dual LP formulations. Let e^* denote the maximum (absolute) difference between the predicted and the observed value of the response variable. Then, the primal LP problem is

minimize e^*
subject to
$$\begin{aligned} X(\beta^+ - \beta^-) + e^*1 &\geqslant y, \\ -X(\beta^+ - \beta^-) + e^*1 &\geqslant -y, \\ e^* &\geqslant 0, \\ \beta^+, \beta^- &\geqslant 0. \end{aligned} \tag{10}$$

Narula and Wellington [92] propose the minimization of the maximum weighted absolute error (MMWAE) criterion that subsumes the MMAE criterion. For the MMWAE criterion we only need to change the constraints in (10) as follows:

$$WX(\beta^+ - \beta^-) + e*1 \geq Wy,$$

$$-WX(\beta^+ - \beta^-) + e*1 \geq -Wy,$$

where W is an $n \times n$ diagonal matrix of weights $w > 0$. Let γ^+ and γ^- be $n \times 1$ vector of dual variables. Then the dual LP problem corresponding to (10) is

maximize $y'(\gamma^+ - \gamma^-)$
subject to
$$X'(\gamma^+ - \gamma^-) = 0,$$
$$1'(\gamma^+ + \gamma^-) \leq 1, \quad (11)$$
$$\gamma^+, \gamma^- \geq 0.$$

2.4. Selection of variables

Irrespective of the criterion used for estimating β it may be desirable, for practical, economic and statistical reasons, to include fewer than k variables, say m variables, in the model. Thus, the problem is to decide the value of m and then to find the "best" set of m variables. For a given value of m, the set of m variables is called the best if the model with these m variables has the smallest criterion value from among all the $\binom{k}{m}$ possible models of m variables. At times, instead of deciding on a specific value of m, we may want to find: (i) the best set of m variables for $m = 1, 2, \ldots, k$; (ii) the best set of m variables for each value of m greater than or equal to m_L (say) and less than or equal to m_U (say); (iii) a "near-best" model for each m, i.e. a model with m variables that does not differ from the best model with m variables by more than a specified amount, $m = 1, 2, \ldots, k$; and (iv) models, called equally-good models, with fewer than k variables that do not differ in the criterion value from the model with k variables by more than a specified amount. We shall consider the problem of variable selection in relation to the MSSE, the MSAE and the MMAE criteria in the following sections.

3. The MSSE criterion

The MSSE criterion, or equivalently the least squares procedure, is the best known and most popular criterion for estimating β in (1). It has dominated the statistical literature for a long time. The popularity and dominance of the criterion can be ascribed, at least partially, to the facts that the theory is simple, well developed and documented. Also, the computer programs for the criterion are easily available. If the errors are independent with mean zero and a common (though unknown) finite variance σ^2, then the criterion gives the best (minimum variance) linear (linear function of the y) unbiased estimator of β. Furthermore, if errors follow a normal distribution, then the estimators are also maximum likelihood estimators. Unfortunately, the MSSE estimators are extremely sensitive to modest amounts of outlier contamination which makes it a very poor estimator in many non-normal, especially thick-tailed distributions. To effectively deal with such situations a number of "robust" procedures have been proposed. The interested reader may refer to Denby and Larsen [46], Hogg [68,69], Hogg and Randles [70], and Huber [71,72].

Since (3) is an unconstrained quadratic programming problem, setting the derivative of (3) with respect to β equal to zero yields, for the minimizing value, the MSSE estimator

$$X'X\hat{\beta} = X'y.$$

If $(X'X)$ is a nonsingular matrix, then $(X'X)^{-1}$ exists and the estimator $\hat{\beta}$ of β is

$$\hat{\beta} = (X'X)^{-1}X'y. \tag{12}$$

By the Gauss–Markov theorem, $\hat{\beta}$ is a minimum variance estimator of β in the class of unbiased estimators that are a linear function of y.

However, in practice it is possible that (i) $X'X$ is ill-conditioned, or (ii) $X'X$ is singular. The first problem, i.e. that $X'X$ is ill-conditioned, occurs if the predictor variables are highly correlated, thus creating the problem of multicollinearity. To overcome this problem, ridge regression (see Hoerl and Kennard [66,67]) has been extensively used since it gives stable estimates in the presence of multicollinearity. A number of other alternatives, such as latent root regression and regression by principal components, have also been proposed (see Gunst [62]). When $X'X$ is ill-conditioned, we may also use a number of numerically stable techniques by Gill and Murray [58], Golub [60], Golub and Saunders [61], Mifflin [86], Schittkowski and Stoer [109] and Stoer [120]. These procedures do not appear to be popular among statisticians.

The second case, i.e. that $X'X$ is singular, arises when either the design matrix X is ill-defined (a linear dependence exists among the predictor variables) or the design matrix X applies to the analysis of a variance model, in which case X has less then k column rank and $X'X$ is singular. In this case (12) does not apply and the generalized inverse of $(X'X)$ should be used.

3.1. Restricted MSSE estimators

In many applications (e.g. Chipman and Rao [41], Judge and Takayama [74]), we need to incorporate additional information in our model in the form of restrictions on the parameters. This problem is known as the restricted MSSE problem.

The restricted MSSE problem can be stated as: find a vector β that minimizes

$$y'y + \beta'X'X\beta - 2\beta'X'y$$

subject to

$$R\beta\rho r, \tag{13}$$

where R is an $s \times k$ matrix of known constants, $s < k$, r is an $s \times 1$ vector of known constants, and ρ an $s \times 1$ vector that denotes the equality and inequality condition. Each ρ_i of $\rho = (\rho_1, \rho_2, \ldots, \rho_s)'$ stands for any of the relations $\leq, =,$ and $\geq$; more precisely, $\rho_i = 0$ means "$=$", $\rho_i = 1$ means "$\leq$", and $\rho_i = -1$ means "$\geq$", $i = 1, \ldots, s$. A closed form solution of this problem does not exist. However (13) can be solved by the quadratic programming algorithms of Gill and Murray [58], Golub and Saunders [61], Mifflin [86], Schittkowski and Stoer [109] and Stoer [120]. The reader may also want to refer to Bibby and Toutenburg [36] and Lawson and Hanson [83] for a more extensive treatment of the subject.

We shall now consider two special cases of linear restrictions: (i) linear equality constraints for which closed form solution exists and (ii) non-negativity constraints on β, i.e. $\beta \geq 0$, for which a number of special purpose algorithms have been developed.

3.1.1. Linear equality restrictions

Sometimes the prior information on the parameters can be stated in the form of an exact linear relation such as $R\beta = r$, where R is an $s \times k$ matrix of known constants, $s < k$, and r an $s \times 1$ vector of known constants. In this situation (13) reduces to

minimize $y'y + \beta'X'X\beta - 2\beta'X'y$
subject to (14)
$R\beta = r$.

This quadratic programming problem can be stated and solved as a classical Lagrangean problem. The solution to (14) that yields the minimizing solution is

$$\beta^* = \hat{\beta} + (X'X)^{-1}R'\left[R(X'X)^{-1}R'\right]^{-1}(r - R\hat{\beta}), \tag{15}$$

which differs from the unrestricted estimator $\hat{\beta}$ by a linear function of the quantity $(r - R\hat{\beta})$. Theil [121] has shown that β^* is a minimum variance estimator within the class of unbiased estimators that are a linear function of y and r.

If $X'X$ is ill-conditioned, as before, a numerically stable algorithm such as those given by Gill and Murray [58], Golub [60], Golub and Saunders [61], Mifflin [86], Schittkowski and Stoer [109] and Stoer [120] can be used.

3.1.2. Non-negativity constraints

When the restrictions on β can be stated as $\beta \geq 0$, then (13) can be written as:

$$\text{minimize } y'y + \beta'X'X\beta - 2\beta'X'y$$
$$\text{subject to} \qquad (16)$$
$$\beta \geq 0.$$

To solve (16), a number of special purpose algorithms have been developed, namely Armstrong and Frome [9], Mantel [84] and Waterman [126]. Mantel's algorithm [84] is a modification of the simplex method of Wolfe [127] for quadratic programming that takes advantage of the special nature of the constraints. The algorithm due to Waterman [126] solves 2^k unrestricted problems, whereas Armstrong and Frome [9] propose a branch-and-bound algorithm that solves the problem without explicitly solving the 2^k unrestricted problems. These three algorithms can be modified to accommodate the upper and lower bounds on β. Lawson and Hanson [83] describe and give a *FORTRAN* program for an algorithm that is based on the Kuhn–Tucker conditions for optimality (see Sposito [113]).

3.2. Selection of variables

Early techniques of variable selection, such as forward selection, backward elimination and stepwise regression, have been well described by Draper and Smith [48]. With the advent of computers, it became possible to compute all possible models. However, researchers exploited the special structure of the problem and developed a number of efficient algorithms; notable among these are algorithms by Furnival [53], Garside [55,56], and Schatzoff, Fienberg and Tsao [108]. From among 2^k possible models, a number of models are clearly inferior and since we are only interested in "good" models, a second wave of algorithms was tried to determine the "best" models by implicit enumeration or branch-and-bound algorithms. Successful attempts in this venture have been by Beale [34], Beale, Kendall and Mann [35], Furnival and Wilson [54], Hocking and Leslie [65] and LaMotte and Hocking [81]. Among these, the leaps-and-bounds algorithm of Furnival and Wilson [54] is the most popular and is currently in the BMDP package as well as the IMSL package. Hocking [64] gives an excellent review of the selection of variables using the MSSE criterion.

4. The MSAE criterion

Although the MSAE criterion was first proposed by Edgeworth [49] in the 1880s, until fairly recently the computational problems associated with the MSAE estimation in any but the simple model effectively prevented the use of the procedure in fitting the regression model. The MSAE criterion results in the maximum likelihood estimator of β in (1) if the errors are independent and follow a Laplace distribution. A number of authors, Appa and Smith [6], Barrodale [18], Blattberg and Sargent [37], Ekblom [51], Kiountouzis [78], Rice and White [103] among others, have observed that the MSAE regression is more resistant to outliers in the data and to thick-tailed error distributions.

For the statistical properties of the MSAE estimators, the reader may refer to Dielman and Pfaffenberger [47].

4.1. Simple linear regression

In the 1880s Edgeworth [50] proposed an algorithm to compute the MSAE estimates of β in a simple linear regression model. Except for an iterative algorithm by Karst [75] in 1958, the major interest in this problem revived in the 1970s. Rao and Srinivasan [101] show that the computational efficiency of Sharpe's [111] algorithm stems from the fact that the dual LP formulation of the problem reduces to a knapsack problem with one parameter and no integer restrictions. They also give another equally efficient algorithm to solve the problem.

Sadovski [107] gives a *FORTRAN* program of Karst's [75] algorithm. However, Sposito [114] points out that the program may not converge in general. Sposito and Smith [116] give an algorithm that will converge. Recently, Armstrong and Kung [15] have published a *FORTRAN* program for the problem that is a specialization of an algorithm by Barrodale and Roberts [21] for the multiple linear regression model. Klingman and Mote [79] formulate and solve the problem as a generalized network problem. They compare the computational efficiency of their algorithm with that of Armstrong and Kung [15]. Sposito, Kennedy and Gentle [115] give a *FORTRAN* program for L_p, $1 \leq p < 2$, that is based on an extension of the iterative procedure of Schlossmacher [110] for the MSAE estimation of β in (1).

4.2. Multiple linear regression

The methods of Rhodes [102] and Singleton [112] extend the proposal of Edgeworth [50] to more than one variable; however, they become extremely unwieldy as the number of variables increases. Only with the advent of computers and development of linear programming has the MSAE criterion become feasible for the multiple linear regression models. In 1955, Charnes,

Cooper and Ferguson [40] proposed the use of linear programming to compute the estimates of β in (1); and in 1959, Wagner [124] suggested a method of obtaining a solution more rapidly by solving the dual problem (7) using LP algorithms designed to accommodate upper bounds directly. Since then a number of algorithms have been proposed.

The primal algorithm of Barrodale and Young [25] was the first to take advantage of the special structure of (8). A number of special purpose algorithms followed in quick succession, namely the descent methods of Davies [45] and Usow [123], an interval programming procedure of Robers and Ben-Israel [104] that has been specialized to solve the dual LP problem (7) by Robers and Robers [105]; and iterative procedures of Abdelmalek [1] and Schlossmacher [110].

In 1973, Barrodale and Roberts [21] introduced a very efficient algorithm to solve the primal LP problem. Their algorithm combines several simplex iterations into one, which has become a special feature of algorithms since then. Their algorithm has become a standard of comparison and compares favorably with the algorithms of Abdelmalek [2,3], Bartels and Conn [26], Bartels, Conn and Sinclair [30] and Spyropoulos, Kiountouzis and Young [118].

In a comparative study, Gilsinn et al. [59] compared a double-precision, dual revised simplex code of Abdelmalek [3], a primal revised simplex code with a partial-sort procedure of Armstrong and Frome [7], a primal simplex code of Barrodale and Roberts [21] and a code of Bartels and Conn [26] that uses a descent method and LU decomposition. In their study the code of Armstrong and Frome [7] outperformed all the other codes.

Computer programs for the estimation of β in (1) for the MSAE criterion appear in Armstrong, Frome and Kung [12], Barrodale and Roberts [22], and Robers and Robers [105]. The computer program of Armstrong, Frome and Kung [12] uses LU decomposition and a modification of the algorithm due to Barrodale and Roberts [21]. The computer code of Barrodale and Roberts [22] is currently included in the IMSL package.

4.3. Restricted MSAE estimators

Most algorithms for the MSAE estimation can be used without any modifications for models with less than full ranks or with linear restrictions on the parameters. However, more efficient procedures have been developed for special design matrices as well as restrictions on the parameters.

For the case of special design matrices, Armstrong, Elam and Hultz [8] give an algorithm for two-way classification models; Armstrong and Frome [10] develop a special purpose algorithm when the model contains some dummy variables; and Armstrong and Frome [11] give two algorithms for one-way and two-way tables.

For restricted MSAE estimation, Armstrong and Hultz [13] solve the MSAE problem by extending the techniques of interval linear programming to include interval constraints on the variables. Bartels, Conn and Sinclair [30] handle the equality constraints by direct elimination and the inequality constraints by including them in the objective function via a penalty function approach. To solve the restricted MSAE problem, Barrodale and Roberts [23,24] generalize their unconstrained algorithm [21] which still allows many intermediate simplex vertices to be bypassed. Bartels and Conn [28] give a descent method which can solve a wide class of problems. Narula and Wellington [88] give an algorithm for MSAE estimation when the linear model is restricted to pass through the mean or the median of each of the variables. The computer program for their algorithm appears in Narula and Wellington [89].

4.4. Selection of variables

Gentle and Hanson [57] propose stepwise procedures for the selection of variables using the MSAE criterion. Roodman [106] gives an implicit enumeration algorithm to find the "best" subset of each size under the MSAE criterion. More recently, Narula and Wellington [94] describe an efficient implicit enumeration algorithm for the selection of variables using the MSWAE criterion. The efficiency of their algorithm lies in the fact that they use primal and dual algorithms to solve the individual problems as well as a special movement through the tree. The computer program for their algorithm appears in Narula and Wellington [97]. Their algorithm has been modified, as in Narula and Wellington [96], to (i) find the best set of m variables for $m = m_L, \ldots, m_U$, where $m_L \geq 1$ and $m_U \leq k$, (ii) find the "near-best" set of variables for any specified m, and (iii) find equally-good models. Another algorithm due to Armstrong and Kung [16] appears in this volume.

5. The MMAE criterion

Of the three criteria, the MMAE criterion has received the least attention. The MMAE estimators are maximum likelihood estimators if the random errors are independent and follow a uniform distribution with a wide range. Rice and White [103] recommend the use of the criterion in the experimental situations where observed response is subject to errors that can be described by a probability distribution having sharply defined extremes with high probability. Barrodale and Young [25] and Stiefel [119] have used the criterion for computing approximations to functions. Barrodale and Phillips [19] found the criterion useful when the response variable contains only small inherent errors. However, Appa and Smith [6] caution against the use of this criterion in certain econometric studies (especially those where error variance may be nonfinite).

For the simple linear regression problem Klingman and Mote [79] formulate and solve the problem as a generalized network problem.

5.1. Multiple linear regression

Kelley [76] gave the first dual simplex algorithm to find the MMAE estimator of β in (1). Since then a number of algorithms have been proposed.

Stiefel [119] describes an exchange method for finding the estimators. The method requires that every $k \times k$ submatrix of X be nonsingular. Osborne and Watson [98] rework Stiefel's exchange method and the simplex algorithms applied to the dual LP formulation. They also give an algorithm free from the major restrictions of the classical approximation theory, including the Haar condition, which they replace by the much weaker condition of nonsingularity of the successive basis matrices. Bartels and Golub [31] give a generalization of the exchange method that requires that the rank of X be k. However, the procedure may fail if there are any linear dependencies.

Barrodale and Young [25] proposed the first special purpose LP algorithm that exploits the special structure of the dual LP problem. Without restrictions on the problem, Barrodale and Phillips [19] have proposed a primal algorithm to solve the dual LP formulation. Their algorithm is more efficient than the one by Barrodale and Young [25], although both algorithms combine several simplex iterations into one. Bartels and Golub [33] give a simplex method that uses LU decomposition and permits "occasional exchange of two equations simultaneously". Abdelmalek [4] gives yet another simplex-based procedure that solves the problem more rapidly than ordinary simplex algorithm.

Besides the foregoing LP based procedures, a number of descent methods have been proposed. Cline [43] proposed the first descent method to obtain the MMAE estimates in (1). His method has been generalized by Bartels and Conn [27], who also incorporate the ideas of Conn [44] where a penalty function technique to solve the problem was proposed. The procedure of Bartels, Conn and Charalambous [29], a descent method, is also very closely related to Cline's algorithm [43]. From computational experience reported in Bartels, Conn and Charalambous [29] and Cline [43], the algorithms due to Barrodale and Phillips [19] and Bartels, Conn and Charalambous [29] appear efficient. Cline [42] points out that the algorithm of Lawson [82] determines the MMAE solution as a limit of weighted MSSE solution on a finite set of points.

The computer programs to determine MMAE estimates appear in Abdelmalek [5], Armstrong and Kung [14], Barrodale and Phillips [20] and Bartels and Golub [32].

5.2. Other problems

Narula and Wellington [92] give an efficient algorithm to estimate β using the MMWAE criterion. For the MMWAE criterion, Narula and Wellington [93] give an efficient implicit enumeration algorithm for the selection of variables. Their algorithm exploits the special structure of the problem and uses the primal and dual algorithms to solve the individual problems as they arise.

6. Conclusions and recommendations

Sections 2 through 5 clearly point out the important role played by optimization techniques in linear regression analysis. For the MSSE criterion, quadratic programming and branch-and-bound algorithms play a significant role. The development of linear programming and the advent of computers made the MSAE and the MMAE criteria computationally viable alternatives to the MSSE criterion. The special purpose algorithms make it possible to solve these problems more efficiently than the ordinary methods of linear programming. In addition, the duality theory and sensitivity analysis of LP, and goal programming formulations give us further insight into the nature of the solution space for these problems. For the three criteria, branch-and-bound procedures have been successfully developed for judiciously selecting the "best" set of variables. For further discussion of the MSAE, the MSSE and the MMAE criteria, the reader may refer to Arthanari and Dodge [17], Harter [63], Kennedy and Gentle [77], Narula and Wellington [95] and Sposito, Smith and McCormick [117].

So far, we have applied a few optimization techniques very successfully to solve the regression analysis problem. Using these techniques, a number of efficient algorithms were developed simultaneously or in a very quick succession. At times the algorithms are variants of the same technique; whereas at other times, algorithms have been developed using altogether different approaches. This leaves the user with "a large menu and no prices". Thus, it is imperative to compare the algorithms. Also, to make the MSAE and the MMAE criteria more appealing we need to develop the tools of statistical inference for these criteria. Furthermore, we have not exhausted all the optimization techniques, techniques such as nonlinear programming, dynamic programming, network analysis, etc. that are available to us. In the future, we may find some of these approaches very useful for solving statistical problems.

Acknowledgment

I would like to thank John F. Wellington for his valuable comments and support during the preparation of this manuscript and Richard S. Sacher for bringing some of the references to my attention.

References

[1] N.N. Abdelmalek, "Linear L_1 Approximation for a Discrete Point Set and L_1 Solutions of Overdetermined Linear Equations", *J. Assoc. Comput. Mach.* 18 (1971) 41–47.
[2] N.N. Abdelmalek, "On the Discrete Linear L_1 Approximation and L_1 Solutions of Overdetermined Equations", *J. Approx. Theory* 11 (1974) 38–53.
[3] N.N. Abdelmalek, "An Efficient Method for the Discrete Linear L_1 Approximation Problem", *Math. Comput.* 29 (1975) 844–850.
[4] N.N. Abdelmalek, "Chebyshev Solution of Overdetermined System of Linear Equations", *BIT* 15 (1975) 117–129.
[5] N.N. Abdelmalek, "A Computer Program for the Chebyshev Solution of Overdetermined System of Linear Equations", *International J. Numerical Methods in Engineering* 10 (1976) 1197–1202.
[6] G. Appa and C. Smith, "On L_1 and Chebyshev estimation", *Math. Programming* 5 (1973), 73–87.
[7] R.D. Armstrong and E.L. Frome, "A Comparison of Two Algorithms for Absolute Deviation Curve Fitting", *J. Amer. Statist. Assoc.* 7 (1976) 328–330.
[8] R.D. Armstrong, J.J. Elam and J.W. Hultz, "Obtaining Least Absolute Value Estimates for a Two-Way Classification Model", *Commun. Statist. B* 6 (1977) 365–381.
[9] R.D. Armstrong and E.L. Frome, "A Branch-and-Bound Solution of a Restricted Least Squares", *Technometrics* 18 (1976) 447–450.
[10] R.D. Armstrong and E.L. Frome, "A Special Purpose Linear Programming Algorithm for Obtaining Least Absolute Value Estimates in a Linear Model with Dummy Variables", *Commun. Statist. B* 6 (1977) 383–398.
[11] R.D. Armstrong and E.L. Frome, "Least-Absolute-Value Estimators for One-Way and Two-Way Tables", *Naval Research Logistic Quarterly* 26 (1979) 79–96.
[12] R.D. Armstrong, E.L. Frome and D.S. Kung, "A Revised Simplex Algorithm for the Absolute Deviation Curve Fitting Problem", *Common. Statist. B* 8 (1979) 175–190.
[13] R.D. Armstrong and J.W. Hultz, "An Algorithm for a Restricted Discrete Approximation Problem in the L_1 Norm, *SIAM J. Numer. Anal.* 14 (1977) 555–565.
[14] R.D. Armstrong and D.S. Kung, "AS135: Mini-max Estimates For a Linear Multiple Regression Problem", *Applied Statist.* 28 (1979) 93–100.
[15] R.D. Armstrong and M.T. Kung, "AS132: Least Absolute Value Estimates for a Simple Linear Regression Problem", *Applied Statist.* 27 (1978), 363–366.
[16] R. Armstrong and M. Kung, "An Algorithm to Select the Best Subset for a Least Absolute Value Problem", *TIMS Studies in the Management Sciences*, this issue.
[17] T.S. Arthanari and Y. Dodge, *Mathematical Programming in Statistics* (John Wiley and Sons, New York, N.Y., 1981).
[18] I. Barrodale, "L_1 Approximation and the Analysis of Data", *Applied Statist.* 17 (1968) 51–57.
[19] I. Barrodale and C. Phillips, "An Improved Algorithm for Discrete Chebyshev Linear Approximation", *Proc. 4th Manitoba Conf. on Numerical Mathematics* (University of Manitoba, Winnipeg, Manitoba, 1974) 177–190.

[20] I. Barrodale and C. Phillips, "Solution of an Over-Determined System of Linear Equations in the Chebyshev Norm", *ACM Transactions on Mathematical Software* 1 (1975) 264–270.
[21] I. Barrodale and F.D.K. Roberts, "An Improved Algorithm for Discrete L_1 Linear Approximation", *SIAM J. Numer. Anal.* 10 (1973), 839–848.
[22] I. Barrodale and F.D.K. Roberts, "Solution of an Overdetermined System of Equations in the L_1-norm", *Commun. of the ACM* 17 (1974) 319–320.
[23] I. Barrodale and F.D.K. Roberts, "Algorithm for Restricted Least Absolute Value Estimation", *Commun. Statist. B.* 6 (1977) 353–363.
[24] I. Barrodale and F.D.K. Roberts, "An Efficient Algorithm for Discrete L_1 Linear Approximation with linear Constraints", *SIAM J. Numer. Analysis* 15 (1978) 603–611.
[25] I. Barrodale and A. Young, "Algorithm for Best L_1 and L_∞ Linear Approximations on a Discrete Set", *Numerische Mathematik* 8 (1966) 295–306.
[26] R.H. Bartels and A.R. Conn, "Linear Constrained Discrete L_1 Problems", Tech. Report 248, Dept. of Math. Sciences, John Hopkins University, Baltimore, Maryland (1976).
[27] R.H. Bartels and A.R. Conn, "A Primal, Penalty Linear Programming Method for Solving Overdetermined Linear Systems in the L_∞ Sense", Tech. Report 282, Dept. of Math. Sciences, The John Hopkins University, Baltimore, Maryland (1977).
[28] R.H. Bartels and A.R. Conn, "Least Absolute Value Regression: A Special Case of Piecewise Linear Minimization", *Commun. Statist. B* 6 (1977), 329–339.
[29] R.H. Bartels, A.R. Conn and C. Charalambous, "On Cline's Direct Method for Solving Overdetermined Linear System in the L_∞ Sense," *SIAM J. Numer. Anal.* 15 (1978) 255–270.
[30] R.H. Bartels, A.R. Conn and J.W. Sinclair, "Minimization Techniques for Piecewise Differentiable Functions. The L_1 Solution of an Overdetermined Linear System", *SIAM J. Numer. Anal.* 15 (1978) 224–241.
[31] R.H. Bartels and G.H. Golub, "Stable Numerical Methods for Obtaining the Chebyshev Solution to an Overdetermined System of Equations", *Commun. of the ACM* 11 (1968) 401–406.
[32] R.H. Bartels and G.H. Golub, "Algorithm 328: Chebyshev Solution to an Overdetermined Linear System", *Commun. ACM* 11 (1968) 428–430.
[33] R.H. Bartels and G.H. Golub, "The Simplex Method of Linear Programming Using LU Decomposition", *Commun. Assoc. Mach.* 12 (1969) 266–268.
[34] E.M.L. Beale, "A Note on Procedure for Variable Selection in Multiple Regression", *Technometrics* 12 (1970) 909–914.
[35] E.M.L. Beale, M.G. Kendall and D.W. Mann, "The Discarding of Variables in Multivariate Analysis", *Biometrika* 54 (1967) 357–366.
[36] J. Bibby and H. Toutenburg, *Prediction and Improved Estimation in Linear Models* (John Wiley and Sons, New York, N.Y., 1977).
[37] R. Blattberg and T. Sargent, "Regression with Non-Gaussian Stable Disturbances: Some Sampling Results", *Econometrica* 39 (1971) 501–510.
[38] A. Charnes and W.W. Cooper, "Goal Programming and Constrained Regression – A Comment", *Omega* 3 (1975) 403–409.
[39] A. Charnes and W.W. Cooper, "Goal Programming and Multiple Objective Optimization, Part I", *European J. of Operations Research* 1 (1977) 39–54.
[40] A. Charnes, W.W. Cooper and R.O. Ferguson, "Optimal Estimation of Executive Compensation by Linear Programming", *Management Science* 1 (1955) 138–150.
[41] J.S. Chipman and M.M. Rao, "The Treatment of Linear Restrictions in Regression Analysis", *Econometrica* 32 (1964) 198–209.
[42] A.K. Cline, "Rate of Convergence of Lawson's Algorithm", *Mathematics of Computation* 26 (1972) 167–176.
[43] A.K. Cline, "A Descent Method for the Uniform Solution to Overdetermined System of Linear Equations", *SIAM J. Numer. Anal.* 13 (1976) 293–303.

[44] A.R. Conn, "Linear Programming Via a Nondifferentiable Penalty Function", *SIAM J. Numer. Anal.* 13 (1976) 145–154.
[45] M. Davies, "Linear Approximation Using the Criterion of Least Total Deviations", *J. Royal Statist. Soc. Series B* 29 (1967) 101–109.
[46] L. Denby and W.A. Larsen, "Robust Regression Estimators Compared via Monte Carlo", *Commun. Statist., Theor. Meth. Part A*, 6 (1977) 335–362.
[47] T. Dielman and R. Pfaffenberger, "LAV (Least Absolute Value) Estimation in Linear Regression: A Review (*TIMS Studies in the Management Sciences*, this issue.
[48] N.R. Draper and H. Smith, *Applied Regression Analysis* (John Wiley and Sons, New York, 1966).
[49] F.Y. Edgeworth, "On Observations Relating to Several Quantities", *Phil. Mag. (5th Series)* 24 (1887) 222–223.
[50] F.Y. Edgeworth "On a New Method of Reducing Observations Relating to Several Quantities", *Phil. Mag. (5th Series)* 25 (1888), 184–191.
[51] H. Ekblom, "L_p Methods for Robust Regression", *BIT*, 14 (1974) 22–32.
[52] A.B. Forsythe, "Robust Estimation of Straight Line Regression by Minimizing pth Power Deviations", *Technometrics* 14 (1972) 159–166.
[53] G.M. Furnival, "All Possible Regression with Less Computations", *Technometrics* 13 (1971) 403–408.
[54] G.M. Furnival and R.W. Wilson Jr., "Regression by Leaps and Bounds,", *Technometrics* 16 (1974) 499–512.
[55] M.J. Garside, "The Best Subset in Multiple Regression Analysis", *Applied Statist.* 14 (1965) 196–200.
[56] M.J. Garside, "Some Computational Procedures for the Subset Problem", *Applied Statist.* 20 (1971) 8–15.
[57] J.E. Gentle and T.A. Hanson, "Variable Selection Under L_1", *Proceedings Statist. Comput. Section ASA* (Washington, D.C., 1977).
[58] P.E. Gill and W. Murray, "Numerically Stable Methods for Quadratic Programming", *Math. Programming* 14 (1978) 349–372.
[59] J. Gilsinn, R.H. Hoffman, F. Jackson, E. Leyendecker, P. Saunders and D. Shier, "Methodology and Analysis For Comparing Discrete Linear L_1 Approximation Codes", *Commun. Statist - Simul. Comput., Part B* 6 (1977) 399–413.
[60] G.H. Golub, "Numerically Stable Methods for Solving Linear Least Squares Problems", *Numer. Math.* 7 (1965) 206–216.
[61] G.H. Golub and M.A. Saunders, "Linear Least Squares and Quadratic Programming", in: J. Abadie, ed., *Linear and Nonlinear Programming* (North-Holland Publishing Co., Amsterdam, 1970).
[62] R.F. Gunst, "Biased Regression: A Ten Year Perspective", Presented at the Annual Amer. Statist. Assoc. Meetings, Houston, Texas, 1980.
[63] H.L. Harter, "The Method of Least Squares and Some Alternatives IV", *International Statist. Review* 43 (1975) 125–190.
[64] R.R. Hocking, "The Analysis and Selection of Variables in Linear Regression", *Biometrics* 32 (1976) 1–49.
[65] R.R. Hocking and R.N. Leslie, "Selection of the Best Subset in Regression Analysis", *Technometrics* 9 (1967) 531–540.
[66] A.E. Hoerl and R.W. Kennard, "Ridge Regression: Biased Estimation for Non-orthogonal Problems", *Technometrics* 12 (1970) 55–67.
[67] A.E. Hoerl and R.W. Kennard, "Ridge Regression: Applications to Non-orthogonal Problems", *Technometrics* 12 (1970) 69–82.
[68] R.V. Hogg, "Adaptive Robust Procedures: A partial Review and Some Suggestions for Future Applications and Theory", *J. Amer. Statist. Assoc.* 69 (1974) 309–323.

[69] R.V. Hogg, "Statistical Robustness: One View of Its Use in Applications Today", *The Amer. Statistician* 33 (1979) 108–115.
[70] R.V. Hogg and R.H. Randless, "Adaptive Distribution-Free Regression Methods and Their Applications", *Technometrics* 17 (1975) 399–407.
[71] P.J. Huber, "Robust Statistics: A Review", *The Annals of Mathematical Statistics* 43 (1972) 1041–1067.
[72] P.J. Huber, "Robust Regression: Asymptotics, Conjectures and Monte Carlo", *The Annals of Statistics* 1 (1973) 799–821.
[73] J.P. Ignizio, *Goal Programming and Extensions* (Lexington Books, Lexington, MA, 1976).
[74] G.G. Judge and T. Takayama, "Inequality Restrictions in Regression Analysis", *J. Amer. Statist. Assoc.* 61 (1966) 167–181.
[75] O.J. Karst, "Linear Curve Fitting Using Least Deviations", *J. Amer. Statist. Assoc.* 53 (1958) 118–132.
[76] J.E. Kelley, "An Application of Linear Programming to Curve Fitting", *SIAM J. Appl. Math.* 6 (1958) 15–22.
[77] W.J. Kennedy Jr. and J.E. Gentle, *Statistical Computing* (Marcel Dekker, Inc., New York, 1980).
[78] E.A. Kiountouzis, "Linear Programming Techniques in Regression Analysis", *Appl. Statist.* 22 (1973) 69–73.
[79] D. Klingman and J. Mote, "Generalize Network Approaches for Solving Least Absolute Value and Tchebycheff Regression Problems", *TIMS Studies in the Management Sciences*, this issue.
[80] R. Koenker and G. Bassett Jr., "Regression Quantiles", *Econometrica* 46 (1978) 33–50.
[81] L.R. Lamotte and R.R. Hocking, "Computational Efficiency in the Selection of Regression Variables", *Technometrics* 12 (1970), 83–93.
[82] C.L. Lawson, "Contributions to the Theory of Linear Least Maximum Approximation", Ph.D. Dissertation, Univ. of California, Los Angeles (1961).
[83] C.L. Lawson and R.J. Hanson, *Solving Least Squares Problems*, (Prentice-Hall, Englewood Cliffs, N.J., 1975).
[84] N. Mantel, "Restricted Least Squares Regression and Convex Quadratic Programming", *Technometrics* 11 (1969) 763–773.
[85] G.F. McCormick and V.A. Sposito, "A Note of L_1 Estimation Based on the Median Positive Quotient", *Applied Statist.* 24 (1975) 347–350.
[86] R. Mifflin, "A Stable Method for Solving Certain Constrained Least Squares Problems", *Math. Programming* 16 (1979) 141–158.
[87] S.C. Narula and J.S. Ramberg, "Letter to the Editor", *The Amer. Statistician* 26 (1972) 42.
[88] S.C. Narula and J.F. Wellington, "An Algorithm for Linear Regression with Minimum Sum of Absolute and Relative Errors Criterion", Tech. Report 76-4, Dept. of Industrial Engineering, SUNY at Buffalo, New York (1976).
[89] S.C. Narula and J.F. Wellington, "AS108: Multiple Linear Regression with Minimum Sum of Absolute Errors", *Applied Statist.* 26 (1977) 106–111.
[90] S.C. Narula and J.F. Wellington, "Prediction, Linear Regression and Minimum Sum of Relative Errors", *Technometrics* 19 (1977) 185–190.
[91] S.C. Narula and J.F. Wellington, "An Algorithm for the Minimum Sum of Weighted Absolute Error Regression", *Commun. Statist. B* 6 (1977) 341–352.
[92] S.C. Narula and J.F. Wellington, "Regression Using Minimization of the Maximum Weighted Absolute Error Criterion", Research Report 37-77-P13, School of Management, Rensselaer Polytechnic Institute, Troy, New York (1977).
[93] S.C. Narula and J.F. Wellington, "Subset Selection Using Minimization of the Maximum Weighted Absolute Error", Research Report 37-77-P14, School of Management, Rensselaer Polytechnic Institute, Troy, New York (1977).

[94] S.C. Narula and J.F. Wellington, "Selection of Variables in Linear Regression Using the Minimum Sum of Weighted Absolute Errors Criterion", *Technometrics* 21 (1979) 299–306.
[95] S.C. Narula and J.F. Wellington, "Multiple Linear Regression: A Review of Two Alternatives to Least Squares Criterion", Research Report 37-79-P10, School of Management, Rensselaer Polytechnic Institute, Troy, New York (1979).
[96] S.C. Narula and J.F. Wellington, "Selection of Variables in Linear Regression: A Pragmatic Approach", Research Report 37-80-P6, School of Management, Rensselaer Polytechnic Institute, Troy, New York (1980).
[97] S.C. Narula and J.F. Wellington, "Variable Selection in Multiple Linear Regression Using the Minimum Sum of Weighted Absolute Errors Criterion", *Commun. Statist.* B10 (1981) 641–648.
[98] M.R. Osborne and G.A. Watson, "On the Best Linear Chebyshev Approximation", *Computer Journal* 10 (1967) 172–177.
[99] R.C. Pfaffenberger and J.J. Dinkel, "Absolute Deviations Curve Fitting: An Alternative to Least Squares", in: H.A. David, ed., *Contributions to Survey Sampling and Applied Statistics* (Academic Press, New York, 1978).
[100] J.O. Ramsay, "A Comparative Study of Several Robust Estimates of Slope, Intercept, and Scale in Linear Regression", *J. Amer. Statist. Assoc.* 72 (1977) 608–615.
[101] M.R. Rao and V. Srinivasan, "A Note on Sharpe's Algorithm for Minimizing the Sum of Absolute Deviations in a Simple Regression Problem", *Management Science* 19 (1972) 222–225.
[102] E.C. Rhodes, "Reducing Observations by the Method of Minimum Deviations", *Phil. Mag. (7th Series)* 9 (1930) 974–992.
[103] J.R. Rice and J.S. White, "Norms for Smoothing and Estimation", *SIAM Review* 6 (1964) 243–256.
[104] P.D. Robers and A. Ben-Israel, "An Interval Programming Algorithm for Discrete Linear L_1 Approximation Problems", *J. Approx. Theory* 2 (1969) 323–336.
[105] P.D. Robers and S.S. Robers, "Algorithm 458: Discrete Linear L_1 Approximation by Interval Linear Programming", *Commun. Assoc. Comput. Mach.* 16 (1973) 629–631.
[106] G. Roodman, "A Procedure for Optimal Stepwise MSAE Regression Analysis", *Operations Research* 22 (1974) 393–399.
[107] A.N. Sadovski, "AS74: L_1-Norm Fit of a Straight Line", *Applied Statist.* 23 (1974) 244–248.
[108] M. Schatzoff, S. Fienberg and R. Tsao, "Efficient Calculations of all Possible Regression", *Technometrics* 10 (1968) 769–779.
[109] K. Schittkowski and J. Stoer, "A Factorization for the Solution of Constrained Linear Least Squares Problems Allowing Subsequent Data Changes", *Numerische Mathematik* 31 (1979) 431–463.
[110] E.J. Schlossmacher, "An Iterative Technique for Absolute Deviations Curve Fitting", *J. Amer. Statist. Assoc.* 68 (1973) 857–859.
[111] W.F. Sharpe, "Mean-Absolute-Deviation Characteristic Lines for Securities and Portfolios", *Management Science* 19 (1971) B1–B13.
[112] R.R. Singleton, "A Method of Minimizing the Sum of Absolute Values of Deviations", *Ann. Math. Statist.* 11 (1940) 301–310.
[113] V.A. Sposito, *Linear and Nonlinear Programming* (Iowa State University Press, Ames, Iowa, 1975).
[114] V.A. Sposito, "A Remark on Algorithm AS74: L_1-Norm Fit of a Straight Line", *Applied Statist.* 25 (1976) 96–97.
[115] V.A. Sposito, W.J. Kennedy and J.E. Gentle, "AS110: L_p-Norm Fit of A Straight Line", *Applied Statist.* 26 (1977) 114–118.
[116] V.A. Sposito and W.C. Smith, "On a Sufficient and a Necessary Condition for L_1 Estimation", *Applied Statist.* 25 (1976) 154-157.

[117] V.A. Sposito, W.C. Smith and G. McCormick, *Minimizing the Sum of Absolute Deviations* (Vandenhoeck and Rupercht, Göttingen, W. Germany, 1978).
[118] K. Spyropoulos, E. Kiountouzis and A. Young, "Discrete Approximation in L_1-norm", *Computer Journal* 16 (1973) 180–186.
[119] E. Stiefel, "Note on Jordan Elimination, Linear Programming and Tchebycheff Approximations", *SIAM J. Numer. Math.* 2 (1960) 1–17.
[120] J. Stoer, "On the Numerical Solution of Constrained Least-Squares Problems", *SIAM J. Numer. Anal.* 8 (1971) 383–411.
[121] H. Theil, "On the Use of Incomplete Prior Information in Regression Analysis", *J. Amer. Statist. Assoc.* 58 (1963) 401–414.
[122] A. Tishler and L. Zang, "An Absolute Deviations Curve Fitting Algorithm for Nonlinear Models", *TIMS Studies in the Management Sciences*, this issue.
[123] K.H. Usow, "On L_1 Approximation II: Computation for Discrete Functions and Discretization Effects", *SIAM, J. Numer. Anal.* 4 (1967) 233–244.
[124] H.M. Wagner, "Linear Programming Techniques for Regression Analysis", *J. Amer. Statist. Assoc.* 54 (1959) 206–212.
[125] R.E. Walls and D.L. Weeks, "A Note on the Variance of a Predicted Response in Regression", *The Amer. Statistician* 23 (1969) 24–26.
[126] M.S. Waterman, "A Restricted Least Squares Problem", *Technometrics* 16 (1974) 135–136.
[127] P. Wolfe, "The Simplex Method for Quadratic Programming", *Econometrica*, 27 (1959) 382–398.

LAV (LEAST ABSOLUTE VALUE) ESTIMATION IN LINEAR REGRESSION: A REVIEW

Terry DIELMAN and Roger PFAFFENBERGER

Texas Christian University

Owing to recent asymptotic distribution results for LAV (least absolute value) estimators in the regression model and the development of efficient algorithms to produce values of the estimators given a regression data set, LAV estimation research and applications are currently attracting considerable attention. In this paper, recent advances in algorithms designed to produce LAV estimates are reviewed; small sample and recently developed large sample (asymptotic) properties of LAV estimators are also reviewed; and estimation procedures based on the large sample properties of the LAV estimators are developed for the regression model and applied to simulated data for illustrative purposes. The role played by optimization for both estimation of model parameters and inferences about these parameters is noted. Avenues of future research, especially pertaining to inference procedures, are suggested.

1. Introduction

In the past few years there has been an increased interest in robust estimation procedures applied to the regression model. The motivation for this interest has arisen largely from the recognition of and the reaction to the following facts: (1) many experimental or survey data sets do not conform to the Gaussian error assumption required for inferences drawn from least squares-fitted regression models, and (2) least squares is sensitive to outliers in the data set. For a fascinating and informative historical account of linear model estimation based on least squares and alternative estimation methods, see Harter's [29] five part series. In part V, Harter reports the research on robust estimation methods in the linear model up to 1975. While there are a number of excellent papers on robust estimation procedures, only a few will be mentioned in this section. In 1964, Huber [33] published what is now considered to be a classic paper on robust estimation of location. Huber's work was subsequently extended to the linear model by Andrews [2], Bickel [15], and Huber [34], among others. In 1977, an entire issue of the *Communications in Statistics – Theory and Methods* was devoted to robust estimation procedures. In this issue, Hogg [32] reviews the *M*-estimation approaches based on maximum likelihood and the *R*-estimation approaches based on ranking schemes.

Received October 8, 1980; revised April 7, 1981.

Harter [28] and Gentle, Kennedy and Sposito [26] address topics related to the LAV (least absolute value) estimation procedure in this series. In the former article, the nonuniqueness of least absolute values estimates of the regression coefficients is considered, and in the latter article properties of LAV estimators in the regression model are explored. In a later *Econometrica* article, Koenker and Bassett [38] introduce a robust estimation method termed, "regression quantiles", a class of estimators that include the LAV estimators as a special case.

In this paper attention is focused on the LAV estimator in the linear regression model. While it is now evident that no single "robust" procedure is best (by mean square error or other criteria), the LAV estimators have attracted considerable attention when the error distribution is "fat-tailed" (e.g. Laplace or Cauchy distributed errors) since least squares is then judged inadequate. The existence of error distributions with long tails in regression applications is becoming widely accepted in the finance and economic literature (e.g. see Fama [25], Mandelbrot [39], Sharpe [50], and Wilson [57]). Recent results to be summarized below provide the necessary "linkage" between the existence of long-tailed error distributions which promote outliers and hence typically poor fits by least squares, and the appropriateness of LAV estimation in such cases.

Use of LAV estimation is supported by noting the following facts.

(1) Bassett and Koenker [14] have recently developed the asymptotic theory for LAV estimators in the regression model. Their paper finally confirms the suspected result (explored previously by Monte Carlo simulation experiments) that for any error distribution for which the median is superior to the mean as an estimator of location, the LAV estimators of the regression coefficients are preferred to the least squares estimators in the sense of having strictly smaller asymptotic confidence ellipsoids for the regression coefficients for a fixed sample size.

(2) Bickel [15], Harvey [30], and Hill and Holland [31], as well as others, recommend the LAV estimator as a starting point (a consistent estimator) for one-step and iteratively weighted multi-step least squares procedures. Hence, in the instances when the median is not the most efficient estimator of location in the error distribution, the LAV estimator may still play an important role by providing an initial consistent estimator for other iterative robust estimation procedures.

(3) Until recently, the computational difficulty in producing the numeric values of the LAV estimates in a regression model-fitting experiment has relegated the LAV estimation procedure to a very secondary role to least squares. Until Charnes, Cooper and Ferguson [19] demonstrated that LAV estimates could be produced by linear programming methods, it was simply not feasible to consider the LAV estimator in most applications. In section 2 of this paper recent advances in computational algorithms to produce LAV

estimates in a regression experiment are reviewed. By the use of the algorithms, it is ordinarily possible to produce LAV estimates in only slightly more computational (computer) time than that required to produce least squares estimates. In addition, it is now possible to procure computer-coded algorithms that will perform best subset variable selection and entry of blocks of variables in a stepwise sense.

(4) Monte Carlo simulation studies of the small-sample properties of the LAV estimators continue to appear in the literature. These studies have confirmed the superiority of the LAV estimators to the least squares estimators in small samples when the error distribution in the regression model has fat-tails. Section 3 of this paper summarizes the principal results reported in these studies.

In summary, the development of the asymptotic theory for LAV estimation and the development of highly efficient computational algorithms for producing LAV estimates is rapidly bringing LAV estimation into the mainstream of regression model-fitting applications.

In the next section of the paper the recent advances in algorithms for the computation of LAV estimates are reviewed. In section 3, in addition to a review of the small sample properties of the LAV estimators, the application of the recent asymptotic results of Bassett and Koenker [14] to hypothesis testing and confidence interval specification is delineated. In section 4 LAV estimation is applied to a regression problem involving simulated data and the use of the section 3 procedures for statistical inference is demonstrated. In section 5 problems requiring further research concerning the LAV estimation approach are discussed.

A final note: LAV estimators have been referred to as MAD (minimum absolute deviation), LAE (least absolute error), MSAE (mean square absolute error), LAR (least absolute residual) or the L_1-norm estimator. The acronym LAV, which is most consistent with the notation for least squares (LS), will be adopted in this paper.

2. Computational algorithms for calculating LAV estimates

Consider the multiple linear regression model,

$$Y_t = \sum_{k=1}^{K} X_{tk} b_k + e_t, \quad t = 1, 2, \ldots, T, \tag{1}$$

where $(Y_t, X_{t1}, X_{t2}, \ldots, X_{tk})$, $t = 1, 2, \ldots, T$, is given and the problem is to find $\hat{b}' = (\hat{b}_1, \hat{b}_2, \ldots, \hat{b}_K)$ so that the sum of the absolute deviations between the predicted Y_t, given by

$$\hat{Y}_t = \sum_{k=1}^{K} X_{tk} \hat{b}_k, \tag{2}$$

and the observed Y_t is minimized.
Thus, the objective is to

$$\underset{\hat{b}}{\text{minimize}} \sum_{t=1}^{T} \left| Y_t - \sum_{k=1}^{K} X_{tk} \hat{b}_k \right|. \tag{3}$$

Wagner [55], among others, has shown that (3) is equivalent to the following linear programming problem:

$$\text{minimize} \sum_{t=1}^{T} (d_t^+ + d_t^-) \tag{4}$$

subject to

$$Y_t - \sum_{k=1}^{K} X_{tk} \hat{b}_k + d_t^+ - d_t^- = 0, \qquad t = 1, 2, \ldots, T,$$

$$d_t^+, d_t^- \geq 0, \qquad t = 1, 2, \ldots, T, \tag{5}$$

$\hat{b}_k$, $k = 1, 2, \ldots, K$ unrestricted in sign,

where d_t^+ and d_t^- are, respectively, the positive and the negative deviations associated with the tth observation.

For later reference, note that the matrix equivalent of eq. (1) is:

$$Y = Xb + e, \tag{6}$$

where $Y' = (Y_1, Y_2, \ldots, Y_T)$ is the $T \times 1$ vector of observations on the dependent variable, $X = (X_1, X_2, \ldots, X_K)$ is the $T \times K$ matrix of observations on the K independent variables and $e' = (e_1, e_2, \ldots, e_T)$ is the vector of random disturbances. The columns of X, X_k, for $k = 1, 2, \ldots, K$, can be written as $X_k' = (X_{k1}, X_{k2}, \ldots, X_{kT})$, where one X_k is a column of ones if it is desired to include a constant in the model.

While it is possible to apply the standard simplex algorithm to either (4) and (5) or its dual, several researchers have demonstrated that modifications to the simplex or to the revised simplex algorithms that take advantage of the special structure of the primal problem in (4) and (5) or its dual could lead to rather substantial reductions in computer execution times to produce the LAV estimates. Barrodale and Roberts [11] devised an algorithm which is a modification of the simplex method applied to the primal problem. Abdelmalek [1] modified the dual simplex algorithm to solve efficiently the dual of (4) and (5). Armstrong and Godfrey [6] demonstrated that the Barrodale and Roberts [11] and the Abdelmalek [1] algorithms are conceptually equivalent, differing only in the manner in which an initial basis is chosen and in the choice of heuristic rules for breaking ties if they occur at any iteration. Sposito, Hand and McCormick [52] have demonstrated that the total CPU time required to obtain LAV estimates can generally be reduced by first computing least squares estimates and using them as starting values in the Barrodale and Roberts [12] algorithm. Armstrong, Frome and Kung [5] have modified the Barrodale and Roberts algorithm, with the result that their algorithm produces significant computer execution time savings over the original B-R algorithm. A FORTRAN listing of their algorithm is given in the appendix of their paper. Recently, Klingman and Mote [37] have begun work on developing a generalized network approach to solving the LAV (L_1-norm) and Tchebycheff (L_∞-norm) estimation problems which holds promise for potentially significant computational efficiencies.

Barrodale and Roberts [10] and Armstrong and Hultz [7] developed algorithms to produce constrained LAV estimates, i.e., solutions to (4) and (5) when linear restrictions are applied to the regression coefficients. Armstrong and Frome [4] have developed a special purpose algorithm for models with dummy variables. Roodman [46] and Wilson [57] have developed stepwise procedures for determining the best subset of m regressors, $m \leq K$ in (1). For a more detailed discussion of best subset selection algorithms see Armstrong and Kung [8]. Although not directly germane to this review, it is interesting to note that LAV algorithms have been developed for the estimation of parameters in one-way and two-way analysis of variance (e.g. see Armstrong and Frome [4]) and for parameter estimation in the nonlinear model (e.g. see El-Attar, Vidyasagar and Dutta [24]). For a more detailed discussion of the development of algorithms see Narula and Wellington [42].

The availability of efficient algorithms to produce LAV estimates in the linear regression model insures the increased used of this robust estimation procedure, particularly in those cases where it can be demonstrated that the error distribution is "fat-tailed".

3. Properties of the LAV regression estimators

3.1. Small sample properties

There have been a number of simulation studies of LAV estimators, most of which compare the small sample efficiency of LAV and least squares estimators for various error distributions. This is done, for example, in Rice and White [45], Blattberg and Sargent [17], Pfaffenberger and Dinkel [43], Rosenberg and Carlson [48], Ashar and Wallace [9], Wilson [56], and Kiountouzis [36]. The results of these simulation studies indicate that LAV estimators are more efficient than least squares estimators in small samples when the residuals have Cauchy, Laplace, Stable Paretian distributions with the characteristic exponent less than or equal to 1.5, or certain contaminated normal distributions.[1] The performance of LAV with normal error distributions is, as would be expected, not as good as that of least squares. Both Ashar and Wallace [9] and Wilson [56] found that the efficiency of the LAV estimators relative to least squares estimators to be about 80% when the least squares assumptions are satisfied.[2]

In other small sample comparisons, Glahe and Hunt [27] and Hunt, Dowling and Glahe [35] provide evidence that the LAV estimator performs well in simultaneous equation models with long-tailed residual distributions. Meyer and Glauber [41] use least squares and LAV estimation to fit a model developed for predicting investment return. The LAV model outperforms least squares with forecasting accuracy measured by Theil's U-coefficient. Cogger [20] also compares forecasting accuracy (based on several criteria) for LAV and least squares estimation in autoregressive models. His results indicated that LAV is about 90% as efficient as least squares when conditions are optimal for the latter and outperforms least squares in other situations tested.

Brecht [18] investigated a model with normal error distributions and unusually large errors in the explanatory variables. LAV estimation proved relatively more efficient than least squares in this situation also.

In articles previously mentioned, Hill and Holland [31] and Ramsay [44] suggest alternative robust estimation procedures that outperform the LAV estimator in certain cases. Hill and Holland suggest the use of LAV estimation to obtain starting values in their robust procedure.

The most extensive simulation study of LAV estimation was performed by Rosenberg and Carlson [47]. They investigated the small sample properties of the LAV estimator in multiple regression models with symmetric error distributions, specifically, normal and contaminated normal distributions. After performing over 100,000 LAV regressions, they reported the following results.

(1) The LAV estimator had a significantly smaller standard error than the least squares estimator for a regression with high-kurtosis disturbances.

(2) The LAV estimators were almost exactly normally distributed in the presence of high-kurtosis disturbances.

(3) The error, $(\hat{b} - b)$, in the LAV estimator was approximately normally distributed with mean zero and covariance matrix $\lambda^2(X'X)^{-1}$, where X is the matrix of K explanatory variables and T observations as in eq. (6), and λ^2/T is

---

[1] A contaminated normal distribution is a residual distribution where the majority of the disturbances are taken from a normal distribution with constant variance except for one or more outliers drawn from a normal distribution with a much larger variance.

[2] Ashar and Wallace's results include models with autocorrelated normal disturbances.

the variance of the median of a sample of size T from the disturbance distribution.

The Rosenberg and Carlson results were first published in 1977 [48]. The results are currently drawing interest since an asymptotic distribution theory for LAV estimation in the regression model has been developed by Bassett and Koenker [14]. The similarity between the Bassett and Koenker results and the Rosenberg and Carlson simulation results is clear: Bassett and Koenker prove that result (3) cited above holds true asymptotically. In the following section, the asymptotic theory is discussed in greater detail.

3.2. Large sample (asymptotic) properties

One of the first asymptotic properties proved for LAV estimators appears in an article by Taylor [53]. The property concerns a certain form of consistency for the estimators for grouped data when the residuals are symmetrically distributed about zero.

The unbiasedness of LAV estimators when the residual distributions are symmetrical is also shown in Taylor's paper. This result holds true when the solution to (4) and (5) is unique. When the solution to (4) and (5) is not unique, certain computational algorithms may lead to biased regression coefficient estimates. Sielken and Hartley [51] describe an algorithm which produces unbiased LAV estimators when the solution to (4) and (5) is nonunique.

The most important findings concerning the asymptotic distribution of LAV estimators are a result of research reported in Bassett [13] and Bassett and Koenker [14]. Their main result is that the sampling distributions of LAV estimators will be asymptotically normal with a specified mean and variance. In addition, they prove that the LAV estimators are consistent. The technical statement of the theorem is given in appendix A of this paper. The theorem requires only very general assumptions and shows that the LAV estimator $\hat{b}$ has a normal distribution asymptotically with mean b (the vector of coefficient values) and covariance matrix $\lambda^2(X'X)^{-1}$ (as suggested by Rosenberg and Carlson), where λ^2/T is again the asymptotic variance of the sample median from random samples of size T taken from the error distribution. An important difference between this result and an analogous result for least squares (see Theil [54, p. 378] for example) is the lack of assumptions made here concerning the mean and variance of the equation disturbances. It is necessary to assume only that the disturbances are independent and identically distributed. This result implies that for any error distribution for which the median is (asymptotically) superior to the mean as an estimator of location, LAV estimation is preferable to least squares. Table 1 compares the asymptotic variances of the mean and the median for several distributions and indicates those cases where the variance of the median is smaller. It is in such cases when the LAV estimator will outperform the least squares estimator in large samples.

Table 1
Comparison of asymptotic variances for the sample mean ($\bar{X}$) and median (m) for selected probability density functions.

Distribution	Var($\bar{X}$)	Var(m)	Superior estimator of location	pdf	Range
Normal	σ^2/T	$\pi\sigma^2/2T$	$\bar{X}$	$f(X) = (2\pi\sigma^2)^{-1/2} \exp[-\frac{1}{2}(X/\sigma)^2]$	$-\infty < X < \infty$
Cauchy	∞	$\pi^2 c^2/4T$	m	$f(X) = 1/[\pi c(1+(X/c)^2)]$ $c > 0$	$-\infty < X < \infty$
Laplace	$1/(2Tc^2)$	$1/(4Tc^2)$	m	$f(X) = (\frac{1}{2}c)\exp[(-|X-b|)/c]$ $c > 0$	$-\infty < X < \infty$
Logistic	$\pi d^2/(12T)$	$\pi d^2/T$	m	$f(X) = \text{sech}^2(X/d)/(2d)$	$-\infty < X < \infty$
Cosine	$(\pi^2-6)/(3T)$	$\pi^2/(4T)$	$\bar{X}$	$f(X) = (1+\cos(X))/2$	$-\pi < X \le \pi$
Triangular	$b^2/6T$	$b^2/(4T)$	$\bar{X}$	$f(X) = (b+X)/b^2$ $f(X) = (b-X)/b^2$	$-b \le X \le 0$ $0 \le X \le b$
Uniform	$a^2/3T$	a^2/T	$\bar{X}$	$f(X) = 1/(2a)$	$-a \le X \le a$
Symmetric stable	∞	finite [a]	m	$f(X)$ does not exist except for the Cauchy and Normal pdfs	$-\infty < X < \infty$

Note: The pdfs are designed so that each distribution is centered at zero. The results apply to more general cases, however.
[a] This result applies to symmetric stable distributions with a characteristic exponent less than 2 (thus, the normal distribution is excluded, but the Cauchy is included). The result is due to Bassett [13] and states that var(m) is finite as long as the sample size, T, satisfies the relation, $T \ge (4/\alpha)+1$, where α is the characteristic exponent. The median therefore has smaller variance than the mean asymptotically.

In addition, the pdfs and ranges for these distributions when they are centered about zero are given in table 1.

One reason for past reluctance of researchers to choose LAV over least squares estimation has been the absence of any theory to provide for inference about the model parameters. With the Bassett and Koenker results in hand, such a theory can be developed for hypothesis testing and confidence interval estimation in the regression model.

3.3. Application of the asymptotic results to inference-making in the regression model

Using the Bassett and Koenker result that $\hat{b}$ is asymptotically normally distributed with mean b and covariance matrix $\lambda^2(X'X)^{-1}$, where λ^2/T is the variance of the median of a sample of T observations from the disturbance distribution, confidence interval formulations and hypothesis-testing procedures may be developed in a rather straightforward fashion. For example, suppose an interval estimator for $r'b$ is desired where r is a vector of constants used to specify a particular linear combination of the regression parameters and b is defined in (1). If $b' = (b_1, b_2, b_3)$, then for $r' = (1, 0, 0)$, the component chosen is b_1; $r' = (1, 1, 0)$ indicates the linear combination $b_1 + b_2$, and so on. The point estimator of $r'b$ is $r'\hat{b}$, where $\hat{b}$ is the LAV estimator of the parameter vector b.

Using the Bassett and Koenker result, the $(1 - \alpha)100\%$ confidence interval for $r'b$ can be written as:

$$r'\hat{b} \pm z_{\alpha/2} \lambda \left[r'(X'X)^{-1} r \right]^{1/2}, \tag{7}$$

where $z_{\alpha/2}$ denotes the appropriate percentile from the standard normal distribution.

For a single component of b, b_i, the $(1 - \alpha)100\%$ confidence interval is:

$$\hat{b}_i \pm z_{\alpha/2} \lambda (X'X)_{ii}^{-1/2}, \tag{8}$$

where $(X'X)_{ii}^{-1/2}$ denotes the square root of the ith diagonal element of the $(X'X)^{-1}$ matrix.

The drawback to this procedure should be obvious: λ is unknown. Thus, an estimate of λ is needed to make the intervals in (7) and (8) operational. One approach to estimating λ in large samples is suggested by Cox and Hinkley [22, pp. 468–470]. First note that, asymptotically

$$\lambda = f'/2 \tag{9}$$

(see Cramer [23, p. 369]), where $f' = 1/f(m)$ and $f(m)$ is the value of the distribution function for the true residuals, e_t, evaluated at the median, m, of the distribution. Estimates of the true residuals can be obtained using the relationship, $\hat{e}_t = Y_t - X_t\hat{b}$, where $\hat{b}$ is the LAV estimate of b. Denoting the ordered residuals by $\hat{e}_{(t)}$, an estimate of f' is:

$$\hat{f}' = \frac{\hat{e}_{(t)} - \hat{e}_{(s)}}{(t-s)/T}. \tag{10}$$

As noted by Cox and Hinkley [22, p. 470], $\hat{f}'$ is a consistent estimator of f'.

The value of t and s in (10) should be symmetric around the index of the median sample residual. Thus,

$$t = [T/2] + v \quad \text{and} \quad s = [T/2] - v, \tag{11}$$

where $[\cdot]$ indicates the integer portion of the number and where v is an integer chosen to spread t and s symmetrically about the index of the median sample residual. The choice of v, or the difference, $t-s$, should be kept fairly small, but little more can be said about this matter. This choice, however, should not cause widely differing values of $\hat{f}'$ in large samples. Additionally, as noted by Cogger [21], $\hat{e}_{(t)}$ and $\hat{e}_{(s)}$ should not be equal to zero.

Using $\hat{f}'$, a consistent estimator of λ is:

$$\hat{\lambda} = \hat{f}'/2. \tag{12}$$

The $(1 - \alpha)100\%$ confidence interval estimator of $r'b$ is now:

$$r'\hat{b} \pm z_{\alpha/2} \hat{\lambda} \left[r'(X'X)^{-1} r \right]^{1/2}. \tag{13}$$

In the remainder of this section some useful formulae are provided for various interval estimates and hypothesis testing techniques. In each case, the procedures and formulae were developed assuming λ known, then the estimator, $\hat{\lambda}$, was substituted. In large samples, such an approach is justified. The small sample behavior of the intervals and test statistics remains a subject for Monte Carlo investigation.[3]

---

[3] Cogger [21] suggests an alternative procedure for inference in small samples in his paper. He also demonstrates, via an example, that confidence intervals and test results can be somewhat unreliable if the techniques suggested in this paper are used in small samples.

3.4. Confidence intervals

(1) Confidence interval for $r'b$:

$$r'\hat{b} \pm z_{\alpha/2}\,\hat{\lambda}\bigl[r'(X'X)^{-1}r\bigr]^{1/2}, \tag{14}$$

(2) Confidence interval for b_i:

$$\hat{b}_i \pm z_{\alpha/2}\,\hat{\lambda}(X'X)_{ii}^{-1/2}, \tag{15}$$

where all terms in (14) and (15) have been previously defined.

(3) Bonferroni joint confidence intervals for setting joint intervals on $k(k \leq K)$ of the regression coefficients with family-wise confidence coefficient $(1-\alpha)$:

$$\hat{b}_i \pm z_{\alpha/2k}\,\hat{\lambda}(X'X)_{ii}^{-1/2}, \tag{16}$$

where $z_{\alpha/2k}$ is the $(1-\alpha/2k)$th percentile of the standard normal distribution.

(4) Confidence interval for a mean response, Y_0. Let $X_0' = (X_{10}, X_{20}, \ldots, X_{K0})$ be one observation of the input vector X_t. Then, the point estimate of the mean response corresponding to X_0, denoted by $\hat{Y}_0$, is given by

$$\hat{Y}_0 = X_0'\hat{b}, \tag{17}$$

and the $(1-\alpha)100\%$ confidence interval for the mean response is:

$$\hat{Y}_0 \pm z_{\alpha/2}\,s(\hat{Y}_0), \tag{18}$$

where

$$s^2(\hat{Y}_0) = \hat{\lambda}^2\bigl(X_0'(X'X)^{-1}X_0\bigr). \tag{19}$$

(5) A prediction interval with $(1-\alpha)100\%$ confidence for a new observation, Y_{new}, corresponding to an observed input vector X_0 is:

$$\hat{Y}_{\text{new}} \pm z_{\alpha/2}\,s(\hat{Y}_{\text{new}}), \tag{20}$$

where

$$s^2(\hat{Y}_{\text{new}}) = \hat{\lambda}^2\bigl(1 + X_0'(X'X)^{-1}X_0\bigr). \tag{21}$$

The confidence interval estimators for Y_0 and for Y_{new} depend upon the asymptotic normality of Y, an assumption that may not be warranted in many applications of LAV estimation. These formulae are suggested since they are analogous to those used in least squares. In addition, since the hypothesis-testing procedures developed below have provided satisfactory results, it is anticipated that these formulae will perform well when compared to the corresponding least squares confidence intervals.

3.5. Hypothesis testing

(1) To test the null hypothesis, $H_0: r'b = h$ versus the alternative, $H_a: r'b \neq h$, with a significance level of α, the rejection and acceptance regions are:
 (a) Reject H_0 if

$$z^* = \frac{r'\hat{b} - h}{\hat{\lambda}(r'(X'X)^{-1}r)^{1/2}} > z_{\alpha/2} \tag{22}$$

or if $z^* < -z_{\alpha/2}$.
 (b) Accept H_0 if $-z_{\alpha/2} \leq z^* \leq z_{\alpha/2}$.

(2) In particular, to test $H_0: b_i = 0$ versus $H_a: b_i \neq 0$ for an individual coefficient:
 (a) Reject H_0 if

$$z^* = \frac{\hat{b}_i}{\hat{\lambda}(X'X)_{ii}^{-1/2}} > z_{\alpha/2} \tag{23}$$

or if $z^* < -z_{\alpha/2}$.
 (b) Accept H_0 if

$$-z_{\alpha/2} \leq z^* \leq z_{\alpha/2}.$$

(3) In a least squares regression an F-test is often used to test the "significance" of the regression; that is, to test the null hypothesis, $H_0: b_2 = b_3 = \ldots = b_K = 0$ (assuming that the intercept term, b_1, has been added to the model) versus the alternative H_a: not all slope coefficients are zero.

An analogous test for LAV estimation is based on the statistic:

$$\frac{(\hat{b}^* - b^*)'X^{*'}X^*(\hat{b}^* - b^*)}{\hat{\lambda}^2}. \tag{24}$$

The asterisks (*) signify that the column of ones in the X matrix and the entry in the b vector corresponding to the constant have been deleted. Under the null hypothesis given above, the statistic in (24) has, asymptotically, a chi-square distribution with $k - 1$ degrees of freedom. The distribution is easily derived since $\hat{b}^* - b^*$ is asymptotically normal with mean zero and covariance matrix $\lambda^2(X'X)^{-1}$ using Corollary 2.2 of Searle [49, p. 58] and the fact that $\hat{\lambda}$ in eq. (12) is a consistent estimator of λ, thus ensuring convergence in distribution.

The rejection and acceptance regions are defined as:

(a) Reject H_0 if

$$C = \frac{(\hat{b}^* - b^*)'X^{*\prime}X^*(\hat{b}^* - b^*)}{\hat{\lambda}^2} > \chi^2_{1-\alpha}. \tag{25}$$

(b) Accept H_0 if

$$C \leq \chi^2_{1-\alpha},$$

where $\chi^2_{1-\alpha}$ denotes the $(1-\alpha)$th percentile of the chi-square distribution with $K-1$ degrees of freedom.

Note that the distributions of the statistics used to construct the confidence intervals and the critical regions for the hypothesis tests in (22) and (23) are assumed to be normal. In small samples we would use a t-table to obtain critical values for the corresponding least squares tests and intervals. We are not guaranteed that this will be an effective procedure when using the LAV fitted models, however, since the theory of Bassett and Koenker only describes the behavior of the sampling distributions of the coefficients in large samples. The shape of the sampling distributions of the LAV fitted coefficients in small samples is a matter to be investigated by Monte Carlo simulation studies.

Regarding the development of hypothesis tests for LAV estimation, it is well known that the common approaches to devising tests such as the Neyman–Pearson (NP) lemma and the likelihood ratio statistic are *optimization* problems (e.g. see Bickel and Doksum [16] for a description of the NP lemma and the likelihood ratio statistic (Chapter 6)). For example, suppose that $X = (X_1, X_2, \ldots, X_n)$ has density $p(x, \theta)$, where $\theta = (\theta_1, \theta_2, \ldots, \theta_K)$ is a vector of unknown parameters, and it is desired to test $H_0: \theta \in \theta_0$ versus $H_a: \theta \in \theta_1$, where θ_0 and θ_1 are specified constant vectors of θ. The *likelihood ratio statistic* for this test is given by:

$$L(x) = \frac{\sup[p(x, \theta): \theta \in \theta_1]}{\sup[p(x, \theta): \theta \in \theta_0]}. \tag{26}$$

Tests that reject H_0 for large values of $L(x)$ are called *likelihood ratio tests*. It may well be that the tests developed in this section are asymptotically equivalent to Neyman–Pearson or likelihood ratio tests. The demonstration that they are or are not is, in fact, an interesting problem for further research. By the use of optimization techniques, it may be possible to use the asymptotic LAV results due to Bassett and Koenker to produce likelihood ratio tests (and interval estimators) where $x = \hat{b}$ and $\theta = b$ in (26).

4. Regression example of LAV estimation

To illustrate the application of LAV estimation to fitting a regression equation to a data set, the following experiment was conducted:

(1) A random sample of size $T = 100$ Cauchy disturbances was generated using the fact that $Y = F(X)$ is uniformly distributed on the interval from 0 to 1, where $F(X)$ is the cumulative distribution function for the Cauchy pdf. A random number generator was used to produce uniform (0, 1) variates and the corresponding Cauchy variates were determined by using $x = F^{-1}(y)$. In the Cauchy pdf the scale parameter c was set to one in the simulation experiment.

(2) The bivariate regression model, $E(Y) = b_1 + b_2 X_1 + b_3 X_2$ was chosen for the experiment, with $b_1 = 0$, $b_2 = 1$, and $b_3 = 1$. The values of X_1 and X_2 were selected as random and independent standard normal variates.

(3) Using the equation $Y = b_1 + b_2 X_1 + b_3 X_2 + e$, where the values of e were determined in step (1), $T = 100$ responses (Y) were generated. The values of X_1, X_2, and Y are shown in appendix B.

(4) Both least squares and LAV estimation techniques were applied to the resulting data.

It should be pointed out that this example is an extreme case since least squares is known to perform poorly with Cauchy distributed data. The example does, however, serve to illustrate the advantages of using the LAV estimator in data sets with "wild points" or extreme outliers.

Table 2
Statistics for X_1, X_2, and Y data.

Variable	Sample size	Mean	Standard deviation	Range
X_1	100	−0.014	0.924	4.752
X_2	100	−0.400	1.006	4.751
Y	100	1.554	11.304	104.776

Table 3
Point estimation results for estimating b_1, b_2 and b_3.

Param- eter	True value	Least squares			LAV		
		Point estimate	Stand. error	z-stat. value	Point estimate	Stand. error	z-stat. value
b_2	1.0	1.95	1.23	1.59	1.10	0.24	4.68
b_3	1.0	0.26	1.13	0.23	0.96	0.22	4.45
b_1	0.0	1.59	1.13	1.41	0.02	0.22	0.09

Table 2 summarizes the statistics for the X_1, X_2, and Y data. Of particular interest is the range of the values of Y. The maximum value was 97.468 and the minimum value was -7.308, giving a range of 104.776 in the sample size $T = 100$. The large range may be attributed to the Cauchy disturbances. Random variates sampled from the Cauchy distribution may vary considerably, with several very large or very small values occurring in a typical sample of size $T = 100$.

In table 3 the point estimation results are summarized for b_1, b_2, and b_3. Results for both least squares estimates and the LAV estimates are compared in this table. For least squares, the standard errors of the coefficients are calculated in the usual manner:

$$\text{standard error } (\hat{b}_{i,\text{OLS}}) = \hat{\sigma}(X'X)_{ii}^{-1/2}, \quad i = 1, 2, 3, \tag{27}$$

where $\hat{\sigma}$ denotes the square root of the residual mean square and $(X'X)_{ii}^{-1/2}$ denotes the square root of the ith diagonal element of the $(X'X)^{-1}$ matrix. The Z-stat values are calculated using

$$Z\text{-stat}(\hat{b}_{i,\text{OLS}}) = \frac{\hat{b}_{i,\text{OLS}}}{\text{stand. error}(\hat{b}_{i,\text{OLS}})}, \quad i = 1, 2, 3. \tag{28}$$

These values can be used to test the individual hypotheses that each coefficient is equal to zero:

$$H_0: b_i = 0, \quad i = 1, 2, 3. \tag{29}$$

For the LAV fitted model, the standard errors can be calculated using

$$\text{stand. error}(\hat{b}_{i,\text{LAV}}) = \hat{\lambda}(X'X)_{ii}^{-1/2}, \quad i = 1, 2, 3, \tag{30}$$

with $\hat{\lambda}$ calculated as shown in eq. (12).

The Z-stats are used to test the hypotheses shown in (29) and are calculated using

$$Z\text{-stat}(\hat{b}_{i,\text{LAV}}) = \frac{\hat{b}_{i,\text{LAV}}}{\text{stand. error}(\hat{b}_{i,\text{LAV}})}, \quad i = 1, 2, 3. \tag{31}$$

Note that rather remarkable differences between the point estimates of b_2 and b_3 for least squares and LAV. The LAV estimates are much closer to the true values of b_2 and b_3, with resulting significant z-test statistic values. Also note the difference in estimating the intercept, b_1, by least squares and LAV. The LAV estimate is very close to zero and the corresponding z-test statistic value

Table 4
Least squares summary for testing $H_0: b_2 = b_3 = 0$ analysis of variance table.

Source	df	ss	ms	F	
Regression	2	327.9	163.9	1.29	$R^2 = 0.006$ [a]
Residual	97	12 322.6	127.0		$F(0.90; 2, 97) = 2.36$
Total	99	12 650.2			

[a] Adjusted for degrees of freedom.

strongly suggests that we should not reject the null hypothesis, $H_0: b_1 = 0$.

Of course, one sample does not guarantee the supremacy of LAV to least squares in a case such as this one. However, asymptotic theory now ensures that in the cases where the median is a more efficient estimator of location than the mean in the error distribution, then LAV estimators will be superior to the least squares estimators. This is the case with Cauchy disturbances. What this simulation example does illustrate is the dramatic differences that can be realized by using LAV estimators rather than least squares estimators in regression model-fitting to Cauchy or other error distributions for which the median is a superior estimator of location when compared to the mean. As mentioned in the introduction, numerous simulation experiments have demonstrated this fact over recent years.

Inference procedures developed in section 3 will now be applied to this example data set.

(1) Test of the null hypothesis, $H_0: b_2 = b_3 = 0$. Table 4 summarizes the least squares results for this hypothesis test. Not surprisingly, the least squares F-test fails to reject the null hypothesis, $H_0: b_2 = b_3 = 0$. Table 5 summarizes the LAV estimation results for this hypothesis test.

In table 5, $\hat{b}*' = (\hat{b}_2, \hat{b}_3)$, $\mathbf{0}' = (0, 0)$, and $X^{*'}X^*$ is the 2×2 matrix of sums of squares and cross products for X_1 and X_2. The LAV fit, tested by the χ^2 procedure developed in section 3, strongly suggests that it is a good representation of the population regression plane.

(2) Confidence interval estimation. To illustrate the application of the

Table 5
LAV summary for testing $H_0: b_2 = b_3 = 0$.

χ^2 Statistic value	df	p-Value
$\dfrac{(\hat{b}-0)'X^{*'}X^*(\hat{b}-0)}{\hat{\lambda}^2} = 41.39$	2	$P(\chi^2 > 41.39) \ll 0.001$

confidence interval formulae to the data set given in this section, intervals are constructed for b_2 and b_3, $r'b$, and Y_0. For these data, $\hat{\lambda} = 2.163$ and $(X'X)^{-1}$ is given by:

$$\begin{bmatrix} 0.0100187 & 0.0001659 & 0.0004058 \\ 0.0001659 & 0.0118317 & 0.0001146 \\ 0.0004058 & 0.0001146 & 0.0099894 \end{bmatrix} \tag{32}$$

(i) 95% confidence intervals on b_2 and b_3. From table 3 and formula (15):

b_2: $\quad 1.102 \pm (1.96)(2.163)(0.0118317)^{1/2}$
$\quad\quad 0.641$ to 1.563;

b_3: $\quad 0.963 \pm (1.96)(2.163)(0.0099894)^{1/2}$
$\quad\quad 0.539$ to 1.387.

(ii) 95% confidence interval on $r'b$, where $r' = (0, 1, -1)$ and $b' = (b_1, b_2, b_3)$. From table 3, formula (14), and (32):

$r'\hat{b} = 0.139; \quad r'(X'X)^{-1}r = 0.0216$

$r'b$: $\quad 0.139 \pm (1.96)(2.163)(0.0216)^{1/2}$
$\quad\quad -0.484$ to 0.762.

Notice that this interval is on the difference between b_2 and b_3. Accordingly, the null hypothesis, $H_0: b_2 = b_3$ could not be rejected based on this confidence interval.

(iii) 95% confidence interval on Y_0, given $X_0' = (0, 1, 1)$. From the formula given in (18) and (19), and from (32):

$\hat{Y}_0 = 2.065; \quad s^2(\hat{Y}_0) = 0.0220$

Y_0: $\quad 2.065 \pm (1.96)(2.163)(0.0220)^{1/2}$
$\quad\quad 1.436$ to 2.694.

5. Summary

With the recently developed asymptotic normal theory for LAV estimators in the linear regression model and the availability of a number of computationally efficient algorithms for calculating LAV estimates, it is reasonable to expect that LAV estimation procedures will be more commonly applied to data sets than before in regression experiments. For the error distributions given in table 1 for which the median is superior to the mean as an estimator of location, LAV estimation is certainly preferred to least squares and should be used in these cases.

In section 3, inference techniques were developed for LAV estimation in the regression model based on the asymptotic normal theory. These techniques are relatively easy to use, as demonstrated in section 4.

It seems evident that future research on LAV estimation in the regression model will be focused on three topics: (i) the development of alternative inference techniques and comparisons to those suggested in this paper; (ii) investigations concerning the rate of convergence in distribution of the LAV estimators to the asymptotic normal distribution; and (iii) tabulation of significance values for test statistics used in hypothesis testing about the regression coefficients in small samples by Monte Carlo simulation methods or by curve-fitting approximations to the actual small-sample distributions of the pertinent test statistics.

Appendix A: Bassett and Koenker theorem

Consider the model $Y = Xb + e$, where the residuals are independently and identically distributed with cumulative distribution function F. Let $\hat{b}$ represent the solution to:

$$\min |Y - X\hat{b}|.$$

Let $[\hat{b}_T]$ denote a sequence of unique solutions to the above problem with the matrix of explanatory variables denoted by X_T. Furthermore, assume that:

(i) F is continuous and has continuous and positive density f at the median, and

(ii) $\lim(T^{-1}X'_T X_T) = Q$, where Q is a positive definite matrix. Then, $T^{1/2}(\hat{b}_T - b)$ converges in distribution to a K-dimensional normal random vector with mean 0 and variance matrix $\lambda^2 Q^{-1}$, where λ^2/T is the asymptotic variance of the sample median from random samples for the distribution F.

Appendix B: Data used in LAV vs. OLS Example

X_1	X_2	Y	X_1	X_2	Y
−0.148	1.333	1.601	0.847	0.572	0.333
0.380	−0.284	97.468	0.077	−1.356	−2.268
−1.002	1.017	−0.089	−1.165	0.759	6.728
−0.930	−0.978	−1.747	−0.034	0.292	−0.256
0.689	0.421	1.649	0.898	−0.175	9.432

X_1	X_2	Y	X_1	X_2	Y
−0.692	−1.140	−1.797	−1.056	−0.353	−2.992
−0.877	−1.571	−3.938	0.346	−0.969	−1.772
−0.165	−0.383	−0.320	1.790	1.772	2.277
−1.103	−0.132	−0.459	0.226	−1.143	−0.827
−0.078	2.526	−1.635	0.333	0.431	−6.472
1.194	−1.641	−0.349	0.954	0.436	1.991
0.284	−0.757	−2.169	0.404	0.506	1.154
0.619	−0.707	−1.262	2.970	0.074	2.650
−0.956	1.373	−0.191	1.101	−0.354	3.809
0.487	0.311	0.221	0.120	0.381	0.817
0.056	1.342	0.103	−1.225	1.128	0.376
−0.303	−0.677	−0.439	1.446	−0.121	2.883
−1.286	−0.698	−2.645	−1.782	−0.311	−2.911
−0.704	−1.439	4.613	0.312	0.314	−0.532
−0.288	−0.260	0.954	0.514	−1.190	−1.320
1.370	1.354	3.928	−0.807	1.107	−0.490
−1.305	0.450	9.289	1.816	−0.684	2.703
−0.221	−1.414	−2.385	1.706	−1.132	13.246
−0.403	1.353	−5.419	0.514	0.507	0.202
−1.046	−0.046	−3.592	−0.162	−0.095	−1.727
0.375	−1.505	−0.611	−0.336	0.103	−0.581
0.075	−1.147	−2.594	−0.279	0.058	0.466
0.183	−2.072	−1.772	−0.681	0.093	−0.516
−0.673	−1.193	−0.908	−0.912	0.329	−1.031
2.151	−1.350	0.188	−1.187	0.012	4.026
−0.044	0.130	1.184	−1.694	−0.566	−2.970
−0.103	−0.748	−1.224	−1.765	−0.059	−2.834
−1.462	0.223	−2.284	1.118	0.866	−3.180
0.496	−0.315	0.355	−0.496	−2.207	−4.185
0.434	−0.538	−0.033	0.614	−0.466	−1.774
0.258	1.719	1.318	−1.368	0.274	−1.223
−1.029	1.504	8.421	1.982	0.712	2.891
0.132	0.812	0.392	−0.695	1.161	3.026
−0.744	−0.167	−1.297	0.826	−0.708	49.092
−0.174	2.138	−0.227	−0.276	−1.485	−1.635
−0.416	−0.212	0.852	−1.424	−1.429	−7.308
1.220	0.813	−2.564	0.634	2.259	5.684
−0.060	0.350	0.389	−0.588	0.512	0.278
−0.182	−0.199	0.065	−0.384	0.051	−4.245
−0.608	−2.225	−0.245	0.783	−0.880	0.099
−0.880	1.468	0.785	0.565	−0.045	2.108
−0.759	0.087	−4.257	0.138	0.831	1.511
0.021	0.944	2.425	0.523	−0.025	1.915
0.390	0.005	1.140	−0.309	0.115	−0.252
0.615	−1.606	−2.315	0.917	−0.281	0.384

References

[1] N. Abdelmalek, "An Efficient Method for the Discrete Linear L_1 Approximation Problem", *Mathematics of Computation* 29 (1975) 844–850.

[2] D.F. Andrews, "A Robust Method for Multiple Linear Regression", *Technometrics* 16 (1974) 523–531.

[3] R.D. Armstrong and E. Frome, "A Special Purpose Linear Programming Algorithm for Obtaining Least Absolute Value Estimators in a Linear Model with Dummy Variables", *Communications in Statistics – Simulation and Computation* B6 (1977) 383–398.

[4] R.D. Armstrong and E. Frome, "Least-Absolute Value Estimators for One-Way and Two-Way Tables", *Naval Research Quarterly* 26 (1979) 79–96.

[5] R.D. Armstrong, E. Frome and D. Kung, "A Revised Simplex Algorithm for the Absolute Deviation Curve-Fitting Problem", *Communications in Statistics – Simulation and Computation* B8 (1979) 175–190.

[6] R.D. Armstrong and J. Godfrey, "Two Linear Programming Algorithms for the Linear Discrete L_1 Norm Problem", *Mathematics of Computation* 33 (1979) 289–300.

[7] R.D. Armstrong and J. Hultz, "An Algorithm for a Restricted Discrete Approximation Problem in the L_1 Norm, *SIAM Journal of Numerical Analysis* 14 (1977) 555–565.

[8] R.D. Armstrong and M.T. Kung, "An Algorithm to Select the Best Subset for a Least Absolute Value Problem,", *TIMS Studies in the Management Sciences*, this issue.

[9] V.G. Ashar and T.D. Wallace, "A Sampling Study of Minimum Absolute Deviations Estimators", *Operations Research* 2 (1963) 747–758.

[10] I. Barrodale and F. Roberts, "An Efficient Algorithm for Discrete L_1 Linear Approximation with Linear Constraints", *SIAM Journal of Numerical Analysis* 15 (1978) 603–611.

[11] I. Barrodale and F. Roberts, "An Improved Algorithm for Discrete L_1 Linear Approximation", *SIAM Journal of Numerical Analysis* 10 (1973) 839–848.

[12] I. Barrodale and F. Roberts, "Solution of an Over-Determined System of Equations in the L_1 Norm", *Communications of the Association of Computer Machinery Journal* 17 (1974) 319–320.

[13] G. Bassett, "Some Properties of the Least Absolute Error Estimator", Unpublished Ph.D. dissertation, Department of Economics, University of Michigan (1973).

[14] G. Bassett and R. Koenker, "Asymptotic Theory of Least Absolute Error Regressions", *Journal of the American Statistical Association* 73 (1978) 618–622.

[15] P.J. Bickel, "One-Step Huber Estimates in the Linear Model", *Journal of the American Statistical Association* 70 (1975) 428–434.

[16] P.J. Bickel and K.A. Doksum, *Mathematical Statistics: Basic Ideas and Selected Topics* (Holden-Day, Inc., San Francisco, 1977).

[17] R.C. Blattberg and T. Sargent, "Regression with Non-Gaussian Stable Disturbances: Some Sampling Results", *Econometrica* 39 (1971) 501–510.

[18] H.D. Brecht, "Regression Methodology with Gross Observation Errors in the Explanatory Variables", *Decision Sciences* 7 (1976) 57–65.

[19] A. Charnes, W.W. Cooper and R.O. Ferguson, "Optimal Estimation of Executive Compensation by Linear Programming", *Management Science* 1 (1955) 138–151.

[20] K. Cogger, "Time Series Analysis and Forecasting with an Absolute Error Criterion", in: S. Makridakis and S.C. Wheelwright, eds., *Forecasting* (North-Holland Publishing Company, Amsterdam, 1979) 189–201.

[21] K. Cogger, "Statistical Inference in Minimum Absolute Deviation Regression", mimeographed (1979).

[22] D.R. Cox and D.V. Hinkley, *Theoretical Statistics* (Chapman and Hall, London, 1974).

[23] H. Cramer, *Mathematical Methods of Statistics* (Princeton University Press, Princeton, 1946).

[24] R. El-Attar, M. Vidyasager and S. Dutta, "An Algorithm for L_1-Norm Minimization with Application to Nonlinear L_1-Approximation", *SIAM Journal of Numerical Analysis* 16 (1979) 76–86.

[25] E.F. Fama, "The Behavior of Stock-Market Prices", *Journal of Business* 38 (1965) 34–105.
[26] J. Gentle, W. Kennedy and V. Sposito, "On Least Absolute Values Estimation", *Communications in Statistics – Theory and Method* A6 (1977) 839–845.
[27] F. Glahe and J. Hunt, "The Small Sample Properties of Simultaneous Equation Least Absolute Estimators Vis-à-Vis Least Squares Estimators", *Econometrica* 38 (1970) 742–753.
[28] H.L. Harter, "Nonuniqueness of Least Absolute Values Regression", *Communications in Statistics – Theory and Methods* A6 (1977) 829–838.
[29] H.L. Harter, "The Method of Least Squares and Some Alternatives", (in five parts), *International Statistical Review* 42 (1974) 147–174, 235–264, 43 (1975) 1–44, 125–190, 269–278.
[30] A.C. Harvey, "A Comparison of Preliminary Estimates for Robust Regression", *Journal of the American Statistical Association* 72 (1977) 910–913.
[31] R.W. Hill and P.W. Holland, "Two Robust Alternatives to Least Squares Regression", *Journal of the American Statistical Association* 72 (1977) 828–833.
[32] R.V. Hogg, "An Introduction to Robust Procedures", *Communications in Statistics – Theory and Method* A6 (1977) 789–794.
[33] P.J. Huber, "Robust Estimation of a Location Parameter", *Annals of Mathematical Statistics* 35 (1964) 73–101.
[34] P.J. Huber, "Robust Regression, Asymptotics, Conjectures, and Monte Carlo", *Annals of Statistics* 1 (1973) 799–821.
[35] J. Hunt, J. Dowling and F. Glahe, "L_1 Estimation in Small Samples with Laplace Error Distributions", *Decision Sciences* 5 (1974) 22–29.
[36] E.A. Kiountouzis, "Linear Programming Techniques in Regression Analysis", *Applied Statistics* 22 (1973) 69–73.
[37] D. Klingman and J. Mote, "Generalized Network Approaches for Solving Least Absolute Value and Tchebycheff Regression Problems", *TIMS Studies in the Management Sciences*, this issue.
[38] R. Koenker and G. Bassett, "Regression Quantiles", *Econometrica* 46 (1978) 33–50.
[39] B. Mandelbrot, "The Variation of Certain Speculative Prices", *Journal of Business* 1 (1963) 394–419.
[40] G. McCormick and V. Sposito, "Using the L_2 Estimator in L_1-Estimation", *SIAM Journal of Numerical Analysis* 13 (1976) 334–343.
[41] J.R. Meyer and R.R. Glauber, "Investment Decisions, Economic Forecasting, and Public Policy", Division of Research Memoir, Graduate School of Business Administration, Harvard University (1964).
[42] S.C. Narula, "Optimization Techniques in Linear Regression: A Review", *TIMS Studies in the Management Sciences*, this issue.
[43] R. Pfaffenberger and J. Dinkel, "Absolute Deviations Curve Fitting: An Alternative to Least Squares", in: H.A. David, ed., *Contributions to Survey Sampling and Applied Statistics* (Academic Press, Inc., New York, 1978) 279–294.
[44] J.O. Ramsay, "A Comparative Study of Several Robust Estimates of Slope, Intercept and Scale in Linear Regression", *Journal of the American Statistical Association* 72 (1977) 608–615.
[45] J. Rice and J. White, "Norms for Smoothing and Estimation", *Society of Industrial and Applied Mathematics Review* 6 (1964) 243–256.
[46] G. Roodman, "A Procedure for Optimal Stepwise MSAE Regression Analysis", *Operations Research* 22 (1974) 393–399.
[47] B. Rosenberg and D. Carlson, "The Sampling Distribution of Least Absolute Residuals Regression Estimates", Working Paper IP-164, Institute of Business and Economic Research, University of California, Berkeley (1971).
[48] B. Rosenberg and D. Carlson, "A Simple Approximation of the Sampling Distribution of Least Absolute Residuals Regression Estimates", *Communications in Statistics – Simulation and Computation* B6 (1977) 421–438.

[49] S. Searle, *Linear Models* (John Wiley and Sons, Inc., New York, 1971).
[50] W. Sharpe, "Mean-Absolute-Deviation Characteristic Lines for Securities and Portfolios", *Management Science* 18 (1971) B1–B13.
[51] R. Sielken and H.O. Hartley, "Two Linear Programming Algorithms for Unbiased Estimation of Linear Models", *Journal of American Statistical Association* 68 (1973) 639–641.
[52] V. Sposito, M. Hand and G. McCormick, "Using an Approximate L_1 Estimator", *Communications in Statistics – Simulation and Computation* B6 (1977) 263–268.
[53] L.D. Taylor, "Estimation by Minimizing the Sum of Absolute Errors", in: Paul Zarembka, ed., Frontiers in Econometrics (New York, 1974) 169–190.
[54] M. Theil, *Principles of Econometrics* (John Wiley and Sons, Inc., New York, 1971).
[55] H. Wagner, "Linear Programming Techniques for Regression Analysis", *Journal of the American Statistical Association* 54 (1959) 206–212.
[56] H.G. Wilson, "Least Squares Versus Minimum Absolute Deviations Estimation in Linear Models", *Decision Sciences* 9 (1978) 322–335.
[57] H.G. Wilson, "Upgrading Transport Costing Methodology", *Transportation Journal* (1979) 49–55.

GENERALIZED NETWORK APPROACHES FOR SOLVING LEAST ABSOLUTE VALUE AND TCHEBYCHEFF REGRESSION PROBLEMS

Darwin KLINGMAN

The University of Texas at Austin

and

John MOTE

Analysis, Research, and Computation, Inc.

This paper provides formulations of the simple linear least absolute value (L_1) and Tchebycheff (L_∞) regression problems as highly structured generalized network models. Computational comparisons are given between a general purpose network optimization code and two specialized optimization codes that were designed to exploit the unique network structure of the L_1 and L_∞ problems. These computational results indicate the overwhelming superiority of the specialized algorithmic implementations.

An underlying relationship between the L_1 and L_∞ network formulations is briefly explored. It is shown that the L_∞ problem can be viewed as a simple relaxation of the L_1 problem.

1. Introduction

The least absolute value (L_1) and Tchebycheff (L_∞) simple linear regression analysis problems have been modeled and solved quite effectively as straightforward linear programs. Charnes, Cooper and Ferguson [8] were the first to show that a primal formulation of the L_1 problem could be solved with the primal simplex algorithm. Later, Wagner [16] suggested that it would be more efficient to solve a dual formulation of the problem since the size of the working basis would be greatly reduced. This conjecture was widely accepted until Barrodale and Roberts [6] showed that a specialized primal simplex algorithm could be used to capitalize on the underlying structure of the primal problem. McCormick and Sposito [11] proposed an improved Barrodale and Roberts algorithm which uses the least squares regression problem to speed convergence. Abdelmalek [1,2] suggested that a dual simplex algorithm could be used to solve a dual formulation of the L_1 problem. The equivalence of the

---

Received July 31, 1980; revised April 6, 1981.

Barrodale and Roberts approach and the Abdelmalek approach was demonstrated by Armstrong and Godfrey [3]. A specialized version of the Barrodale and Roberts procedure was developed by Armstrong and Kung [5] to solve the unweighted simple linear L_1 regression problem. Their approach is currently regarded as the most efficient one for this class of problems.

Like the L_1 problem, most of the suggested solution procedures for the L_∞ regression problem have been based on the efficient simplex algorithm [4,7,14,15].

We present alternative formulations of these two problems as highly structured generalized networks. While most of the previous researchers have solved a primal problem with a primal algorithm, or equivalently a dual problem with a dual algorithm, our approach follows Wagner's suggestion [16] and applies a primal algorithm to a dual formulation of the problem. Not only is this approach unique in that the resulting problem is a generalized network, but the computational results presented in section 5 suggest that this approach is quite efficient.

Although not presented here, generalized network formulations can also be developed for a variety of extensions to the weighted L_1 regression problem. Among these are

- simple upper and/or lower bounds on the intercept and/or slope,
- general linear constraints on the slope and intercept,
- multiple intercepts,
- multiple slopes, and
- unbalanced weights.

2. Weighted L_1 regression analysis

The weighted L_1 simple regression analysis problem is examined in this section. Given a set of paired observations, $\{(x_k, y_k)\}$, and a corresponding set of n positive weights, $\{w_k\}$, the weighted L_1 regression problem is to select a linear regression equation that minimizes the weighted sum of the absolute deviations. It is well known that this problem can be formulated as a linear programming problem. One such formulation is:

$$\text{minimize} \sum_{k=1}^{n} w_k D_k$$

subject to

$$-A - x_k B + D_k \geq -y_k, \quad \text{for } k = 1, 2, \ldots, n,$$
$$A + x_k B + D_k \geq y_k, \quad \text{for } k = 1, 2, \ldots, n.$$

This formulation, hereafter referred to as the primal problem, involves $2n$ constraints and $2+n$ structural variables (the intercept, slope, and n deviation variables). Clearly, there are more attractive formulations of the primal, but this one was selected because it yields a dual problem that will be used in section 4 to examine the relationship between the L_1 and L_∞ regression problems.

The straightforward dual transformation of the L_1 problem is:

$$\text{maximize} - \sum_{k=1}^{n} y_k \alpha_k + \sum_{k=1}^{n} y_k \beta_k \tag{1}$$

subject to

$$-\sum_{k=1}^{n} \alpha_k + \sum_{k=1}^{n} \beta_k = 0, \tag{2}$$

$$-\sum_{k=1}^{n} x_k \alpha_k + \sum_{k=1}^{n} x_k \beta_k = 0, \tag{3}$$

$$\alpha_k + \beta_k = W_k, \quad \text{for } k=1,2,\ldots,n, \tag{4}$$

$$\alpha_k \geq 0, \quad \beta_k \geq 0. \tag{5}$$

This problem can be further transformed by eliminating the β_k variables using (4), scaling (1) and (3) by $1/2$, and scaling (2) by $-1/2$. The resulting problem is:

$$\text{minimize} \sum_{k=1}^{n} y_k \alpha_k - \frac{\sum_{k=1}^{n} w_k y_k}{2} \tag{6}$$

subject to

$$-\sum_{k=1}^{n} \alpha_k = \frac{-\sum_{k=1}^{n} w_k}{2}, \tag{7}$$

$$\sum_{k=1}^{n} x_k \alpha_k = \frac{\sum_{k=1}^{n} w_k x_k}{2}, \tag{8}$$

$$0 \leq \alpha_k \leq w_k. \tag{9}$$

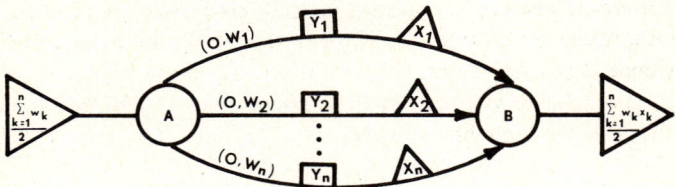

Fig. 1. L_1 Generalized network.

This formulation of the dual problem is a capacitated generalized network problem with two nodes and n arcs. Fig. 1* illustrates the special network topology of this problem.

All arcs for this network problem are directed from node A to node B. The kth arc has a lower bound of zero, upper bound of w_k, multiplier of x_k, and objective function coefficient of y_k.

For reasons that will become apparent, the node associated with constraint (7) will be labeled A and the node associated with constraint (8) will be labeled B. Node A has a net supply of $\sum_{k=1}^{n} w_k/2$ units and node B has a net demand of $\sum_{k=1}^{n} w_k x_k/2$ units. It is interesting to note that the net gain through the network is equal to the weighted mean of the observations of the independent variable:

$$\sum_{k=1}^{n} w_k x_k / \sum_{k=1}^{n} w_k.$$

Any basic solution to this network problem consists of two basic arcs, a set of nonbasic arcs with capacity flow levels. If U is an index set of arcs that are nonbasic at their upper bounds, then

$$b_A = \frac{-\sum_{k=1}^{n} w_k}{2} + \sum_{k \in U} w_k$$

is the net demand (negative supply) at node A after the non basic arcs have been taken into consideration. Likewise,

$$b_B = \frac{\sum_{k=1}^{n} w_k x_k}{2} - \sum_{k \in U} w_k x_k$$

* Ed. note: For a brief review of generalized network notation, applications and solution techniques, see F. Glover and D. Klingman, "Network Application in Industry and Government", *AIIE Transactions* 9 (1977) 363–376.

is the net demand at node B. The corresponding flows on the basic arcs can be easily determined using these net demands. If k_1 and k_2 are the indices of the two basic arcs, then their flows are given by:

$$\alpha_{k_1} = \frac{b_B + b_A x_{k_2}}{x_{k_1} - x_{k_2}}; \quad \alpha_{k_2} = \frac{b_B + b_A x_{k_1}}{x_{k_2} - x_{k_1}}. \tag{10}$$

Here it must be assumed that $x_{k_1} \neq x_{k_2}$. This is equivalent to assuming that the basis matrix has full row rank.

A basic solution is a feasible solution if the flows given by (10) satisfy the bounds (9). The selection of an initial feasible solution is greatly simplified by adding an artificial arc to the generalized network problem. It is convenient to number the artificial arc 0. Like the n real arcs, the artificial arc is directed from node A to node B. Given an initial selection of U, the multiplier for the artificial arc is defined as

$$x_0 = -b_B/b_A.$$

The artificial arc is given an infinite objective function coefficient in order to ensure that its flow is zero at optimality.

Using this artificial arc, the choice of an initial feasible basis is very simple. Let $k_1 = 0$ denote that the artificial arc is the first basic arc. For the second basic arc, any arc $k_2 \notin U$ can be selected provided that $x_{k_1} \neq x_{k_2}$. The flows on these two basic arcs are given by (10). Owing to the specific choice of x_0 that was made, the flows on the two basic arcs are simply $\alpha_{k_1} = -b_A$ and $\alpha_{k_2} = 0$.

The specialized revised simplex algorithm for solving generalized network problems uses node potentials in order to simplify the selection of entering arcs. These node potentials are defined as:

$$\pi_A = \frac{x_{k_2} y_{k_1} - x_{k_1} y_{k_2}}{x_{k_1} - x_{k_2}}; \quad \pi_B = \frac{y_{k_1} - y_{k_2}}{x_{k_1} - x_{k_2}}.$$

Using these node potentials, a given basic feasible solution is an optimal solution if

$$\begin{aligned} -\pi_A + x_k \pi_B &\leq y_k, \quad \text{if } \alpha_k = 0, \\ -\pi_A + x_k \pi_B &\geq y_k, \quad \text{if } \alpha_k = w_k. \end{aligned} \tag{11}$$

If for some arc, k_e, (11) is not satisfied, then the arc is pivot eligible. If such an entering arc is found, then the leaving arc can be determined by the standard minimum ratio test. This pivoting process continues until (11) is satisfied for all nonbasic arcs.

The status of the α_k variables in the optimal solution to the generalized network problem provides some useful insight into the relationship between the original problem data, $\{(x_k, y_k)\}$, and the optimal regression equation. From duality theory it is known that the basic variables in the optimal solution to the dual problem correspond to the binding constraints in the primal problem. This means that the optimal regression line passes directly through the two data points corresponding to the optimal basis of the generalized network problem. This result can be used to derive the formulae for the slope and intercept of the optimal regression equation. The slope of the line through points (x_{k_1}, y_{k_1}) and (x_{k_2}, y_{k_2}) is:

$$B = \frac{y_{k_1} - y_{k_2}}{x_{k_1} - x_{k_2}} = \pi_B, \tag{12}$$

and the intercept of the line is:

$$A = y_{k_1} - Bx_{k_1} = \frac{x_{k_1} y_{k_2} - x_{k_2} y_{k_1}}{x_{k_1} - x_{k_2}} = -\pi_A. \tag{13}$$

In light of this, the optimality conditions (11) can be re-stated as

$$A + Bx_k \leq y_k, \quad \text{if } \alpha_k = 0,$$
$$A + Bx_k \geq y_k, \quad \text{if } \alpha_k = w_k.$$

In other words, all observations that lie above (below) the optimal regression line have flow values that are nonbasic at zero (upper bound). This relationship between the sign of the error term and the nonbasic status of the arcs for problems (10)–(13) can be used to develop heuristics for the initial selection of the set U. This is examined further in the section on computational testing.

3. L_∞ Regression analysis

In this section we will briefly examine the unweighted L_∞ simple regression analysis problem. Given a set of n paired observations $\{(x_k, y_k)\}$, the L_∞ regression problem is to determine a linear regression equation that minimizes the largest absolute deviation. Like the L_1 problem, the L_∞ problem can be stated as a linear programming problem. The following formulation, referred to as the primal problem, has $2n$ constraints and three structural variables (the intercept, slope, and maximum deviation):

minimize D

subject to

$$-A - x_k B + D \geq -y_k, \quad \text{for } k = 1, 2, \ldots, n,$$
$$A + x_k B + D \geq y_k, \quad \text{for } k = 1, 2, \ldots, n.$$

The dual of the L_∞ problem is given as:

$$\text{maximize} \quad -\sum_{k=1}^{n} y_k \alpha_k + \sum_{k=1}^{n} y_k \beta_k \tag{14}$$

subject to

$$-\sum_{k=1}^{n} \alpha_k + \sum_{k=1}^{n} \beta_k = 0, \tag{15}$$

$$-\sum_{k=1}^{n} x_k \alpha_k + \sum_{k=1}^{n} x_k \beta_k = 0, \tag{16}$$

$$\sum_{k=1}^{n} \alpha_k + \sum_{k=1}^{n} \beta_k = 1, \tag{17}$$

$$\alpha_k \geq 0, \quad \beta_k \geq 0, \quad \text{for } k = 1, 2, \ldots, n. \tag{18}$$

The following formulation is equivalent to (14)–(18) and therefore can be used to solve the original problem.

$$\text{minimize} \quad \sum_{k=1}^{n} y_k \alpha_k + \sum_{k=1}^{n} \left(\frac{-y_k}{x_k}\right) \beta_k \tag{19}$$

subject to

$$-\sum_{k=1}^{n} \alpha_k = -\tfrac{1}{2}, \tag{20}$$

$$\sum_{k=1}^{n} x_k \alpha_k - \sum_{k=1}^{n} \beta_k = 0, \tag{21}$$

$$\sum_{k=1}^{n} \left(\frac{1}{x_k}\right) \beta_k = \frac{1}{2}, \tag{22}$$

$$\alpha_k \geq 0, \quad \beta_k \geq 0, \quad \text{for } k = 1, 2, \ldots, n. \tag{23}$$

This new formulation is obtained by replacing eq. (15) with one-half times eq. (15) less one-half times eq. (17), replacing eq. (17) with one-half times eq. (15), plus one-half times eq. (17), scaling the objective function (14) by $-1/2$, and scaling β_k variable by $1/x_k$. The scaling of each β_k implicitly assumes that each x_k is positive. Clearly, since all x_k can be translated by an arbitrary

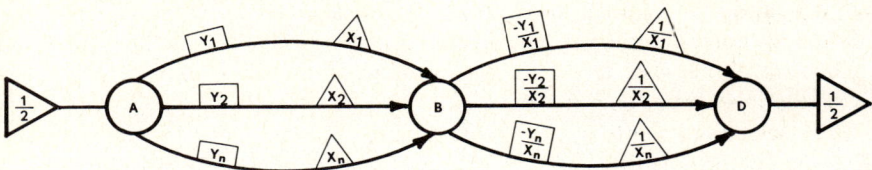

Fig. 2. L_∞ Generalized network.

constant, this positivity requirement is not critical.

This formulation of the dual problems (19)–(23) is an uncapacitated generalized network problem with three nodes and $2n$ arcs. Fig. 2 illustrates the special network topology of this problem.

The nodes corresponding to constraints (20), (21), and (22) will be labeled A, B, and D, respectively. Node A has a net supply of $1/2$ units and node D has a net demand of $1/2$ units.

The arcs corresponding to the α_k variables are directed from node A to node B. The multipliers on these arcs are x_k and the objective function coefficients are y_k. The arcs corresponding to the β_k variables are directed from node B to node D. The multipliers on these arcs are $1/x_k$ and the objective function coefficients are $-y_k/x_k$. All $2n$ arcs have lower bounds of zero.

Since problem (19)–(23) is a generalized network problem, it can be solved using a specialized revised simplex algorithm. Since the problem is uncapacitated, at most three arcs will have nonzero flows when a simplex approach is used to solve the problem. That is, only the three arcs associated with a given basis can have nonzero flows. All nonbasic arcs have zero flow.

Following through steps similiar to those in section 2, it is possible to determine simple closed-form expressions for the flows and node potentials associated with a given basis. These will not be repeated here, but may be obtained from the authors in [10].

Given a feasible basic solution, it is necessary to examine the nonbasic arcs in order to either select an entering arc or to verify optimality. Expressed in terms of the three-node potentials, the optimality conditions for the L_∞ generalized network problem are:

$$\pi_D \leq x_k \pi_B - y_k \leq \pi_A, \quad \text{for } k = 1, 2, \ldots, n.$$

If $x_k \pi_B - y_k > \pi_A$, then α_k is a candidate to enter the basis. If $x_k \pi_B - y_k < \pi_D$, then β_k is a candidate to enter the basis. The *most* pivot eligible arc can easily be identified by solving the following two subproblems.

$$\max_k z^\alpha = x_k \pi_B - y_k, \tag{24}$$

$$\min_k z^\beta = x_k \pi_B - y_k. \tag{25}$$

If $z^\alpha = \pi_A$ and $z^\beta = \pi_D$, then the current solution is optimal. If $z^\alpha - \pi_A \geq \pi_D - z^\beta$, then the α_{k_e} corresponding to the observation that solved (24) is the most pivot eligible arc. Otherwise, the β_{k_e} corresponding to the observation that solved (25) is the most pivot eligible arc.

After selecting an entering arc, the standard simplex minimum ratio test can be used to determine the leaving arc. However, due the special structure of the L_∞ generalized network problem, the minimum ratio test can be performed entirely with logical, instead of arithmetic, operations [10].

After solving the generalized network problem, it is necessary to translate the optimal solution to this problem, back into an optimal solution for the original L_∞ regression problem. By simply reversing the steps of the transformation of the dual linear programming problem, the following intercept and slope of the L_∞ regression equation are obtained:

$$A = \frac{-(\pi_A + \pi_D)}{2}; \quad B = \pi_B.$$

The maximum absolute error is given by:

$$D = \frac{\pi_A - \pi_D}{2}.$$

4. Relationship between the L_1 and L_∞ regression problems

The underlying relationship between the L_1 and L_∞ regression problems will be considered here. This relationship is best illustrated by examining the dual linear programming formulations of the problems.

The dual linear programming formulation of the weighted L_1 regression problem was developed in section 2. The unweighted, or equal weighted, L_1 regression problem is obtained by setting $w_k = 1/n$ for each observation. The unweighted L_1 regression problem is therefore equivalent to:

$$\text{maximize} - \sum_{k=1}^{n} y_k \alpha_k + \sum_{k=1}^{n} y_k \beta_k \qquad (26)$$

subject to

$$-\sum_{k=1}^{n} \alpha_k + \sum_{k=1}^{n} \beta_k = 0, \qquad (27)$$

$$-\sum_{k=1}^{n} x_k \alpha_k + \sum_{k=1}^{n} x_k \beta_k = 0, \qquad (28)$$

$$\alpha_k + \beta_k = 1/n, \qquad \text{for } k = 1, 2, \ldots, n, \tag{29}$$
$$\alpha_k \geq 0, \qquad \beta_k \geq 0, \qquad \text{for } k = 1, 2, \ldots, n. \tag{30}$$

In section 3 the dual linear programming formulation of the unweighted L_∞ regression problem was shown to be:

$$\text{maximize} - \sum_{k=1}^{n} y_k \alpha_k + \sum_{k=1}^{n} y_k \beta_k \tag{31}$$

subject to

$$-\sum_{k=1}^{n} \alpha_k + \sum_{k=1}^{n} \beta_k = 0, \tag{32}$$

$$-\sum_{k=1}^{n} x_k \alpha_k + \sum_{k=1}^{n} x_k \beta_k = 0, \tag{33}$$

$$\sum_{k=1}^{n} \alpha_k + \sum_{k=1}^{n} \beta_k = 1, \tag{34}$$

$$\alpha_k \geq 0, \qquad \beta_k \geq 0, \qquad \text{for } k = 1, 2, \ldots, n. \tag{35}$$

Clearly, the dual L_∞ problem, (31)–(35) is a relaxation of the dual L_1 problem, (26)–(30). This relaxation is obtained by simply summing the n constraints of form (29). This yields a constraint of form (34). Since the dual L_∞ problem is a relaxation of the dual L_1 problem, this suggests that the dual L_∞ problem should be easier to solve than the corresponding dual L_1 problem. The computational results presented in the next section strongly support this hypothesis.

5. Computational testing

Two computer implementations of a primal simplex algorithm were developed for solving the generalized network formulations of the L_1 and L_∞ regression problems. Computational comparisons are presented here between these codes (LONE and LINF), the L_1 simple regression code of Armstrong and Kung (SIMLP), and a commercially available large-scale generalized network code (NETG). The SIMLP code [5] employs a specialized simplex algorithm that uses a multiple pivoting technique to improve its efficiency. Armstrong and Kung have demonstrated that their code is from ten to one hundred times faster than the code of Sadovski [12], which is based on the algorithm of Edgeworth [9]. Furthermore, it has been demonstrated by Sposito [13] that the Sadovski code may fail to converge.

The computational testing reported here was carried out on two different computers, a CDC 6600 and PRIME 550. All of the test problems were generated by adding an error term to a fixed linear relationship between the independent and dependent variables. The observations of the independent variables as well as the error terms were randomly sampled from pre-specified uniform distributions. Computational testing, not presented here, indicates that none of the solution algorithms is adversely affected by either the range of the independent variable, the magnitude of the error terms, or the fixed relationship between the independent and dependent variables.

Two of the most crucial determinants of the relative efficiency of network-based simplex codes are (1) the choice of the starting basis (i.e. the rule for constructing the initial basic feasible solution) and (2) the choice of the pivot strategy (i.e. the rule for selecting entering arcs). As expected, both of these aspects have major ramifications for the solution efficiency of the weighted L_1 regression code, LONE. Two start rules were developed and tested as well as two standard pivot rules. A brief discussion of each follows.

In section 2 it was shown that an initial feasible (artificial) basis can be constructed by (1) selecting a set of arcs U to be nonbasic at their upper bounds, (2) constructing a basic artificial arc that transfers the remaining units of supply from node A to node B, and (3) selecting a real arc to complete the full row rank basis. The most important step of this start procedure is the initialization of the set U. The first start rule tested simply initializes U as the empty set. The second start rule that was tested was motivated by an observation based on duality theory. This observation is that $\hat{y}_k \geq y_k (\hat{y}_k \leq y_k)$ if the optimal network solution has $\alpha_k = 0 (\alpha_k = w_k)$. The heuristic start rule based on this observation is referred to as the advanced L_2 start. Quite simply, the heuristic is to use the weighted least squares regression equation to assign arcs to the set U. That is, U initially contains those arcs associated with positive least squares residuals. The solution times (in c.p.u. seconds on a CDC 6600) and the number of pivots taken for both start rules are presented in table 1. It is clear from the table that the advanced start is far superior to the simple empty start.

The second aspect of the network-based L_1 regression code that was studied was the choice of pivot rule. Numerous rules have been suggested in the network literature for the selection of entering arcs. Two of the most straightforward rules are to pivot on (1) the most pivot eligible arc, and (2) the first pivot eligible arc found. The philosophy underlying these rules is that the most eligible rule requires a lot of effort to identify the entering arc, but is expected to produce the fewest total pivots. On the other hand, the first eligible rule requires little effort to identify an entering arc, but is expected to require more total pivots. Both pivot strategies were tested in order to determine the relative

Table 1
Start rule.

Number of observations	Empty U		Advanced L_2	
	Solution time	Pivots	Solution time	Pivots
100	0.174	102	0.016	8
200	0.671	203	0.049	11
300	1.474	301	0.148	29
400	2.604	400	0.166	23
500	4.025	498	0.291	32

Note: LONE with most eligible pivot rule. Mean times and pivots reported for multiple problems of each size on a CDC 6600.

trade-off between the amount of work required and the number of pivots performed. Table 2 presents the solution times and pivots for these rules.

The impact of the pivot rule is clear. The first eligible rule requires from two to four times as many pivots as the most eligible rule; however, the first eligible rule is consistently faster than the most eligible.

Based on the computational testing of the start rules and the pivot rules, the best network-based weighted L_1 regression code, referred to as LONE, uses the advanced L_2 start and the first eligible pivot rule. In order to properly assess the performance of this approach, LONE was compared to non-network L_1

Table 2
Pivot strategy rule.

Number of observations	Most pivot eligible		First pivot eligible	
	Solution time	Pivots	Solution time	Pivots
50	0.016	11	0.016	28
100	0.016	8	0.016	16
150	0.032	9	0.015	31
200	0.049	11	0.022	33
250	0.114	24	0.053	60
300	0.148	29	0.063	72
350	0.179	28	0.065	61
400	0.166	23	0.088	88
450	0.300	37	0.102	107
500	0.291	32	0.108	123

Note: LONE with L_2 advanced start. Mean times and pivots reported for multiple problems of each size on a CDC 6600.

Table 3
LONE, SIMLP and LINF comparisons.

Number of observations	LONE		SIMLP		LINF	
	Solution time	Pivots	Solution time	Multiple pivots	Solution time	Pivots
50	0.016	28	0.010	3	0.004	3
100	0.016	16	0.023	4	0.008	4
150	0.015	31	0.038	6	0.010	5
200	0.022	33	0.049	4	0.015	4
250	0.053	60	0.076	4	0.026	5
300	0.063	72	0.095	5	0.022	5
350	0.065	61	0.129	5	0.035	6
400	0.088	88	0.139	4	0.050	6
450	0.102	107	0.180	5	0.046	6
500	0.108	123	0.213	6	0.043	6

Note: Mean times and pivots reported for multiple problems of each size on a CDC 6600.

code, SIMLP. As shown in table 3, LONE is roughly twice as fast as SIMLP.

Also reported in table 3 are the solution statistics for the generalized network L_∞ regression code, LINF. As hypothesized in section 4, the L_∞ problem is indeed easier to solve than the corresponding L_1 problem. This is due to the fact that the L_∞ dual linear programming problem can be interpreted as a relaxation of the L_1 dual linear programming problem.

Table 4
LONE, LINF and NETG comparisons.

Number of observations	L_1 problems		L_∞ problems	
	LONE	NETG	LINF	NETG
100	0.06	1.74	0.01	0.21
200	0.10	5.70	0.02	0.29
300	0.16	11.55	0.03	0.53
400	0.18	20.29	0.04	0.55
500	0.24	29.55	0.05	0.74
600	0.29	42.84	0.06	0.90
700	0.32	56.64	0.05	0.91
800	0.37	75.21	0.08	0.99
900	0.43	92.72	0.09	1.21
1000	0.45	113.33	0.09	1.34

Note: Mean times reported for multiple problems of each size on PRIME 550.

After developing and testing LONE and LINF on the CDC 6600, it was decided to compare them with the state-of-the-art generalized network code, NETG. This testing, which was carried out on a PRIME 550, is reported in table 4. As expected, the two specialized regression codes are substantially faster than the general purpose network code.

The FORTRAN regression codes LONE and LINF can be obtained by contacting either author.

References

[1] N.N. Abdelmalek, "An Efficient Method for the Discrete Linear L_1 Approximation Problem", *Mathematics of Computation* 29 (1975) 844–850.

[2] N.N. Abdelmalek, "On the Discrete Linear L_1 Approximation and L_1 Solutions of Overdetermined Linear Equations", *Journal of Approximation Theory* 11 (1974) 38–53.

[3] R.D. Armstrong and J.P. Godfrey, "A Comment on Two Linear Programming Algorithms for the Linear Discrete L_1 Norm Problem", *Mathematics of Computation* 33 (1979) 289–300.

[4] R.D. Armstrong and D.S. Kung, "A Dual Method for Discrete Chebychev Curve Fitting", Working Paper 78-9, Graduate School of Business, The University of Texas at Austin (February 1978).

[5] R.D. Armstrong and M.T. Kung, "Least Absolute Value Estimates for a Simple Linear Regression Problem", *Applied Statistics* 27 (1978) 363–366.

[6] I. Barrodale and F.D. Roberts, "An Improved Algorithm for Discrete Linear Approximation", *SIAM Journal of Numerical Analysis* 10 (1973) 839–848.

[7] I. Barrodale and A. Young, "Algorithms for Best L_1 and L_∞ Linear Approximation on a Discrete Set", *Numerische Mathematik* 8 (1966) 295–306.

[8] A. Charnes, W.W. Cooper and R. Ferguson, "Optimal Estimation of Executive Compensation by Linear Programming", *Management Science* 2 (1955) 138–151.

[9] F.Y. Edgeworth, "On a New Method of Reducing Observations Relating to Several Quantities", *Philosophical Magazine (5th Ser.)* 25 (1888) 184–191.

[10] D. Klingman and J. Mote, "Generalized Network Approaches for Solving Least Absolute Value and Tchebycheff Regression Problems", CCS Report 365, Center for Cybernetic Studies, The University of Texas at Austin (December 1979).

[11] G. McCormick and V. Sposito, "Using the L_2-Estimation in the L_1-Estimation", *SIAM Journal of Numerical Analysis* 13 (1976) 337–343.

[12] A. Sadovski, "L_1-Norm Fit of a Straight Line", *Applied Statistics* 23 (1974) 244–248.

[13] V. Sposito, "A Remark on Algorithm AS74. L_1-Norm Fit of a Straight Line", *Applied Statistics* 25, 96–97.

[14] E. Stiefel, "Note on Jordan Elimination, Linear Programming, and Tchebyscheff Approximation", *Numerische Mathematik* 2 (1960) 1–17.

[15] E. Stiefel, "Methods – Old and New – For Solving the Tchebycheff Approximation Problem", *SIAM Journal of Numerical Analysis* 1 (1964) 164–176.

[16] H.W. Wagner, "Linear Programming Techniques for Regression Analysis", *Journal of the American Statistical Association* 54 (1959) 206–242.

AN ALGORITHM TO SELECT THE BEST SUBSET FOR A LEAST ABSOLUTE VALUE REGRESSION PROBLEM

R.D. ARMSTRONG
University of Georgia

and

M.T. KUNG
McMaster University

This paper considers the problem of obtaining the best subset of variables under a least absolute value criterion. The model is the classic linear regression model with m explanatory variables and a dependent variable. The importance of the explanatory variables is measured by obtaining the minimum sum of absolute deviations when only k of the m explanatory variables are included in the model. An algorithm is presented to obtain the "best" subset of size k, $k = 1, \ldots, m$.

Several algorithms to solve the best subset problem are available when the criterion for evaluation is least squares. However, recently statisticians have become increasingly aware of the limitations of least squares and have popularized "robust-resistant" estimation techniques. Least absolute value is such a technique. Special purpose computer codes which utilize the simplex algorithm of linear programming are used to solve the least absolute value regression problem.

This paper incorporates two of these specialized codes within a branch-and-bound algorithm to solve the best subset problem. The advantages and disadvantages of the two codes, one primal and one dual, will be discussed. Also, a description of the branch-and-bound implementation and the results of computational testing will be given.

1. Introduction

In the experimental design models using regression analysis, it is common practice to examine different mathematical model formulations. As stated by Draper and Smith [7], there are two opposing criteria for establishing a useful and efficient regression equation. They are as follows.

(a) To make the equation useful for predictive purposes, it is advantageous to include as many independent variables as possible so that reliable fitted values can be determined.

(b) Because of the costs, in terms of money, labor, and time, involved in obtaining information about a large number of independent variables and subsequently monitoring them, it is preferable to formulate the regression

---

Received June 24, 1980; revised April 13 and July 30, 1981.

equation with as few independent variables as possible.

The compromise between these extremes leads to the notion of selecting the best subset of independent variables, or the best regression equation.

The problem of selecting variables has been well discussed in the literature in relation to the minimization of the sum of squared error criterion. The well-known procedures include various methods of selection from all possible regressions, backward elimination, forward selection and the stagewise regression procedure.

In recent years, the least absolute value criterion has been recommended in certain cases as an alternative to the least squares (see [6,15]). For investment models (see [12]) and for economic models with errors having nonfinite variance (see [8]), the least absolute value criterion provides a more robust estimator than the least squares criterion. Kennedy and Gentle [10] and Narula [14] provide reviews of solution procedures. It is the purpose of this paper to inspect algorithmic procedures to obtain the best subset of independent variables for a linear regression model under a least absolute value criterion.

2. The best subset regression problem

To estimate parameters for a multivariate linear regression model, the problem for the least absolute value criterion is stated as follows.

Given a set of n observational measurements $(y_i, x_{i1}, x_{i2}, \ldots, x_{im})$, $i = 1, 2, \ldots, n$, determine the values of the parameters, $\beta_j, j \in J$, which will

$$\text{minimize}_{\beta_j} \sum_{i=1}^{n} \left| y_i - \sum_{j \in J} x_{ij} \beta_j \right|, \tag{1}$$

where $J \supseteq \{1, 2, \ldots, m\}$ is the index set of the parameters in the model.

Charnes, Cooper and Ferguson [4] have shown that the optimal values of the parameters in (1) can be obtained via a linear programming formulation. Employing their result here, problem (1) is equivalent to:

Minimize $\sum_{i=1}^{n} (P_i + N_i)$

subject to

$$\begin{aligned} \sum_{j \in J} x_{ij} \beta_j + P_i - N_i &= y_i, \quad i = 1, 2, \ldots, n, \\ P_i \geq 0, \quad N_i \geq 0, \quad & i = 1, 2, \ldots, n, \end{aligned} \tag{2}$$

where P_i and N_i are, respectively, the positive and negative deviation associated with the ith observation.

The "best" subset of a given number of independent variables, k, where $k \leq m$, is one which yields the minimum objective value of all possible subsets of k variables from among the set of m variables under consideration. The statistical problem of choosing the most appropriate value for k in an applied model is not discussed in this paper.

There are $m!/[k!(m-k)!]$ ways of choosing k variables from a set of m variables. In theory, each combination can be enumerated and solutions obtained and analyzed to determine the best subset of k variables. However, total enumeration is generally not computationally efficient. Implicit enumeration procedures have been developed to find the best subset of k variables without examining all possible subsets. Beale, Kendall and Mann [3], LaMotte and Hocking [11], and Furnival and Wilson [9] developed procedures to achieve this purpose under the least squares criterion. Two algorithms using the least absolute value criterion have been reported. Roodman [16] gives a partial enumerative search procedure using an upper bounding simplex algorithm to solve the dual of (2) and binary decision variables to specify the subset of regressors being assigned in the regression problem at each stage. Narula and Wellington [13] use an implicit enumerative procedure that employs both a primal and a dual simplex algorithm along with a pre-optimality test which may indicate suboptimality before a simplex iteration.

The algorithm proposed in this paper uses a branch-and-bound technique to find the best subset regression for different values of k. This algorithm has features which are not present in the Roodman and Narula–Wellington algorithms. It uses a binary tree for enumerative purposes. Also, a rule for selecting a parameter to be restricted at each branch is introduced. The Roodman and Narula–Wellington algorithms do not choose the parameter to restrict based on information from the previous subproblems. We implement two linear programming algorithms to solve the least absolute value regression problems. The first is a dual approach developed by Armstrong and Kung [2], while the other is a primal method by Armstrong, Frome and Kung [1]. Both approaches utilize information obtained from the solution of previous regressions to provide an advanced starting solution for the least absolute value regression problem currently considered. The dual approach guarantees a dual feasible starting solution for each subproblem after the first subproblem which includes all the variables. This is not true of the Roodman or Narula–Wellington algorithms. Like the Narula-Wellington algorithm, however, bounding tests are also considered.

The algorithm is presented in three parts. First, the branch-and-bound framework is outlined. Next, the special characteristics of this algorithm when using a dual linear programming method are given. In the final part, the implementation of the primal simplex approach within the branch-and-bound algorithm is described.

3. The branch-and-bound framework

Enumerative algorithms are usually easier to understand when they are related to a *tree* composed of *nodes* and *branches*, rather than presented using only mathematical terms. Here, a node corresponds to a least absolute value regression problem containing a specified set of parameters with each parameter corresponding to a variable in the model. This subproblem may be stated and solved as a linear programming problem of the form given by (2). The initial subproblem contains all the regression parameters in the model. After a subproblem is solved, the associated node is either *fathomed* or two *descendants* are created from it. Fathoming occurs when it can be ascertained that no regression problems of interest exist in any descendants of the node. If two descendants are created, they differ in the states of the parameters, where a parameter is forced out of the model in one node and the same parameter is required to be included in the model in the other node. The criterion for selecting the parameter to be restricted is the following.

From a list of free parameters (that is, the parameters that are not fixed to be in or out of the model), it is advantageous to select for restriction the parameter which, when removed from the model, gives the least change in the optimal objective value. Thus, the best solutions should be examined earlier in the algorithm. Other subproblems which yield inferior solutions need not be solved. To us, an intuitively appealing rule is to select the free parameter whose removal from the model will result in the smallest objective change during the first dual simplex iteration of the subsequent problem:

$\beta_r = \{\beta_j, j \in F$, and the first iteration objective change is minimum when

$$\times \beta_j = 0\}, \tag{3}$$

where F is the index set of the free parameters. This rule is based on the supposition that the first iteration reflects the overall objective change.

Once a parameter, β_r, is chosen to be restricted, one of the two descendant nodes deletes β_r from the problem, while the other node forces β_r to be included in the problem. Once a specification is established in a parent problem, it must also be satisfied in every descendant that follows. The restriction of parameters and the creation of more branches and nodes continue until there are not free parameters.

A solution tree for a four-parameter problem is used for illustrative purposes. The complete structure of the tree is shown in fig. 1. It is assumed that the parameters to be fixed based on (3) are in the following hierarchical fashion: β_1, β_3, β_4, and β_2. The nodes in fig. 1 show the parameters included in the model. At each node, the right-hand branch indicates the deletion of a specified parameter, and this parameter remains in the model on the left branch.

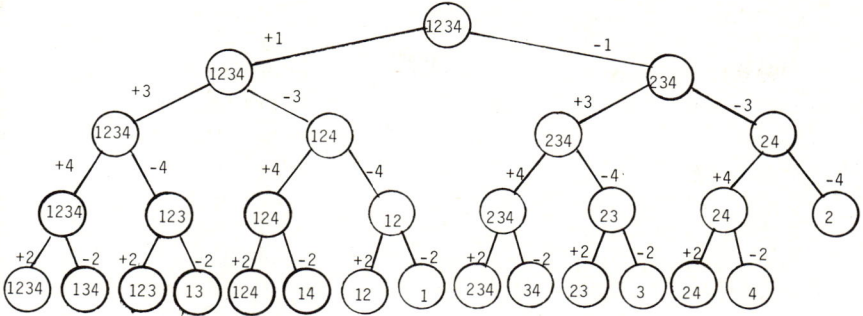

Fig. 1. The complete tree for a four-variable problem. The numbers in each node indicate the indexes of the parameters of a subproblem. The negative number on the branch indicates the parameter to be taken out of the model, while the positive number on the branch states that the parameter is required to be in the model.

As seen in fig. 1, more than one node corresponds to problems with parameters (1234), (234), (124), (123), (23), (24), and (12) in the model. Solving a problem each time the associated node is encountered would result in a series of needless repetitive calculations. It is therefore important to construct and traverse a tree in a way that requires the least amount of effort and reduces redundant computations.

The search procedure for selecting the best subset of all sizes is described here. The selection of the best subset of q, $q+1,\ldots,m$, parameters is a straightforward generalization of this procedure. The computer code developed by the authors does handle the more general case.

In the implicit enumeration procedure, generally not all the subproblems need be solved to optimality. For a current subproblem consisting of k parameters, if the objective value of the optimal solution to this subproblem is greater than the best objective value of previous subproblems of h parameters, where $h < k$, the descendant nodes from the subproblem with k parameters will not yield improved solutions. Thus, these nodes need not be examined. As will be discussed below, other methods exist for determining when a given subset will not yield a better fit than a previously examined subset with the same size.

Define z_k^u to be the upper bound on the objective value of an optimal solution with k parameters in the subproblem. Initially, every parameter is included in the model, namely, $J = \{1, 2,\ldots,m\}$ and the value of z_k^u, $k = 1.2,\ldots,m$, is set to infinity.

The tree is inspected using a last-in-first-out (LIFO) branching rule. The subproblem chosen to be solved next is called the *current candidate problem*. When two descendants are created from the parent problem, the subproblem with a parameter removed from the parent problem becomes the current candidate problem. When no further progress can be made descending a branch, the algorithm backtracks up the tree and chooses the most recently created subproblem for inspection. Because of the LIFO branching rule, the

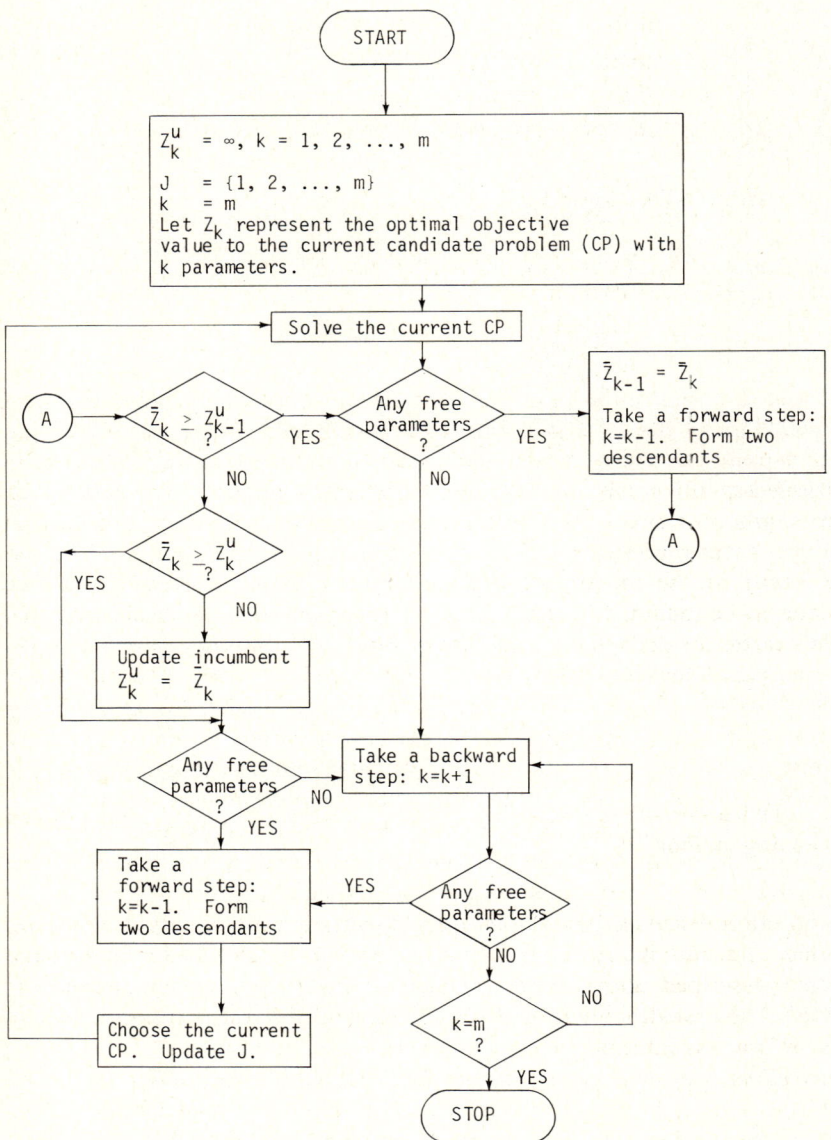

Fig. 2. Flowchart for the branch-and-bound algorithm to obtain the best subset regression.

current candidate problem is created with a minimal amount of effort and the tree can be described with parameter length arrays. Two arrays, IPAR and ISTAT, are utilized to define the current subproblem. The array IPAR is defined as follows:

$$\text{IPAR}(i) = \begin{cases} -k, & \text{if the } k\text{th parameter is forced out of the model at level } i, i = 1, 2, \ldots, m; \\ k, & \text{if the } k\text{th parameter is required to be included in the model at level } i, i = 1, 2, \ldots, m. \end{cases}$$

The other array ISTAT has the following functions:

$$\text{ISTAT}(k) = \begin{cases} 0, & \text{when the } k\text{th parameter is free}; \\ 1, & \text{when the } k\text{th parameter is forced in the model}; \\ -1, & \text{when the } k\text{th parameter is forced out of the model}. \end{cases}$$

At any stage of the algorithm the partial assignment of subsets of parameters corresponds to a list of candidate problems. Once a candidate problem (CP) is selected, it is solved via a linear programming algorithm ([1] or [2]). The current solution at any stage is used to indicate a starting procedure for the next stage. A forward step consists of selecting a parameter based on (3) and fixing it out of the model. A backward step consists of requiring a free parameter to be included in the model. The complete tree has been inspected when all entries of the ISTAT array are positive.

A conceptual flowchart of the branch-and-bound procedure is illustrated in fig. 2.

4. The application of the branch-and-bound algorithm using a dual linear programming method

This section discusses how the dual linear programming method developed by Armstrong and Kung [2] is implemented within the branch-and-bound algorithm described above. Two strategies employing the dual method are inspected. The first strategy is the use of a reoptimization start, and the second one is an implementation of a more powerful bounding test. The dual of problem (2) is:

$$\text{maximize } \sum_{i=1}^{n} \pi_i y_i$$

subject to

$$\sum_{i=1}^{n} \pi_i x_{ij} = 0, \quad j \in J,$$
$$\pi_i \leq 1, \quad i = 1, 2, \ldots, n, \qquad (4)$$
$$\pi_i \geq -1, \quad i = 1, 2, \ldots, n.$$

The two fundamental procedures to solve linear programming problems are the primal and dual algorithms (see [5]). Because of the symmetry of linear programs (the dual of the dual is the primal), it is sometimes difficult to distinguish the two algorithms. The dual algorithm applied to (4) is the same as the primal algorithm applied to (2), and vice versa. A dual method will be termed to be an algorithm that maintains a feasible solution to (4) and strives to obtain a feasible solution to (2). A primal algorithm maintains a feasible solution to (2) and strives to obtain a feasible solution to (4). A detailed description of these two algorithms are found in [1] and [2].

At each stage of the dual algorithm the value of the parameters are the simplex multipliers for (4). These multipliers can be calculated as $\beta^* = Y_B B^{-1}$, where β^* has dimension $m(J)$, the cardinality of J, Y_B is a vector of dimension $m(J)$ corresponding to the basic components of Y, where $Y = (y_1, y_2, \ldots, y_n)^T$, and B^{-1} is the current basic inverse of dimension $m(J)$ by $m(J)$.

4.1. Reoptimization start

The optimal basic solution to a parent problem is stored and used as a start for the immediate descendant which has a parameter removed from the parent problem. Thus, the dimension of a basis for this descendant is one less than a basis for the parent problem. The process of initializing the basis and obtaining the starting solution for the descendant is as follows. Let IB represent the index set of the basic variables and NB represent the index set of the nonbasic variables. Consider individually the constraints $\beta_j = 0$ for all β_j not forced in or out of the model by some previous restriction. Perform minimum ratio tests to determine, for each possible new restriction, the objective change during the first iteration and the basic π_s to leave the basis furing this interation. Choose β_r using (3). The observations associated with β_r are removed from the problem and π_s is removed from the basis creating a new basis of dimension one less than that of the immediate predecessor:

$$\pi_s = \{\pi_j \text{ removed from basis when constraint } \beta_r = 0 \text{ is added}\}. \qquad (5)$$

The new solution is dual feasible and the linear programming solution process can begin.

The variable leaving the basis will be set to the bound prescribed by the ratio test. The remaining nonbasic variables are set equal to their value in the optimal solution of the parent problem. The values for the basic variables are assigned to satisfy the constraint equations. This start enables the algorithm to determine an initial solution to the subproblem which should be a reasonable approximation of the optimal solution. The computational experience shows the efficiency of this start when compared to an initialization procedure does not utilize information obtained during the solution of subproblems considered previously.

4.2. Bounding test

In addition to the bound check described in the general branch-and-bound scheme, an additional test to be performed during each phase 2 iteration is introduced. The purpose of this bounding test is to eliminate needless calculations when the best solution of the k-parameter subproblem cannot be improved.

In the dual method, dual feasible solutions are available at each iteration. For a current subproblem consisting of k parameters, if the objective value of a basic feasible solution, Z_k, is greater-than-or-equal to Z_k^u which is the objective value of the best k-parameter regression found thus far, the current subproblem need not be solved. This procedure evaluates a node (or a subproblem) without solving it to optimality. As described in [13], this bounding test is carried out prior to a simplex pivot. Thus, the amount of computation is reduced substantially.

5. The application of the branch-and-bound algorithm using a primal simplex method

Another linear programming method to evaluate the subproblems in the branch-and-bound procedure is a primal simplex approach to problem (2). Since this is a primal method, only the final basic solution is feasible for (4). Hence, the bounding test utilizing the basic feasible solutions in the dual method cannot be applied. Only the reoptimization start employing the primal method will be described here.

The reoptimization process is very similar to the start procedure discussed in the dual approach. The main difference is the method to obtain the values of the nonbasic variables for the initial solution of the descendant problem.

If the basic variable, π_s, is selected to become nonbasic based on (5), the index s will be removed from IB and added to NB. The initial basis of full rank for the descendant problem, say $\hat{B}$, can be attained by means of the operation described above in section 4.1. However, the values of the nonbasic variables need to be computed to guarantee feasibility for (2). Their values are based on the sign of their reduced costs. The reduced costs of the nonbasic variables are given by:

$$\bar{y}_j = y_j - \beta^* X_j, \quad j \in NB, \tag{6}$$

where X_j is the jth row of X which is the observational matrix. The dimension of X is n by $m(J)$.

From the reduced costs, the initial values of the nonbasic variables for the descendant problem are:

$$\pi_j = \text{sign}(\bar{y}_j), \quad j \in NB. \tag{7}$$

As indicated in section 4.1, the reoptimization start has its advantage in finding the initial solution to the descendant based on the results of the parent problem. The efficiency of this advanced start will be indicated in the computational tests reported in the next section.

6. Computational experience

The branch-and-bound algorithms using the primal [1] and the dual [2] methods have been programmed in FORTRAN. All of the original information, including the observational matrix, is preserved by the program during execution. All the problems were solved on the CDC Dual Cyber 170/750 computer at the University of Texas at Austin Computation Center using an FTN compiler. The computer jobs were executed during periods when the machine load was approximately the same. The reported times are total execution times in CPU seconds and the number of iterations are updates of the basis.

All the observations for the test problems have been drawn from various uniform and normal distributions using a random number generator. The tolerance value for zero was set at 1.0E-8. The number of parameters may not exceed 20 and the number of observations may not exceed 300. The user can easily extend these limitations by changing the dimensions on the appropriate working arrays in the program. The matrix of observations was required to have full column rank although a less than full rank condition is easily handled with the linear programming framework.

The branch-and-bound algorithm employing the dual linear programming approach with the reoptimization start was compared to an initialization procedure which does not utilize information obtained during the solution of subproblems considered previously. Thus, the dual algorithm will require a phase 1 procedure when the reoptimization start is not used. Three sets of data consisting of 50 observations on 6, 8, and 10 parameters, respectively, were drawn from a random number generator. The computational results indicated that the reoptimization start enables problems to be solved approximately ten to twenty times faster than the version of the algorithm without the advanced start procedure. Thus, all further comparisions are made with algorithms that include the advanced start.

The second phase of testing evaluated the implementation of the bounding test within the dual approach. The testing involved three codes for comparison purposes. The first code is the primal version of the branch-and-bound algorithm with the feature of reoptimization start. The second code, TDUAL, includes the strategies of the reoptimization start and the bounding test in the dual version of the best subset algorithm. The third is a version of TDUAL without the option of the bounding test. Several different sizes of observations on 6, 8, and 10 parameters, respectively, were randomly drawn. The computational time and iteration count uniformly indicated that TDUAL is five to ten times faster than the other two versions.

The final phase of computational testing involved comparing TDUAL with a code SUBSET written by Narula and Wellington and based on the algorithm of [13]. SUBSET and TDUAL do differ slightly in the available options. These differences are the following.

(1) Minimum sum of weighted absolute error and minimum sum of relative error are available as alternative criterion in SUBSET. This is not available directly from TDUAL but can be obtained by scaling the x and y values.

(2) TDUAL allows any variable to be fixed in every model and SUBSET allows an intercept parameter to be included in every model. TDUAL requires a column of 1's be entered for the intercept term and SUBSET does not.

(3) SUBSET does allow the upper bound on the number of parameters considered to be different from m and TDUAL does not.

TDUAL and SUBSET are written completely in standard FORTRAN.

A summary of the computational testing on TDUAL and SUBSET is given in tables 1 and 2. All the solutions obtained by the two codes matched out to seven significant places. No attempt was made to evaluate the stability of the codes. As can be seen, the timing difference between SUBSET and TDUAL codes become more apparent as n increases in size. This can be attributed to rule (3) for choosing the parameter to restrict and the use of the dual algorithm with the advanced start to solve the subproblems at each node.

All the results in tables 1 and 2 are for algorithms which guarantee the optimal regression from each subset. It is possible (see [11,14]) to obtain time–accuracy tradeoffs by considering near-optimal models. This type of modification is easy to implement in the previously described algorithm (a single line of FORTRAN is changed in the code). The fathoming is based on a function of the current incumbent other than the objective value. The optimal model is not guaranteed, but savings in solution time can be significant.

Table 1
Computational comparison of TDUAL and SUBSET obtaining the best subset for $k = 1, 2, \ldots, m$.

Number of observations (n)	Number of parameters (m)					
	$m=6$		$m=8$		$m=10$	
	TDUAL	SUBSET	TDUAL	SUBSET	TDUAL	SUBSET
100	0.193	0.25	0.781	1.12	3.03	5.39
	(128)	(270)	(385)	(1351)	(1035)	(2965)
150	0.381	0.789	1.16	1.79	3.80	7.37
	(249)	(486)	(659)	(839)	(1240)	(3081)
200	0.564	1.16	1.35	3.04	7.28	17.83
	(350)	(533)	(598)	(1124)	(3200)	(5735)
250	0.855	2.11	2.61	5.70	8.70	22.52
	(468)	(849)	(1293)	(1789)	(3460)	(5952)
300	1.33	2.85	2.97	6.23	10.46	22.20
	(643)	(1019)	(1202)	(1703)	(3763)	(4906)

Note: Three problems were solved in each combination of m and n. The upper entry in each row is the mean CPU time in seconds and the lower entry in parentheses is the number of iterations.

Table 3 gives the results of solving a set of test problems with a requirement that the regression be within 90%, 95% and 98% of the optimal. Even though only a certain percentage of optimality is guaranteed, the optimal solution was frequently obtained because of rule (3) to choose the parameter to remove from a subproblem. For example, when guaranteeing 95% of optimality the optimal solution was obtained over 90% of the time.

Table 2
Time and iterations for best subset obtained for $k = q, \ldots, m$.

q	TDUAL	SUBSET
1	12.544	47.322
	(2711)	(4992)
3	11.390	44.621
	(2167)	(4368)
5	6.268	19.938
	(588)	(1895)
7	2.006	4.420
	(107)	(476)
9	0.343	2.335
	(25)	(233)

Note: The upper number in each row is the CPU time in seconds and the lower number in parentheses is the number of iterations required. Number of parameters, 10; number of observations, 200.

Table 3
Computational comparison of TDUAL guaranteeing various percentages of optimality.

Number of parameters	Percentage of optimality guaranteed			
	90	95	98	100
6	0.495	0.718	0.828	1.33
	(178)	(318)	(376)	(643)
8	0.834	0.964	1.60	2.97
	(172)	(242)	(590)	(1202)
10	2.61	3.04	5.13	10.46
	(250)	(512)	(1661)	(3763)
12	11.11	11.43	12.94	35.85
	(246)	(389)	(1238)	(7941)

Note: The upper entry in each row is the mean CPU time in seconds and the lower entry in parentheses is the number of iterations.

7. Conclusion

In this paper a branch-and-bound algorithm to select the best subset of variables in linear multiple regression problems under the least absolute value criterion is presented. The algorithm is implemented with a selection rule to restrict a particular parameter, a fathoming test and a last-in-first-out (LIFO) branching rule for the inspection of the tree. Versions of the algorithm applying a dual as well as a primal simplex technique were formulated and tested. A reoptimization start procedure is implemented in both the primal and dual version of the algorithm. In the dual version an additional bounding test is employed.

As indicated from the computational results, the advanced start procedure saves considerable computational time. With the addition of the bounding test in the dual approach, the dual version of the algorithm is consistently faster than the primal method, especially on problems where a large number of parameters are to be examined. In general, the branch-and-bound algorithm utilizing the dual approach with the advanced start and bounding test was found to be the most efficient method for finding the best subset of variables for least absolute value problems.

A computer code version of the algorithm is available from the authors for academic purposes.

Acknowledgment

The authors wish to thank Professors S.C. Narula and J.F. Wellington for providing a copy of their best subset program for computational comparisons.

References

[1] R.D. Armstrong, E.L. Frome and D.S. Kung, "A Revised Simplex Algorithm for the Absolute Deviation Curve Fitting Problem", *Communications in Statistics* B8 (1979) 175–190.
[2] R.D. Armstrong and M.T. Kung, "A Dual Algorithm, to Solve Linear Least Absolute Value Approximation", Research Report, CCS 370, Center for Cybernetic Studies, The University of Texas, Austin, Texas (1980).
[3] E.M.L. Beale, M.G. Kendall and D.W. Mann, "The Discarding of Variables in Multivariate Analysis", Biometrica 54 (1967) 357–366.
[4] A. Charnes, W.W. Cooper and R.O. Ferguson, "Optimal Estimation Executive Compensation by Linear Programming", Management Science 1 (1955) 138–151.
[5] A. Charnes and W.W. Cooper, Management Models and Industrial Applications of Linear Programming, vols. I and II (John Wiley & Sons, Inc., New York, 1961).
[6] T. Dielman and R. Pfaffenberger, "LAV (Least Absolute Value) Estimation in Linear Regression: A review", TIMS Studies in the Management Sciences, this issue.
[7] N.R. Draper and H. Smith, Applied Regression Analysis (John Wiley & Sons, Inc., New York, 1966).
[8] E.F. Fama and R. Roll, "Some Properties of Symmetric Stable Distributions", *Journal of the American Statistical Association* 63 (1968) 817–836.
[9] G.M. Furnival and R.W. Wilson, "Regression by Leaps and Bounds", *Technometrics* 16 (1974) 499–512.
[10] W.J. Kennedy and J.E. Gentle, *Statistical Computing* (Marcel Dekker, New York 1980).
[11] L.R. La Motte and R.R. Hocking, "Computational Efficiency in the Selection of Regression Variables", *Technometrics* 12 (1970) 83–93.
[12] J.R. Meyer and R.R. Glauber, "Investment Decisions, Economic Forecasting and Public Policy", Division of Research Memoir, Graduate School of Business Administration, Harvard University, Cambridge, Massachusetts (1964).
[13] S.C. Narula and J.F. Wellington, "Selection of Variables in Linear Regression Using the Minimum Sum of Weighted Absolute Errors Criterion", *Technometrics* 1 (1979) 299–306.
[14] S.C. Narula, "Optimization Techniques in Linear Regression: A Review", *TIMS Studies in the Management Sciences,* this issue.
[15] J.R. Rice and J.S. White, "Norms for Smoothing and Estimation", *SIAM Review* 6 (1964) 243–256.
[16] G. Roodman, "A Procedure for Optimal Stepwise MSAE Regression Analysis", *Operations Research* 22 (1974) 393–399.

AN ABSOLUTE DEVIATIONS CURVE-FITTING ALGORITHM FOR NONLINEAR MODELS

Asher TISHLER

Tel Aviv University and University of Southern California

and

Israel ZANG *

Tel Aviv University and University of British Columbia

We present a simple and efficient method for the absolute deviations (AD) estimation problem. The method is applicable to models which are either linear or nonlinear in the estimated parameters. It solves the original problem via some smooth approximation which replaces the original estimation problem in arbitrarily small neighborhoods of the points of discontinuous differentiability caused by the absolute value operator. Unlike the original AD problem, this smooth approximation can be minimized using efficient gradient techniques. It also contains a single parameter controlling the accuracy of the approximation. Consequently, the original AD estimation problem is obtained from the approximate one, as the controlling parameter approaches its limit. The choice of this parameter determines a priori the length of the uncertainty interval in the mean absolute error for the solution of the original AD problem. The interval itself, having the above pre-specified length, is obtained from solution of the approximate problem. It is therefore sufficient, in general, to solve the approximate problem only once. Numerical examples demonstrating the efficiency of the method are presented.

1. Introduction

In this paper we develop an algorithm for nonlinear estimation problems, in which the sum of absolute deviations is the minimization criterion. This problem, known as the absolute deviations (AD) curve-fitting problem, has been the subject of several studies in the last few years. Since the problem has discontinuous first-order partial derivatives, it is considered to be difficult to solve, especially when the fitted curve is nonlinear in its parameters.

From the statistical point of view, it is well known that the properties of the estimated parameters highly depend upon the underlying distribution of the

---

Received November 4, 1980; revised March 19, 1981.

* The authors are indebted to a referee of this paper for helpful remarks, to R. Arad for valuable discussions, and to Y. Hoch for carrying out the computer experiments. Research was partially supported by the Israel Institute for Business Research at Tel Aviv University and by a UBC research grant 23-9641.

error terms in the model. Recently, Basset and Koenker [4] proved, under some general assumptions, that the AD estimator for a general linear model is consistent and asymptotically normal. They also discussed the conditions under which the AD estimator is superior to the least squares estimator. Rice and White [16] presented a numerical study of the effectiveness of different norms in estimation models, and discussed the advantage of the AD estimator for certain distributions. The usefulness and applicability of this model were also discussed by Taylor [21], Barrodale and Roberts [3], Fama and Roll [7] and Sharpe [19].

Efficient algorithms for the AD problem exist for the case where the fitted curve is linear in the estimated parameters (see, for example, [6,12,13,20]). In particular, Barrodale and Roberts [3] greatly simplified the use of the AD estimator in the linear case by applying a modification of the simplex method to the primal formulation of the problem as a linear program. There are several algorithms [2,14,18] for AD estimation, which may be applied in case the curve is nonlinear in the estimated parameters. We mention, in particular, the one suggested by Abdelmalek [1], which determines the best L_1 (AD) approximation as the limit of best L_p approximations as $p \to 1^+$.

In this paper we suggest a new algorithm for nonlinear AD estimation. As in [1], our algorithm approximates the original problem by a continuously differentiable one which can be solved using efficient gradient (e.g. quasi-Newton) techniques. The accuracy of this approximation is determined by a single parameter, β. Unlike in [1], our approximation replaces the original problem only in some arbitrarily small neighborhoods of the points of discontinuous differentiability. Moreover, one of our main results, corollary 3.1, shows that it is possible to choose a priori a value for the parameter β in a way that guarantees that the minimal value of the objective function of the approximate problem is within a definite predetermined distance of the optimal value of the objective of the original problem. As we shall show below, this implies a priori tight bounds on the mean absolute error.

The method we suggest here is very simple to use. It is easy to program and it only requires an unconstrained minimization routine using first derivatives which is today available in every computer library. (A quasi-Newton method is especially recommended.) This last property makes our method applicable even to linear AD problems, in case there are many observations and no special LP code is available. However, although no comparison has been made, we expect that for linear AD models our method will be less efficient than special LP codes, such as that suggested by Barrodale and Roberts [3].

The outline of this article is as follows: section 2 presents our approximations and method. Preservation of convexity by these approximations and some convergence properties are established in section 3. Then, in section 4, we present three nonlinear numerical examples with several parameters, where the number of observations varies between 40 and 3,000. The numerical results are very satisfactory.

2. The method

The absolute deviation curve-fitting problem can be stated as follows. Given N observations on the dependent variable y_t and the independent variables $x_t = (x_{1t}, x_{2t}, \ldots, x_{mt})'$, where $t = 1, \ldots, N$, we want to determine the vector of k parameters $\theta = (\theta_1, \theta_2, \ldots, \theta_k)'$ which minimizes the function $G(\theta)$ given by

$$G(\theta) = \sum_{t=1}^{N} |u_t| = \sum_{t=1}^{N} |y_t - f(x_t, \theta)|, \tag{1}$$

where u_t is a random error, and we assume that $N > k$ holds. The function $f(x, \theta)$, by which we specify the functional dependence between x and y, can be linear or nonlinear in x and/or θ. However, for practical purposes, we assume that f possesses everywhere continuous first-order partial derivatives with respect to θ, for every x_t, $t = 1, \ldots, N$. Consequently, the sole difficulty in finding the "best" vector of parameters, θ^*, is caused by $G(\theta)$ not being continuously differentiable owing to the presence of the absolute value operator in its specification. This property prohibits the use of efficient gradient techniques in the estimation process.

In a recent paper [22] it was suggested to smooth out derivative discontinuities which are introduced into piecewise regression models by the presence of "max" (or "min") operators. The basic idea in [22] is to smooth (or approximate) the "max" operator,

$$q(r) = \max(0, r), \tag{2}$$

by the once continuously differentiable approximation:

$$q_1(\beta, r) = \begin{cases} 0, & \text{if } r \leq -\beta, \\ (r + \beta)^2/4\beta, & \text{if } -\beta \leq r \leq \beta, \\ r, & \text{if } \beta \leq r, \end{cases} \tag{3}$$

or by the twice continuously differentiable approximation suggested in [23] and given by:

$$q_2(\beta, r) = \begin{cases} 0, & \text{if } r \leq -\beta, \\ -r^4/16\beta^3 + 3r^2/8\beta + r/2 + 3\beta/16, & \text{if } -\beta \leq r \leq \beta, \\ r, & \text{if } \beta \leq r, \end{cases} \tag{4}$$

where β is a positive parameter determining the accuracy of the approximations. It is quite easy to see that these approximations replace the original "max" operator only for $-\beta \leq r \leq \beta$. Consequently, we have that as $\beta \to 0$, both $q_1(\beta, r)$ and $q_2(\beta, r)$ approach $q(r)$ as a limit. For some more properties of these approximations see [22] and [23].

The applicability of the above approximations to our problem evolves out of the equation

$$|r| = \max(0, r) + \max(0, -r). \tag{5}$$

Applying $q_1(\beta, r)$ to both "max" operators in (5), we obtain the following once continuously differentiable approximation to the absolute value operator:

$$A_1(\beta, r) = \begin{cases} -r, & \text{if } r \leq -\beta, \\ (r^2 + \beta^2)/2\beta, & \text{if } -\beta \leq r \leq \beta, \\ r, & \text{if } \beta \leq r. \end{cases} \tag{6}$$

Similarly, applying $q_2(\beta, r)$ in (5) results in a twice continuously differentiable approximation to the absolute value function given by:[1]

$$A_2(\beta, r) = \begin{cases} -r, & \text{if } r \leq -\beta, \\ -r^4/8\beta^3 + 3r^2/4\beta + 3\beta/8, & \text{if } -\beta \leq r \leq \beta, \\ r, & \text{if } \beta \leq r. \end{cases} \tag{7}$$

The absolute value function and its approximations (6) and (7) are shown in fig. 1. It can be easily seen that $|r|$ is approximated only in the interval where $-\beta \leq r \leq \beta$ holds, and which can be made arbitrarily small by reducing β. Furthermore,

$$\lim_{\beta \to 0} A_j(\beta, r) = |r|, \quad j = 1, 2, \tag{8}$$

holds. We now approximate problem (1) by replacing the absolute value function with its approximations. The approximate problem will be:

$$\min_{\theta} G_j(\beta, \theta), \quad j = 1 \text{ or } 2, \tag{9}$$

where

$$G_j(\beta, \theta) = \sum_{t=1}^{N} A_j(\beta, y_t - f(x_t, \theta)), \quad j = 1, 2. \tag{10}$$

---

[1] See Zang [23] for similar approximations to the "max" operators that have higher order continuous derivatives. These approximations can be used to generate approximations to the absolute value function with the same properties.

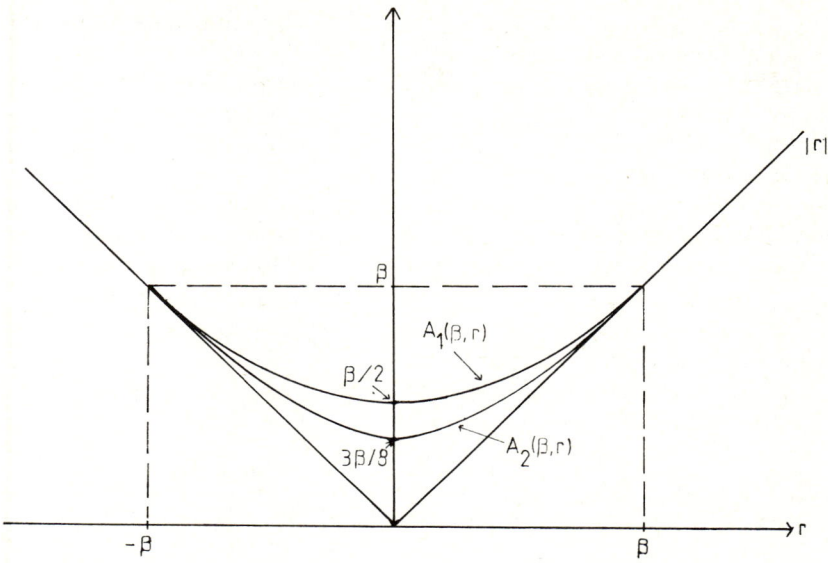

Fig. 1. The absolute value function and its approximations.

The function $G_j(\beta, \theta)$ can be made arbitrarily close to $G(\theta)$. This is simply carried out by reducing β, since by (8)

$$\lim_{\beta \to 0} G_j(\beta, \theta) = G(\theta), \quad j = 1, 2, \tag{11}$$

must hold. Moreover, $G_1(\beta, \theta)$ is once continuously differentiable. This property makes possible the use of efficient gradient techniques, such as quasi-Newton methods [15], for the solution of problem (9), which is impossible for problem (1). In the case where $f(x_t, \theta)$ is twice continuously differentiable with respect to θ for every x_t, $t = 1, \ldots, \bar{n}$, then so will $G_2(\beta, \theta)$ be. Consequently, it is possible to find the minimum of this function using second-order techniques such as Newton's method, or that suggested by Goldfeld, Quandt and Trotter [10].

The approximation (3) can be obtained from an approximation to the "max" operator which was suggested by Bertsekas [5].[2] This approximation contains two parameters and, unlike q_1, it replaces the "max" operator everywhere except for a single point ($r = 0$). Thus, following our derivation, (6) can also be obtained from the approximation suggested in [5]. Because of the

---

[2] Setting $y = 1/2$ in eq. (5) in [5], substituting $1/2c = \beta$ and adding a constant $\beta/4$ to the resulting approximation (a vertical translation) gives $\tilde{q}_1$.

additive nature of $G_1(\beta, \theta)$, the constant $\beta/4$ which translates the approximation vertically (see footnote 2) can be neglected. Thus, our procedure that uses A_1 can be obtained as an application of the method suggested by Bertsekas [5]. We would also like to point out that the idea of using some "piecewise" regression criteria, as we do by introducting (6) and (7), has been applied for robust regression. See, for example, Huber [11].

Note also that the approximations suggested here are more exact than the approximation suggested by Abdelmalek [1]; by controlling β, our approximations replace the original problem only in some arbitrarily small neighborhoods of the points where $G(\theta)$ has discontinuous derivatives. Everywhere else $G_j(\beta, \theta) = G(\theta)$, and consequently the original problem remains unchanged. On the other hand, the approximation suggested in [1] replaced $G(\theta)$ everywhere, and therefore it is less accurate. Moreover, even in a given neighborhood where G is replaced, our approximation operates only on observations for which the error is smaller than β. Consequently, all other observations contribute exactly the same weight to both approximate (G_j) and original (G) objective functions. These observations, having large residuals, contribute most of the weight to both objective functions, and our method leaves them unchanged everywhere.

It is now possible to find the minimum of $G(\theta)$ via minimizing $G_j(\beta, \theta)$ ($j = 1$ or 2) for some small enough value of β. One can even do better, by using the following simple algorithmic scheme.

Stage 1. Assume θ^1 and β^1 are given. Solve

$$\min_{\theta} G_j(\beta^1, \theta), \qquad j = 1 \text{ or } 2, \tag{12}$$

starting from θ^1. Let θ^2 be the solution point to problem (12). Take $\beta^2 < \beta^1$. Go to stage 2.

Stage l. l = 2, 3,... . Given θ^l and β^l, solve

$$\min_{\theta} G_j(\beta^l, \theta), \qquad j = 1 \text{ or } 2, \tag{13}$$

starting from θ^l. Let θ^{l+1} be the solution point to problem (13). In the case where

$$\|\theta^{l+1} - \theta^l\| < \epsilon \tag{14}$$

holds, where ϵ is a small predetermined tolerance, then declare θ^{l+1} as an optimal solution and stop. In any other case, take $\beta^{l+1} < \beta^l$ and go to stage $l + 1$.

In the next section (corollary 3.2) we establish convergence of the above iterative process. However, for practical purposes, it will generally suffice to use only one iteration, choosing a value of β which is substantially smaller than the expected mean absolute error (MAE). This recommendation will be later reinforced by corollary 3.1. In our experiments, which are described in section 4, we used the value of 0.01 for β, which gave satisfactory results. The user of the method should be warned, however, not to choose too small a value for β; for such a value $G_j(\beta, \theta)$ becomes too close to $G(\theta)$, which has discontinuous first-order partial derivatives. This might cause some numerical difficulties in the solution process. This last remark becomes clearer in light of the discussion at the end of section 3; we show there that the method suggested here can be viewed as a penalty function method [8] for a particular constrained nonlinear programming problem associated with problem (1). It is well known that penalty functions become numerically ill-conditioned as the penalty parameter (β in this case) approaches its limit. This explains, from a slightly different point of view, the need to avoid working with too small values of β.

In light of the above discussion, one may ask what happens in the case where the chosen β is too large. We claim that in using the approximation (6) there is, in some sense, an upper limit to the inaccuracy obtained by such a choice. Suppose that β is so large such that

$$|u_t| = |y_t - f(x_t, \theta)| \leq \beta, \quad t = 1, \ldots, N,$$

in a neighborhood of an optimal θ. Then it is easy to see from (10) and (6) that $2\beta G_1(\beta, \theta) - N\beta^2$ is the sum of squared errors. Consequently, since β is a constant, there exists a threshold value β', depending upon the data and the specific equation estimated, such that for all $\beta \geq \beta'$ problem (9) is likely to produce the least squares estimators.

3. Some properties of the algorithm

In this section we establish some properties of the approximate objective function $G_j(\beta, \theta)$. First we show that it retains some convexity properties in the case where these properties are possessed by $G(\theta)$:

Theorem 3.1. Let $|y_t - f(x_t, \theta)|$ be convex functions of θ for $t = 1, \ldots, N$ (and consequently $G(\theta)$ is convex). Then for every $\beta > 0$, $G_j(\beta, \theta)$ are convex functions of θ.

Proof. It is sufficient to establish the convexity of $A_j(\beta, y_t - f(x_t, \theta))$. First we note that by (6) and (7) we have:

$$A_j(\beta, y_t - f(x_t, \theta)) = A_j(\beta, |y_t - f(x_t, \theta)|), \quad j = 1, 2. \tag{15}$$

Furthermore, the functions $A_j(\beta, r)$ are monotone increasing for $r \geq 0$. It is also quite simple to show that $A_j'(\beta, r)$ are monotone increasing for $j = 1, 2$, and consequently $A_j(\beta, r)$ are convex functions of r. The proof follows now directly from these two properties, eq. (15), and theorem 5.1 in Rockafellar [17].

Note that in the case where G_j is nonconvex, several nonglobal local minima may exist. In this case the solution obtained depends on the starting point θ^1, and it is recommended to solve the problem repeatedly using different values for θ^1. Next we establish bounds on the difference between $G_j(\beta, \theta)$ and $G(\theta)$. The derivation of these bounds is based upon the inequalities

$$0 \leq A_1(\beta, r) - |r| \leq \beta/2 \tag{16}$$

and

$$0 \leq A_2(\beta, r) - |r| \leq 3\beta/8, \tag{17}$$

which can be easily shown to hold since for $r = 0$ the difference between $A_j(\beta, r)$ and the absolute value function is maximal. As a direct outcome we have:

Theorem 3.2. The inequalities

$$0 \leq G_1(\beta, \theta) - G(\theta) \leq N\beta/2 \tag{18}$$

and

$$0 \leq G_2(\beta, \theta) - G(\theta) \leq 3N\beta/8 \tag{19}$$

hold.

Using the above theorem we obtain

Corollary 3.1 Let θ^* be a global minimum point of $G(\theta)$, and let θ_j be global minimum points of $G_j(\beta, \theta)$, $j = 1, 2$, respectively. Then

$$0 \leq G_1(\beta, \theta_1) - G(\theta^*) \leq N\beta/2 \tag{20}$$

and

$$0 \leq G_2(\beta, \theta_2) - G(\theta^*) \leq 3N\beta/8 \tag{21}$$

hold.

Proof. The proof follows directly from theorem 3.2 and theorem 1 in Geoffrion [9].

A similar result that gives a bound which is less sharp than (20) but for a more general problem than (1) was obtained by Bertsekas [5] (see his corollary 2.3).

We consider corollary 3.1 as our most important result. In practice it shows that is is possible to determine a priori the length of the uncertainty interval in $G(\theta^*)/N$ by the choice of the value for β. For example, let $j = 1$, then by (20):

$$G_1(\beta, \theta_1)/N - \beta/2 \leq G(\theta^*)/N \leq G_1(\beta, \theta_1)/N \tag{22}$$

must hold, and consequently the length of the uncertainty interval in the mean absolute error (MAE), given by $G(\theta^*)/N$, is $\beta/2$. In the same manner, an uncertainty interval of $3\beta/8$ for the MAE is obtained whenever $j = 2$ is used. This important property supports our previous recommendation that in general one iteration of the algorithmic scheme may be sufficient to determine satisfactory parameters, since it provides a proper choice of β.

However, it may happen that the objective function $G(\theta)$ is flat in a neighborhood of θ^* (due to multicollinearity). In such cases a value of β, giving reasonable a priori bounds on the length of the uncertainty interval of the MAE, must not produce a vector of parameters θ_j which is close enough to θ^*. This, however, is a situation where other methods may produce unsatisfactory results as well, and in our case a further decrease of β, according to the algorithmic scheme given in section 2, is needed.

In practice the upper bounds in (20) and (21) are substantially lower, as we show in the next theorem. We first denote

$$T(\theta) = \{t: 1 \leq t \leq N, y_t - f(x_t, \theta) = 0\} \tag{23}$$

and

$$T_\beta(\theta) = \{t: 1 \leq t \leq N, |y_t - f(x_t, \theta)| < \beta\}. \tag{24}$$

We also let $M(\theta)$ and $M_\beta(\theta)$ be the number of elements in $T(\theta)$ and $T_\beta(\theta)$, respectively, and observe that

$$\lim_{\beta \to 0} T_\beta(\theta) = T(\theta) \tag{25}$$

holds. Since in practice $M(\theta)$ is much smaller than N, so will $M_\beta(\theta)$ be for small enough values of β. We also let

$$N^*(\beta) = \max_\theta M_\beta(\theta). \tag{26}$$

Clearly, $N^*(\beta) \leq N$ holds. Furthermore, the above discussion implies that $N^*(\beta)$ is generally much smaller than N, for sufficiently small β. We may now prove

Theorem 3.3. The inequalities

$$0 \leq G_1(\beta, \theta_1) - G(\theta^*) \leq N^*(\beta) \beta/2 \tag{27}$$

and

$$0 \leq G_2(\beta, \theta_2) - G(\theta^*) \leq 3N^*(\beta)\beta/8 \tag{28}$$

hold, where θ_1, θ_2 and θ^* are as defined in corollary 3.1.

Proof. We observe that by (6) and (7):

$$A_j[\beta, y_t - f(x_t, \theta)] = y_t - f(x_t, \theta), \quad j = 1 \text{ or } 2, \tag{29}$$

holds for all $t \notin T_\beta(\theta)$. Therefore, the difference between $G_j(\beta, \theta)$ and $G(\theta)$ is determined by the terms for which $t \in T_\beta(\theta)$. Thus, as in theorem 3.2, inequalities

$$0 \leq G_1(\beta, \theta) - G(\theta) \leq M_\beta(\theta) \beta/2 \tag{30}$$

and

$$0 \leq G_2(\beta, \theta) - G(\theta) \leq M_\beta(\theta) \beta/8 \tag{31}$$

must hold, which, by (26), implies that the inequalities

$$0 \leq G_1(\beta, \theta) - G(\theta) \leq N^*(\beta) \beta/2 \tag{32}$$

and

$$0 \leq G_2(\beta, \theta) - G(\theta) \leq 3N^*(\beta) \beta/8 \tag{33}$$

must hold as well. Inequalities (27) and (28) follow now from (32), (33), and theorem 1 in [9].

The fact that the bounds (27) and (28) are probably much better in practice for small values of β, is demonstrated by the four numerical problems, presented in the next section. For $N = 3{,}000$, using $j = 1$ and a Cauchy distribution for the error term, we obtained the values of 16, 24 and 22 for $M_{0.01}(\theta_1)$ for the three problems, respectively. Unfortunately, it is impossible to use these better bounds, since they depend upon problem specifications which are unknown a priori ($N^*(\beta)$), whereas the bound of $\beta/2$ if $j = 1$, or $3\beta/8$ for $j = 2$, on the error in the MAE, are problem independent.

Let us now establish convergence of the algorithmic scheme suggested at the end of section 2.

Corollary 3.2. Let $\{\beta^l\} \to 0$ be a sequence of monotonically decreasing positive numbers and assume that θ_j^l is the solution to

$$\min G_j(\beta^l, \theta), \quad j = 1, 2. \tag{34}$$

Also let θ' and θ'' be any accumulation points of the sequences $\{\theta_1^l\}$ and $\{\theta_2^l\}$, respectively. Then

$$G(\theta') = G(\theta'') = G(\theta^*) \tag{35}$$

holds.

Proof. The proof follows directly from (20), (21), and (11).

We conclude this section by deriving an interpretation of our method, which ties up with the well-known concept of penalty functions for constrained optimization (see Fiacco and McCormick [8]). Again, letting θ^* be the vector of parameters that minimizes (1), we may formulate the original problem as:

$$\left.\begin{array}{l}\min \sum\limits_{t \notin T(\theta^*)} \sigma_t[y_t - f(x_t, \theta)] \\ \text{subject to} \\ y_t - f(x_t, \theta) = 0, \quad \forall t \in T(\theta^*),\end{array}\right\} \tag{36}$$

where $T(\theta)$ is given by (23) and σ_t is given by [3]

$$\sigma_t = \text{sgn}(y_t - f(x_t, \theta)). \tag{37}$$

The quadratic loss penalty function ([8]) to problem (36) can be expressed by:

$$\phi(\beta, \theta) = \sum_{t \notin T(\theta^*)} \sigma_t[y_t - f(x_t, \theta)] + \frac{1}{2\beta} \sum_{t \in T(\theta^*)} [y_t - f(x_t, \theta)]^2, \tag{38}$$

where β is a penalty parameter which forces the minimum of ϕ to approach θ^* as $\beta \to 0$. since $|y_t - f(x_t, \theta^*)| > 0$ for all $t \notin T(\theta^*)$, we have by continuity that

---
[3] Note that σ_t, $t \notin T(\theta^*)$ is, by continuity, constant in the neighborhood of θ^*.

there exists a neighborhood $B(\theta^*)$ of θ^* and a positive number β' such that

$$\left.\begin{array}{l} \beta' \leq \min_{t \notin T(\theta^*)} \{|y_t - f(x_t, \theta)|\} \\ \text{and} \\ \beta' \geq \max_{t \in T(\theta^*)} \{|y_t - f(x_t, \theta)|\} \end{array}\right\} \quad (39)$$

hold for all $\theta \in B(\theta^*)$. Therefore, if we now compute $G_1(\beta', \theta)$ for some $\theta \in B(\theta^*)$ and neglect all the constant terms $\beta'/2$ which appear in (6) in the zone where $-\beta' \leq r \leq \beta'$ holds[4], it can be easily shown that we obtain $\phi(\beta', \theta)$. To be more specific, we have

$$\phi(\beta', \theta) = G_1(\beta', \theta) - N(\theta^*)\beta'/2, \quad (40)$$

where $N(\theta^*)$ was defined after (24).

Thus, for a sufficiently but not too small β and in some neighborhood of θ^*, minimizing G_1 is equivalent to minimizing the penalty function ϕ. However, the subset $T(\theta^*)$ is unknown a priori and this precludes a direct use of the above penalty approach. Consequently, our approach can be regarded as an efficient mechanism for locating first the subset $T(\theta^*)$, and then the point θ^* (via a penalty-like approach).

Note that it is also possible to apply to problem (1) penalty function schemes which are different from (38). However, these approaches require the reformulation of the original problem as a constrained nonlinear minimization problem, and they will not result with the a priori bounds given by corollary 3.1.

4. Numerical examples

To test the method developed in section 2, we solved three numerical examples. In all cases the x_{it} and u_t were arbitrarily chosen (see details below). Then, using a given set of predetermined parameters $\hat{\theta}$, we computed the "observed" y_t's. In each example we performed the estimation for five sample sizes ($N = 40, 100, 500, 1,000, 3,000$) and three distributions of the u_t's (the Cauchy, normal and uniform distributions), denoted by $C(0)$, $N(0, \sigma^2)$ and $U(-a, a)$, respectively.

---

[4] Note that constant terms can be neglected without affecting the minimization.

Example 1 [14]:

$$y_t = \frac{\sum_{i=1}^{4} \theta_i x_{it}}{1 + \sum_{i=5}^{6} \theta_i x_{it}} + u_t, \tag{41}$$

where $\hat{\theta} = (1.5, 1.0, 2.7, 0.75, -0.35, 0.7)'$, and the x_i were uniformly distributed in the ranges $(65, 75)$, $(10, 20)$, $(0, 30)$, $(-10, 90)$, $(0, 2)$, and $(0, 4)$, respectively. The distributions of the u_t's were: $C(0)$, $N(0, 12^2)$, and $U(-15, 15)$.

Example 2:

$$y_t = \theta_1 x_{1t}^2 + \theta_2 x_{2t}^2 + \frac{\theta_3}{\theta_1 + \theta_2} x_{1t} x_{2t} + u_t, \tag{42}$$

where $\hat{\theta} = (1.5, 3.0, 2.25)'$, and the x_i were uniformly distributed in the ranges $(3,9)$ and $(3,5)$, respectively. The distributions of the u_t's were: $C(0)$, $N(0, 8^2)$ and $U(-12, 12)$.

Example 3 (C.E.S. production function):

$$y_t = \theta_1 \left[\theta_2 x_{1t}^{-\theta_3} + (1-\theta_2) x_{2t}^{-\theta_3} \right]^{-\theta_4/\theta_3} + u_t, \tag{43}$$

where $\hat{\theta} = (1.5, 0.6, 1.1, 0.75)'$, and

$$\begin{pmatrix} x_{1t} \\ x_{2t} \end{pmatrix} \sim N\left[\begin{pmatrix} 800 \\ 350 \end{pmatrix}, \begin{pmatrix} 10{,}000 & 5{,}500 \\ 5{,}500 & 8{,}000 \end{pmatrix} \right]. \tag{44}$$

The distributions of the u_t were: $C(0)$, $N(0, 6^2)$ and $U(-8, 8)$.

For each function, sample size, and distribution we estimated the parameters three times. The first and second times we used our method with $j=1$ and $j=2$, setting $\beta=0.01$ in both cases. Then we used least squares. For all examples the starting point θ^1 was chosen[5] as $\hat{\theta}+0.1$. The minimization was carried out by subroutine VA13A of Harwell subroutine library, which is a quasi-Newton method [15]. The computations were carried out on a CDC 6600 computer located at Tel Aviv University.

[5] Somewhat more distant θ^1 may affect the method (as well as all other methods) to converge to local minima other than the global one. This may happen to examples 1 and 3 which are nonconvex minimization problems. We preferred to avoid this phenomenon.

Table 1
Estimated parameters and summary statistics for example 1.

			θ_1	θ_2	θ_3	θ_4	θ_5	θ_6	RMSE	MAE	MY	NF	CPU
True parameters $\hat{\theta} \to$			1.5	1.0	2.7	0.75	−0.35	0.7					
N	Distribution	Norm of estimator	Estimated parameters										
40	C	1	1.478	1.236	2.791	0.761	−0.349	0.718	5.87	2.90	115.3	59	1.32
		2	1.241	1.893	2.638	0.809	−0.360	0.682	5.19	3.52	115.3	26	0.28
	NO	1	1.602	1.009	3.234	0.744	−0.348	0.807	11.87	8.94	115.5	82	1.88
		2	1.619	0.833	3.069	0.771	−0.350	0.778	11.76	9.09	115.5	25	0.29
	U	1	1.499	0.602	2.948	0.573	−0.365	0.685	8.92	6.90	114.2	80	1.82
		2	1.572	0.411	3.064	0.682	−0.344	0.699	8.33	7.09	114.2	23	0.26
100	C	1	1.500	0.972	2.699	0.760	−0.352	0.699	5.93	2.86	116.7	62	3.36
		2	1.273	1.755	2.672	0.754	−0.358	0.673	5.65	3.31	116.7	28	0.68
	NO	1	1.599	1.243	3.063	0.716	−0.344	0.788	12.28	9.86	117.9	58	3.14
		2	1.566	1.024	3.035	0.724	−0.343	0.741	12.13	10.10	117.9	22	0.54
	U	1	1.624	1.194	2.659	0.841	−0.323	0.777	9.17	7.44	116.2	53	2.89
		2	1.520	1.181	2.801	0.819	−0.336	0.748	8.96	7.58	116.2	21	0.55

500	C	1	1.515	0.913	2.679	0.752	−0.350	0.695	29.46	5.36	116.0	52	13.72
		2	1.554	0.513	3.110	0.865	−0.337	0.726	29.27	6.57	116.0	32	3.52
	NO	1	1.577	0.931	2.684	0.681	−0.349	0.697	11.79	9.30	116.2	43	11.17
		2	1.554	1.081	2.701	0.714	−0.346	0.709	11.77	9.34	116.2	29	3.18
	U	1	1.485	1.084	2.689	0.762	−0.352	0.708	8.96	7.79	115.4	62	16.12
		2	1.484	1.043	2.685	0.755	−0.351	0.697	8.94	7.81	115.4	23	2.55
1000	C	1	1.506	0.965	2.688	0.747	−0.350	0.696	24.33	4.65	115.4	51	26.74
		2	1.503	0.839	2.805	0.840	−0.343	0.712	24.23	5.30	115.4	21	4.64
	NO	1	1.506	1.135	2.708	0.711	−0.350	0.703	12.12	9.64	115.5	73	37.94
		2	1.495	1.149	2.725	0.736	−0.348	0.709	12.12	9.66	115.5	26	5.66
	U	1	1.529	0.833	2.652	0.772	−0.352	0.705	8.82	7.66	115.1	50	25.80
		2	1.494	0.964	2.672	0.768	−0.351	0.699	8.81	7.67	115.1	22	4.83
3000	C	1	1.503	0.978	2.702	0.750	−0.350	0.699	86.69	6.88	114.6	40	62.01
		2	1.506	0.736	2.902	0.729	−0.346	0.692	86.67	7.40	114.6	22	14.95
	NO	1	1.520	0.996	2.749	0.743	−0.350	0.710	12.21	9.72	115.1	55	84.98
		2	1.508	1.021	2.757	0.750	−0.349	0.710	12.21	9.73	115.1	35	23.02
	U	1	1.510	0.908	2.695	0.763	−0.353	0.702	8.68	7.50	115.0	56	86.84
		2	1.497	0.985	2.682	0.753	−0.352	0.699	8.68	7.50	115.0	23	15.04

Table 2
Estimated parameters and summary statistics for example 2.

N	Distri-bution	Norm of estimator	θ_1	θ_2	θ_3	RMSE	MAE	MY	NF	CPU
True parameters $\hat{\theta} \rightarrow$			1.5	3.0	2.25					
			Estimated parameters							
40	C	1	1.452	2.950	2.784	5.75	2.95	113.2	61	0.50
		2	1.464	2.961	2.648	5.75	2.97	113.2	23	0.15
	NO	1	1.181	2.736	4.580	7.77	6.07	113.3	89	0.74
		2	1.171	2.729	4.746	7.74	6.17	113.3	31	0.20
	U	1	1.144	2.357	5.061	6.91	5.75	112.3	89	0.73
		2	1.272	2.720	4.136	6.80	5.90	112.3	33	0.21
100	C	1	1.493	3.044	2.244	5.93	2.86	113.4	49	0.89
		2	1.505	3.115	1.977	5.93	2.87	113.4	22	0.28
	NO	1	1.271	2.753	4.097	8.25	6.74	114.1	76	1.41
		2	1.358	2.946	3.425	8.15	6.89	114.1	26	0.32
	U	1	0.994	2.126	5.815	7.24	6.19	112.9	83	1.50
		2	1.071	2.238	5.652	7.14	6.24	112.9	37	0.45

500	C	1	1.477	2.974	2.514	29.46	5.37	122.2	29	2.48
		2	1.385	2.504	4.051	29.42	5.89	122.2	30	1.61
	NO	1	1.555	3.174	1.504	7.88	6.24	122.1	32	2.71
		2	1.521	3.090	1.994	7.87	6.24	122.1	21	1.13
	U	1	1.375	2.679	3.762	7.16	6.23	121.6	41	3.50
		2	1.432	2.834	3.108	7.15	6.24	121.6	27	1.47
1000	C	1	1.487	2.979	2.412	24.33	4.66	119.0	28	4.75
		2	1.307	2.456	4.445	24.29	4.92	119.0	32	3.44
	NO	1	1.557	3.202	1.346	8.09	6.42	119.1	37	6.16
		2	1.519	3.109	1.853	8.09	6.43	119.1	19	2.04
	U	1	1.432	2.810	3.061	7.06	6.14	118.8	40	6.72
		2	1.460	2.921	2.679	7.05	6.15	118.8	23	2.44
3000	C	1	1.495	2.993	2.310	86.69	6.88	118.0	37	18.54
		2	2.316	4.242	−11.021	86.59	8.83	118.0	43	13.56
	NO	1	1.529	3.089	1.816	8.14	6.49	118.4	35	17.51
		2	1.506	3.047	2.096	8.14	6.49	118.4	28	9.06
	U	1	1.477	2.979	2.457	6.95	6.01	118.3	44	22.07
		2	1.486	2.987	2.374	6.95	6.01	118.3	22	7.00

Table 3
Estimated parameters and summary statistics for example 3.

		θ_1	θ_2	θ_3	θ_4	RMSE	MAE	MY	NF	PU
True parameters $\hat{\theta} \rightarrow$		1.5	0.6	1.1	0.75					
N	Distribu-tion	Norm of estimator	Estimated parameters							
40	C	1	1.303	0.729	1.510	0.763	5.62	2.85	161.2	90	3.27
		2	0.939	0.739	1.320	0.811	5.47	3.23	161.2	45	1.55
	NO	1	1.532	0.903	3.670	0.732	6.16	4.70	161.3	103	3.82
		2	1.809	0.784	2.380	0.713	5.97	4.85	161.3	47	1.85
	U	1	0.828	0.698	1.208	0.832	4.70	3.58	160.6	117	4.35
		2	0.985	0.705	1.351	0.807	4.39	3.70	160.6	44	1.51
100	C	1	1.482	0.603	1.086	0.752	5.93	2.87	162.9	52	4.59
		2	1.093	0.674	1.148	0.793	5.82	3.05	162.9	53	4.44
	NO	1	1.462	0.824	2.713	0.744	6.28	5.10	163.4	116	10.26
		2	1.444	0.666	1.443	0.753	6.15	5.21	163.4	49	4.05
	U	1	1.034	0.543	0.532	0.808	4.87	4.10	162.5	114	10.29
		2	1.075	0.637	0.975	0.798	4.77	4.16	162.5	62	5.13

500	C	1	1.449	0.663	1.393	0.752	29.45	5.36	163.4	54	23.66
		2	1.414	0.695	1.527	0.755	29.45	5.49	163.4	52	21.59
	NO	1	1.371	0.749	1.885	0.757	5.90	4.65	163.0	83	36.88
		2	1.374	0.671	1.423	0.760	5.89	4.66	163.0	52	21.67
	U	1	1.242	0.645	1.194	0.776	4.75	4.15	162.6	65	29.84
		2	1.298	0.643	1.178	0.770	4.75	4.15	162.6	52	21.35
1000	C	1	1.456	0.633	1.248	0.753	24.31	4.65	162.1	51	44.38
		2	1.395	0.708	1.628	0.756	24.31	4.69	162.1	57	45.34
	NO	1	1.554	0.698	1.660	0.740	6.07	4.82	162.2	62	53.54
		2	1.569	0.639	1.343	0.741	6.07	4.83	162.2	39	30.86
	U	1	1.195	0.627	0.984	0.782	4.68	4.06	162.0	64	55.75
		2	1.300	0.609	0.978	0.770	4.67	4.07	162.0	50	40.64
3000	C	1	1.497	0.601	1.102	0.750	86.69	6.88	162.2	39	102.35
		2	1.173	0.640	1.087	0.784	86.69	7.05	162.2	43	100.57
	NO	1	1.517	0.595	1.084	0.749	6.11	4.87	162.6	47	111.77
		2	1.511	0.602	1.119	0.749	6.11	4.87	162.6	35	77.66
	U	1	1.458	0.647	1.323	0.752	4.63	4.00	162.6	43	106.53
		2	1.425	0.624	1.167	0.757	4.63	4.01	162.6	32	75.66

The results of the experiments are contained in tables 1–3. Since the performance of both approximations $j = 1$ and $j = 2$ was almost identical, we report only the results obtained for $j = 1$. These results apear in the rows for which the norm of the estimator is 1. The results obtained for the least squares solution are contained in the rows where the norm of the estimator is 2. Besides giving the values of the true and estimated parameters, we also report the RMSE, the mean absolute error (MAE), the mean of the dependent variable (MY), the number of function calls in the optimization process (NF), and the CPU time used for the optimization. The three distributions considered are denoted by C, NO and U, respectively

In the experiments carried out, convergence to the global optimum was always achieved, although it is possible, in the nonlinear models, to reach a local nonglobal optimum.

The results reported in the tables are very accurate. Using corollary 3.1 it can be shown that the MAE for the solution of the problem (1) must lie within $\beta/2$ (which is 0.005) below the reported MAE.

To demonstrate the performance of the algorithmic scheme of section 2 and the discussion concerning the choice of β, we show in table 4 the results of an experiment carried out with the above scheme, using example 1, $N = 40$, $j = 1$, and errors generated according to the Cauchy distribution. In table 4, NL denotes the number of observations which satisfy $-\beta \leq r \leq \beta$ in (6) at the optimal solution. Note that for $\beta = 20$, $NL = 40 = N$, and thus the least squares parameters are obtained. This coincides with the discussion contained in the final paragraph of section 2.

The experiments reported in tables 1–3 were designed mainly to evaluate the numerical properties of the new AD method developed in this paper. The results point out that, as expected, the AD estimator is superior to the least squares (hereafter LS) estimator for the Cauchy distribution. For the normal distribution, the opposite is true. Finally, the LS estimator is somewhat better than the AD for the uniform distribution. All these results are in agreement with the statistical properties of the above distributions (see Rice and White [16]). Another conclusion that can be inferred from the tables is that for large samples it makes little difference which norm is minimized, if the "true" error distribution is either normal or uniform. If the "true" distribution of the random errors is Cauchy, then even in large samples the AD method is superior to the least squares method. Our conjecture is that one may make a major error using the LS estimator when the AD method should be used (long-tailed distributions), but the contrary is not necessarily true, at least in large samples.

Also note that, for the Cauchy distribution, the estimated parameters become more accurate as the number of observations increases. The increasing of the RMSE and MAE for $N = 500$, as compared to $N = 100$ (or $N = 40$) for the Cauchy distribution, should not be misinterpreted. It occurs because some

Table 4
Results obtained for different values of β.

	θ_1	θ_2	θ_3	θ_4	θ_5	θ_6	RMSE	MAE	NL	CPU
True parameters $\hat{\theta} \to$	1.5	1.0	2.7	0.75	−0.35	0.7				
β	Estimated parameters									
20	1.241	1.893	2.638	0.809	−0.360	0.682	5.19	3.52	40	0.65
10	1.398	1.417	2.723	0.769	−0.356	0.700	5.50	3.07	39	0.67
5	1.461	1.209	2.763	0.756	−0.355	0.707	5.78	2.99	32	0.72
1	1.487	1.146	2.782	0.762	−0.350	0.712	5.84	2.93	20	0.84
0.01	1.478	1.236	2.791	0.761	−0.349	0.718	5.87	2.90	6	1.32

very large u_t's (as large as 500) are found in the sample size of 500 and not for 100 or less.

The execution times, obtained using our AD method, were quite reasonable. However, it is quite obvious that the LS method is in general faster, in terms of CPU, than our method. This is because each evaluation of $G_1(\beta, \theta)$ requires N calls for a routine computing values for $A_1(\beta, r)$, which takes more time than the computation of N squares. Yet, for models where the effort involved in computing $f(x, \theta)$ becomes dominant, such as in the highly nonlinear example 3, the time difference decreases, and it becomes almost negligible for the large sample Cauchy distribution.

References

[1] N.N. Abdelmalek, "Linear L_1 Approximation for a Discrete Point Set and L_1 Solutions of Overdetermined Linear Equations", *Journal of the Association for Computing Machinery* 18 (January 1971) 41–47.
[2] R.D. Armstrong and E.L. Frome, "A Comparison of Two Algorithms for Absolute Deviation Curve Fitting", *Journal of the American Statistical Association* 71 (June 1976) 328–330.
[3] I. Barrodale and F.D.K. Roberts, "An Improved Algorithm for Discrete L_1 Linear Approximation", *SIAM Journal of Numerical Analysis* 10 (October 1973) 839–848.
[4] G. Basset, Jr. and R. Koenker, "Asymptotic Theory of Least Absolute Error Regression", *Journal of the American Statistical Association* 73 (September 1978) 618–622.
[5] D.P. Bertsekas, "Nondifferentiable Optimization Via Approximation", *Mathematical Programming Study* 3 (November 1975) 1–25.
[6] T. Dielman and R. Pfaffenberger, "LAV (Least Absolute Value) Estimation in Linear Regression: A Review", *TIMS Studies in the Management Sciences,* this volume.
[7] E.F. Fama and R. Roll, "Some Properties of Symmetric Stable Distributions", *Journal of the American Statistical Association* 63 (September 1968) 817–836.
[8] A.B. Fiacco and G.P. McCormick, *Nonlinear Programming: Sequential Unconstrained Minimization Techniques* (John Wiley and Sons, Inc., New York, 1968).
[9] A.M. Geoffrion, "Objective Function Approximations in Mathematical Programming", *Mathematical Programming* 13 (August 1977) 23–37.
[10] S.M. Goldfeld, R.E. Quandt and H.F. Trotter, "Maximization by Quadratic Hill-Climbing", *Econometrica* 34 (July 1966) 541–551.
[11] P.J. Huber, "Robust Estimation of a Location Parameter", *Annals of Mathematical Statistics* 35 (1964) 73–101.
[12] D. Klingman and J. Mote, "Generalized Network Approaches for Solving Least Absolute Value and Tchebycheff Regression Problems", *TIMS Studies in the Management Sciences,* this volume.
[13] S.C. Narula, "Optimization Techniques in Linear Regression: A Review", *TIMS Studies in the Management Sciences,* this volume.
[14] M.R. Osborn and G.A. Watson, "On an Algorithm for Discrete Nonlinear L_1 Approximation", *The Computer Journal* 14 (May 1971) 184–188.
[15] M.J.D. Powell, "A View of Unconstrained Optimization", in: L.C.W. Dixon, ed., *Optimization in Action* (Academic Press, 1976) pp. 117–152.
[16] J.R. Rice and J.S. White, "Norms for Smoothing and Estimation", *SIAM Review* 6 (July 1964) 243–256.

[17] R.T. Rockafellar, *Convex Analysis* (Princeton University Press, Princeton, N.J., 1970).
[18] E.J. Schlossmacher, "An Iterative Technique for Absolute Deviations Curve Fitting", *The Journal of the American Statistical Association* 68 (December 1973) 857–859.
[19] W.F. Sharpe, "Mean-Absolute-Deviation Characteristic Lines for Securities and Portfolios", *Management Science* 18 (October 1971) B1–B13.
[20] R.L. Sielken, Jr. and H.O. Hartley, "Two Linear Programming Algorithms for Unbiased Estimation of Linear Models", *The Journal of the American Statistical Association* 68 (1973) 639–641.
[21] L.D. Taylor, "Estimation by Minimizing the Sum of Absolute Errors", in: P. Zarembka, ed., *Frontiers in Econometrics* (Academic Press, 1974) pp. 169–190.
[22] A. Tishler and I. Zang, "A New Maximum Likelihood Algorithm for Piecewise Regression", *The Journal of the American Statistical Association* 76 (1981) 980–987.
[23] I. Zang, "A Smoothing-Out Technique for Min-Max Optimization", *Mathematical Programming* 19 (1980) 61–67.

SOME ALGORITHMS FOR CONCAVE AND ISOTONIC REGRESSION

Chien-Fu WU
University of Wisconsin, Madison

In many practical applications of regression analysis the functional relationship between the dependent and independent variables is implicitly defined by some constraints on the variables. Concave regression and isotonic regression arise from concavity and monotonicity constraints, respectively. We propose a slightly modified version of Kruskal's "up-and-down blocks" algorithm for obtaining the least squares solution of the isotonic regression. We then characterize the least squares solution of the concave regression problem and propose an algorithm for finding it. Another algorithm, which gives approximate solutions to the least squares solution, is also proposed. The latter algorithm can be implemented on a calculator and without much programming skill, which renders it a useful tool for concave regression analysis. The relationships of our algorithms to those proposed by Hildreth [5], Dent [2], Dent et al. [3] and Holloway [6] are discussed. The algorithms are illustrated using two examples.

1. Introduction

To study the relationship between dependent and independent variables, one may not have enough knowledge to assume a particular parametric form for the relationship being investigated. For example, the relationship between corn yield and the level of nitrogen may be described by a quadratic curve, a power curve or an exponential curve, each of which has a very different implication [5]. In this situation economists may be more willing to make the weaker assumption that the marginal productivity of corn decreases as the level of nitrogen increases. The underlying curve describing this relationship is thus concave. Other nonparametric assumptions on economic models include the concavity of utility functions and the homogeneity of some demand and production relations [3,5]. The problem of estimating the functional relationship subject only to concavity constraints also arises in the analysis of enzyme kinetic data [7].

The problem was first formulated by Hildreth [5] for the estimation of marginal productivity curves. By adopting the least squares method of estimation, he formulated this as a quadratic programming problem and gave an algorithm for solving it. Dent [2] and Holloway [6] furthered this work by embedding the problem in the more general framework of quadratic program-

---

Received September 22, 1980; revised March 27, 1981.

ming with linear inequality constraints. All these methods lead to the least squares solution but require considerable programming skill. Dent et al. [3] took a different approach. They proposed an algorithm which gives approximate solutions but can be hand calculated.

In this paper we consider both approaches and give algorithm 2 and algorithm 3 for each approach. To motivate these algorithms, we first review the isotonic regression problem and propose in section 2 a version of the "pool-adjacent-vilators" algorithm, which is slightly more efficient than the "up-and-down blocks" algorithm due to Kruskal [8]. In section 3 we give a characterization of the least squares solution for the concave regression problem, and propose algorithm 2 which is analogous to the "pool-adjacent-violators" algorithm in section 2. Sufficient conditions for the convergence of algorithm 2 to the least squares solution are given in section 4. Algorithm 2 involves a quadratic programming problem on a subset of variables and may not be easy to implement. In section 5 we propose algorithm 3 which is similar to the "pool-adjacent-violators" algorithm of section 2 except that one involves weighted averages, the other weighted least squares. Like the algorithm due to Dent et al. [3], algorithm 3 gives approximate solutions but is easy to implement. Conditions under which algorithm 3 converges to the least squares solution are also discussed in section 5. The algorithms are illustrated on two examples in section 6.

2. A modified version of "up-and-down blocks" algorithm for isotonic regression

In the isotonic regression problem, m_i independent measurements $y_{j,i}$ taken at x_i have mean $u_i = u(x_i)$ and variance σ^2, where $x_1 < \ldots < x_k$. The problem is to estimate u_i subject to the monotonicity constraints $u_1 \leq u_2 \leq \ldots \leq u_k$. For examples and justifications, see Barlow et al. [1]. Since

$$\sum_{i=1}^{k} \sum_{j=1}^{m_i} (y_{j,i} - u_i)^2 = \sum_{i=1}^{k} m_i (\bar{y}_i - u_i)^2 + \sum_{i=1}^{k} \sum_{j=1}^{m_i} (y_{j,i} - \bar{y}_i)^2$$

with

$$\bar{y}_i = m_i^{-1} \sum_{j=1}^{m_i} y_{j,i},$$

the least squares estimate of $\boldsymbol{u} = \{u_i\}_{i=1}^{k}$ under the monotonicity assumption is obtained by solving

$$\min_{\boldsymbol{u}} \sum_{i=1}^{k} m_i (u_i - \bar{y}_i)^2, \tag{1}$$

subject to $u_1 \leq u_2 \leq \ldots \leq u_k$. Since this is a quadratic program with linear inequality constraints, the solution $\{\hat{u}_i\}_{i=1}^k$ of (1) is obtained from the Kuhn–Tucker–Lagrange conditions [9, p. 87], which can be shown to be equivalent to the following:

$$\sum_{j=1}^{i} m_j \hat{u}_j = \sum_{j=1}^{i} m_j \bar{y}_j - \hat{\lambda}_i, \quad \hat{\lambda}_i \geq 0, \quad 1 \leq i \leq k-1, \tag{2a}$$

$$\sum_{j=1}^{k} m_j \hat{u}_j = \sum_{j=1}^{k} m_j \bar{y}_j, \tag{2b}$$

$$\hat{\lambda}_i = 0, \quad \text{if } \hat{u}_{i+1} > \hat{u}_i, \tag{2c}$$

$$\hat{u}_1 \leq \hat{u}_2 \leq \ldots \leq \hat{u}_k, \tag{2d}$$

where $\hat{\lambda}_i$ are uniquely determined by (2a)–(2d). The usual interpretation of the isotonic regression estimator [1, p.9] as the greatest convex minorant of the cumulative sum diagram of $(\bar{y}_i)$ with weights (m_i) follows easily from (2a)–(2d). Our approach thus provides another proof of this geometric characterization.

Based on (2a)–(2d), we give a version of the "pool-adjacent-violators" algorithm for obtaining $\hat{u}$, which is a slight variant of Kruskal's "up-and-down blocks" algorithm. We call the monotonicity constraint on $(i, i+1)$ *inactive*, *active*, and *violated*, respectively, iff $u_i - u_{i+1}$ is $<, =, >0$. If the constraint on $(i, i+1)$ is violated, then there exists a unique block B of points containing i and $i+1$ such that:

(i) any two consecutive points in B are related by a violated or active constraint, and
(ii) the two extreme (boundary) points in B are related to their neighboring points outside B by inactive constraints. (3)

In table 1 the constraint $u_2 - u_3 \leq 0$ is violated, the constraint $u_4 - u_5 \leq 0$ is active, and the block of points as defined above is $x_i = 2, 3, 4, 5$.

Table 1

x_i	1	2	3	4	5	6
u_i	0	2	1	0	0	3

Algorithm 1
Step 0: $\hat{u}_i = \bar{y}_i$, $i = 1, \ldots, k$.
Step 1: Check if there is any violated constraint. If none, stop; otherwise, denote the points in a violated constraint by i_0, $i_0 + 1$, go to step 2.
Step 2: Let $B = (i_0 - p, \ldots, i_0 + q)$ be the block of points containing i_0, $i_0 + 1$ as defined in (3). Replace $\{\hat{u}_i\}_{i \in B}$ by its weighted average $\sum_{i \in B} \hat{u}_i / \sum_{i \in B} m_i$. Update $\{\hat{u}_i\}_{i=1}^k$ and the "status" of each constraint, go to step 1.

In at most $k - 1$ steps, all the violated constraints will be eliminated. It can be shown further that (2a)–(2c) are preserved at each iteration. Therefore algorithm 1 terminates in at most $k - 1$ iterations and the final $\{\hat{u}_i\}_{i=1}^k$ are the desired solution $\hat{u}$. The difference between our algorithm 1 and the algorithms considered in [1, §2.3], including the "up-and-down blocks" algorithm, is in the formation of consecutive points into a block. Block B in algorithm 1 contains in one iteration as many consecutive points as allowed according to the Kuhn–Tucker–Lagrange conditions (2), while the other algorithms considered in [1] only merge points of the form $(i_0 - r, \ldots, i_0, i_0 + 1, \ldots, i_0 + s)$ into one block where $r \leq p$, $s \leq q$ and $\hat{u}_{i_0-r} = \ldots = \hat{u}_{i_0} > \hat{u}_{i_0+1} = \ldots = \hat{u}_{i_0+s}$. Such a block is necessarily a subset of block B defined in (3). Therefore the saving of iteration steps in algorithm 1 can be substantial. This is especially significant when k, the number of parameters, is large, a situation typically encountered in multidimensional scaling.

3. Least squares solutions in concave regression: characterization and algorithm

In the concave regression problem, the assumptions on $y_{j,i}$ are the same as in section 2 except that $u_i = u(x_i)$ now satisfy the concavity constraints

$$\frac{u_2 - u_1}{x_2 - x_1} \geq \frac{u_3 - u_2}{x_3 - x_2} \geq \ldots \geq \frac{u_k - u_{k-1}}{x_k - x_{k-1}}. \tag{4}$$

The least squares estimate of u under the concavity assumption is obtained by solving

$$\min_{u} \sum_{i=1}^{k} m_i (u_i - \bar{y}_i)^2 \tag{5}$$

subject to condition (4)

By the Kuhn–Tucker–Lagrange condition and some algebraic manipulation, the solution $\hat{u} = (\hat{u}_i)_{i=1}^k$ of (5) is characterized by the following equations:

$$\hat{U}'_i = Y'_i - \hat{\lambda}_i, \quad \hat{\lambda}_i \geq 0, \quad i = 1, \ldots, k-2, \tag{6a}$$
$$\hat{U}'_i = Y'_i, \quad\quad\quad\quad\quad\quad\quad i = k-1, k, \tag{6b}$$

$$\hat{\lambda}_i = 0, \quad \text{if} \frac{\hat{u}_{i+1} - \hat{u}_i}{x_{i+1} - x_i} > \frac{\hat{u}_{i+2} - \hat{u}_{i+1}}{x_{i+2} - x_{i+1}}, \tag{6c}$$

$$\frac{\hat{u}_2 - \hat{u}_1}{x_2 - x_1} \geq \ldots \geq \frac{\hat{u}_k - \hat{u}_{k-1}}{x_k - x_{k-1}}, \tag{6d}$$

where

$$\hat{U}'_i = \sum_{j=1}^{i} (x_{i+1} - x_j) m_j \hat{u}_j = \sum_{r=1}^{i} \Delta x_r U_r,$$

$$Y'_i = \sum_{j=1}^{i} (x_{i+1} - x_j) m_j \bar{y}_j = \sum_{r=1}^{i} \Delta x_r Y_r,$$

$$U_r = \sum_{j=1}^{r} m_j \hat{u}_j, \quad Y_r = \sum_{j=1}^{r} m_j \bar{y}_j; \quad \Delta x_r = x_{r+1} - x_r,$$

and $\hat{\lambda}_i$ are uniquely determined by (6a)–(6d).

No geometric characterization like the greatest convex minorant in isotonic regression can be drawn from (6). In the isotonic case, there are minimax formulae for the least squares estimator of u in terms of the weighted averages of $\{\bar{y}_j\}$. For concave regression, it is tempting to try the same minimax formulae by simply replacing the weighted averages of $\{\bar{y}_j\}$ by the weighted least squares fitted values using $\{\bar{y}_j\}$ with weights $\{m_j\}$. See the lemma for detail. We have tried this on the Hildreth's data in table 4 but failed to produce anything close to the least squares solution. No closed form for the least squares solutions of the concave regression problem has yet been found [3,4].

Before giving our algorithm for concave regression, we need to define several terms. We call the concavity constraint on $(i, i+1, i+2)$ *inactive* (strictly concave), *active* (linear), and *violated* (strictly convex), respectively, iff

$$\frac{u_{i+1} - u_i}{x_{i+1} - x_i} - \frac{u_{i+2} - u_{i+1}}{x_{i+2} - x_{i+1}} \text{ is } >, =, < 0.$$

If the constraint on $(i, i+1, i+2)$ is violated, then there exists a unique block B of points containing $i, i+1,$ and $i+2$ such that:

(i) any two consecutive points in B are related by at least one violated or active constraint, and
(ii) the two extreme (boundary) points in B are related to their neighboring points outside B only by inactive constraints. (7)

Table 2

x_i	1	2	3	4	5	6	7	8	9
u_i	8	6	5	4	2	3	2	0	1
$\frac{u_{i+1}-u_i}{x_{i+1}-x_i}$	-2	-1	-1	-2	1	-1	-2	1	

Since each constraint involves three consecutive points, any two consecutive points except at the two ends can be related by two different constraints. In (3), any constraint involving two consecutive points in B is either active or violated, while in (7) some constraints involving three consecutive points in B may be inactive. In table 2 the constraint on $x_i = 1, 2, 3$ is violated and the block B of points as defined in (7) is $x_i = 1, 2, 3, 4, 5, 6$. Within B the constraint on $x_i = 3, 4, 5$ is inactive but points 3, 4 are related by one active constraint and points 4, 5 are related by one violated constraint.

Algorithm 2
Step 0: $\hat{u}_i = \bar{y}_i$, $i = 1, \ldots, k$.
Step 1: Check if there is any violated constraint. If none, stop; otherwise, denote the points in a violated constraint by i_0, $i_0 + 1$, $i_0 + 2$, go to step 2.
Step 2: Let B be the block of points containing i_0, $i_0 + 1$, $i_0 + 2$ as defined in (7). Replace $\{\hat{u}_i\}_{i \in B}$ by the solution $\{\hat{\mu}_i\}_{i \in B}$ of the constrained minimization problem:

$$\min_{\mu} \sum_{i \in B} m_i(\mu_i - \hat{u}_i)^2, \ \mu = (\mu_i)_{i \in B} \tag{8}$$

subject to the concavity constraints (4) on μ. Update $\{\hat{u}_i\}_{i=1}^k$ and the "status" of each constraint, go to step 1.

Step 2 of algorithm 2 is analogous to step 2 of algorithm 1 since the weighted average of $\hat{u}_i$ over B in algorithm 1 is the solution of the corresponding optimization subproblem, i.e. the concavity constraints in (8) are replaced by monotonicity constraints. The solution of (8) has no closed form, because some of the constraints involving three consecutive points in B may be inactive. But when all the constraints involved in B are either violated or active (then the piecewise linear graph formed by $(x_i, u_i)_{i \in B}$ is convex), the solution of (8) is obtained by fitting a weighted least squares line to $(u_i)_{i \in B}$, which is stated as the following

Lemma. Suppose $\{a_i\}_1^n$ satisfies

$$\frac{a_2 - a_1}{x_2 - x_1} \leq \frac{a_3 - a_2}{x_3 - x_2} \leq \ldots \leq \frac{a_n - a_{n-1}}{x_n - x_{n-1}}. \tag{9}$$

Then the solution to

$$\min_z \sum_{i=1}^n m_i(z_i - a_i)^2 \text{ subject to } \frac{z_2 - z_1}{x_2 - x_1} \geq \ldots \geq \frac{z_n - z_{n-1}}{x_n - x_{n-1}} \quad (10)$$

is

$$\hat{z}_i = \bar{a} + \hat{\beta}(x_i - \bar{x}), \quad i = 1, \ldots, n,$$

obtained by fitting the weighted least squares line with weights $\{m_i\}_1^n$ to "data" $\{a_i\}_1^n$, where

$$\bar{a} = \sum_{i=1}^n m_i a_i / \sum_{i=1}^n m_i,$$

$$\bar{x} = \sum_{i=1}^n m_i x_i / \sum_{i=1}^n m_i,$$

and

$$\hat{\beta} = \sum_{i=1}^n m_i(a_i - \bar{a})(x_i - \bar{x}) / \sum_{i=1}^n m_i(x_i - \bar{x})^2.$$

Proof. Since problem (10) is the projection of $\{a_i\}_{i=1}^n$ onto the convex set defined by the linear inequalities in (10), its solution, denoted $\{\hat{a}_i\}_{i=1}^n$, exists and is unique. Let $G(G^*)$ be the piecewise linear graph formed by connecting $\{(x_i, a_i)\}_{i=1}^n (\{(x_i, \hat{a}_i)\}_{i=1}^n)$. Then G is convex and G^* is concave. If G^* is a linear graph, then $\hat{a}_i = \bar{a} + \hat{\beta}(x_i - \bar{x})$, according to the definition of the weighted least squares. Suppose now G^* is not linear, we will show that $\{\hat{a}_i\}_{i=1}^n$ is not the solution to (10), thus contradicting its definition. If the intersection of G and G^* contains at least two points, Q_1 and Q_2, then from the convexity of G and the concavity of G^* the points on the straight line connecting Q_1 and Q_2 are closer to $\{a_i\}_{i=1}^n$ than $\{\hat{a}_i\}_{i=1}^n$ are. If the intersection of G and G^* consists of one point or is empty, then by the separating hyperplane theorem [9, p.40] there exists a straight line which separates the convex graph G and the concave graph G^*. Again the points on this line are closer to $\{a_i\}_{i=1}^n$ than $\{\hat{a}_i\}_{i=1}^n$ are.

According to the lemma the optimization subproblem (8) in algorithm 2 has a simple solution by using least squares fitting if none of the constraints involved in B is inactive. For example, the points in table 2 can be divided into two blocks, $B_1 = \{1, 2, \ldots, 6\}$ and $B_2 = \{7, 8, 9\}$, according to (7). The lemma is applicable to B_2 but not to B_1.

4. Convergence conditions

We conjecture that algorithm 2 eliminates all the violated constraints and terminates in a finite number of steps, though we are not able to prove so. Assuming that all the violated constraints are eliminated at the end, we proceed to prove the convergence of $\{u_i\}$ in algorithm 2 to the optimal solution by verifying (6a)–(6c). For $n \geq \ell + 1$,

$$U'_n = U'_\ell + (x_{n+1} - x_\ell) U_\ell + \sum_{i=\ell+1}^{n} \Delta x_i \sum_{j=\ell+1}^{i} m_j \hat{u}_j. \tag{11}$$

If step 2 is applied to the first ℓ points, then from relation (6) as applied to these points, $U'_\ell = Y'_\ell$, $U_\ell = Y_\ell$ and the third term of (11) is unchanged. If there are two disjoint blocks of points B_1 and B_2, the updating of points in B_1 by solving (8) will not affect the points in B_2. By applying (6a) and (6b) to each block of points, $U'_i \leq Y'_i$, for all i, and $U'_i = Y'_i$, for $i = k - 1, k$, hold true in each iteration. It remains to prove (6c). Let $U'_{i,s}$ be the U_i value at the s th iteration, $U'_{i,0} = Y'_i$. From the above discussion, $U'_{i,s}$ is nonincreasing in s. If $U'_{i,s} < U'_{i,s-1} \leq U'_{i,0} = Y'_i$ at the s th iteration, then in order to satisfy (6c), the constraint on $(i, i+1, i+2)$ cannot be inactive when the algorithm terminates. According to (6a) and (6c), $U'_{i,s} < U'_{i,s-1}$ only when the constraint on $(i, i+1, i+2)$ is active or violated at the $(s-1)$th iteration and becomes active at the s th iteration. A set of sufficient conditions for convergence to optimality is now in order.

Theorem. The $\{u_i\}_{i=1}^k$ in algorithm 2 converges to the least squares solution of the concave regression problem (5), provided
  (i) the algorithm eliminates all the violated constraints in finite number of steps, and
  (ii) if a violated or active constraint becomes active after one iteration, it should not be inactive at the end.

Conditions (i) and (ii) are satisfied by two important special cases.

Corollary. The convergence of algorithm 2 to optimality in the main theorem is guaranteed, if (a) *or* (b) is satisfied:
  (a) In each iteration no constraint involving any three consecutive points in B of Step 2 is inactive.
  (b) Each constraint can be updated at most once.

That (b) implies (i) and (ii) is obvious. Under (a), problem (8) is solved by least squares fitting and all the constraints involved in block B become active (linear) according to the lemma. After each iteration, only inactive constraints involving one or two points in B may become violated; in the next iteration a larger block B' containing B and the points in the violated constraints will be formed, because all the constraints in B are active and the violated constraints have points in B. Since in each iteration at least one violated constraint is removed (and never comes back), the algorithm terminates in $k-2$ steps under (a).

5. An approximate algorithm using least squares fitting

Algorithms for statistical purpose, especially in small data sets, should be simple to implement. Algorithm 1 for isotonic regression falls in this category. One may carry out the computations on a calculator. The problem with algorithm 2 is in solving the optimization problem (8), which may require another subroutine like those developed in [2,5,6] unless condition (a) of the corollary is met. It is sometimes desirable to have an algorithm which is easy to implement but gives only approximate solutions.

Denote the $k-2$ constraints by $C_1, C_2, \ldots, C_{k-2}$. If a constraint C_j is violated, there is the *largest* block of constraints $(C_{j-f}, \ldots, C_{j+g})$ containing C_j such that all C_i, $j-f \leq i \leq j+g$, are either violated or active. The block C of points involved in C_i, $j-f \leq i \leq j+g$, is contained in the block B of points defined in (7) and the inclusion may be strict. In table 2 the block C_1, containing the violated constraint on (1, 2, 3), is (1, 2, 3, 4); the block C_2, containing the violated constraint on (4, 5, 6), is (4, 5, 6). C_1 and C_2 have point 4 in common and are both properly included in $B = (1, 2, 3, 4, 5, 6)$. From the lemma, the solution for minimization problem (8), when restricted to C, has a simple form in terms of the weighted least squares fitting. Formally we state.

Algorithm 3
Steps 0, 1: Same as in algorithm 2.
Step 2: Let C be the block of points defined above. Replace $\{\hat{u}_i\}_{i \in C}$ by $\{\hat{\mu}_i\}_{i \in C}$, where $\hat{\mu}_i = \bar{u} + \hat{\beta}(x_i - \bar{x})$, $\bar{u} = \Sigma_{i \in C} m_i \hat{u}_i / \Sigma_{i \in C} m_i$, $\bar{x} = \Sigma_{i \in C} m_i x_i / \Sigma_{i \in C} m_i$, $\hat{\beta} = \Sigma_{i \in C} m_i (\hat{u}_i - \bar{u})(x_i - \bar{x}) / \Sigma_{i \in C} m_i (x_i - \bar{x})^2$. Update $\{\hat{u}_i\}_{i=1}^k$ and the "status" of each constraint, go to step 1.

Under assumption (a) of the corollary, the block C of points in algorithm 3 coincides with the block B of points in algorithm 2 and the two algorithms are identical. Convergence to optimality of algorithm 3 is thus guaranteed via the corollary. Other than this, there is no general sufficient condition for convergence to optimality. As pointed out in the paragraph before algorithm 3, two different blocks, C_1 and C_2 as defined there, may have one point in common and so relation (14) may not hold. The updating of $\{\hat{u}_i\}$ in C_1 may affect that in C_2 and (6a) cannot be verified in general for algorithm 3.

Despite this defect, algorithm 3 has three appealing features.

(i) It terminates in at most $k-2$ steps (for reasons see the end of section 4).

(ii) It can easily be implemented without much programming skill and software support.

(iii) It gives estimates of $\{u_i\}$ which are useful for practical purpose. A similar (but not as efficient) algorithm was proposed by Dent et al. [3]. A Monte Carlo comparison [3, section 5] of their approximate algorithm and Hildreth's exact algorithm [5] indicates that the loss of statistical efficiency due to the use of their method is marginal.

When serious programming is not contemplated and high precision not required, we recommend the use of algorithm 3. Our recommendation is best defended by quoting Dent et al. [3, p. 27]:

On an absolute basis, little gain is provided by the QP solution [Hildreth's method], and given the considerably greater computational costs involved, it is not clear that this effort is worthwhile. The QLS estimator [similar to our algorithm 3], which can be calculated by hand, appears to be a valuable approximation to the QP solution and of considerable merit in its own right. By construction it will be consistent and is no more biased than the QP estimator.

6. Illustrations

Example 1. Holloway [6, pp. 403–404] gave this example to illustrate his general optimization algorithm for convex estimation. To fit into our concavity framework, we change the signs of his $\{y_i\}$ and $\{\hat{u}_i\}$ and give them in table 3. The only violated constraint is on $x_i = 4, 6, 9$. According to our algorithms 2 or 3, we fit the least squares line (with equal weight $m_i = 1$) to the three points $(x_i, y_i) = (4, -2), (6, -6)$ and $(9, -4)$ and obtain the fitted values $\hat{u}_i = -3.26$, -3.90 and -4.84, given in table 3. Since all the concavity constraints are now satisfied, we get the same estimates as in [6]. Note that both (a) and (b) of the Corollary are satisfied.

Example 2. Hildreth [5] applied his quadratic programming (QP) estimator to study the relationship (assuming concavity) between $\bar{y}_i$, average corn yield (in bushels), and x_i, nitrogen fertilizer (in lb). From table 4, the three constraints

Table 3

x_i	2	4	6	9	10
y_i	-10	-2	-6	-4	-8
$\dfrac{y_{i+1}-y_i}{x_{i+1}-x_i}$	4	-2	$\tfrac{2}{3}$	-4	
$\hat{u}_i$	-10	-3.26 [a]	-3.90 [a]	-4.84 [a]	-8

[a] Obtained by least squares fitting ($m_i = 1$ for all i).

Table 4

x_i	0	20	40	60	80	120	160	180
m_i	27	9	8	10	9	19	10	8
$\bar{y}_i$	22.94	41.58	65.46	58.81	81.74	82.15	96.59	94.01
$\frac{\bar{y}_{i+1} - \bar{y}_i}{x_{i+1} - x_i}$	0.93	1.19	−0.33	1.15	0.01	0.36	−0.13	
$\hat{u}_i$	22.94	41.58	60.13	67.34	74.55	84.47	94.39	94.01
$\hat{u}_i^{(1)}$	22.94	41.58	59.65 [a]	68.11 [a]	76.57 [a]	82.15	96.59	94.01
$\hat{u}_i^{(2)}$	22.94	41.58	59.65	68.11	74.24 [a]	84.36 [a]	94.49 [a]	94.01

[a] Obtained by weighted least squares fitting with weight m_i.

on $x_i = (0, 20, 40)$, $(40, 60, 80)$ and $(80, 120, 160)$ are violated. According to algorithm 2, we solve the optimization problem (8) on the first seven points (the block B of points in step 2 of the algorithm consists of the first seven points) by the standard quadratic programming methods [3,5,6]. On further analysis it can be shown that the $\{\hat{u}_i\}$ in table 4 are obtained by fitting two straight lines to the $\bar{y}_i$ values over $x_i = (40, 60, 80)$ and $x_i = (80, 120, 160)$ with a joint at $x_i = 80$, and leaving the other $\bar{y}_i$ values intact. Therefore algorithm 2 coincides with Hildreth's method on these data except that it only involves seven variables.

Algorithm 3, when applied to the data, gives approximate solutions. We first select the violated constraint on $x_i = (40, 60, 80)$ and replace the three $\bar{y}_i$ values by their weighted least squares fitted values 59.65, 68.11 and 76.57, obtaining $\hat{u}_i^{(1)}$ at the first iteration. Now only the constraint on $x_i = (80, 120, 160)$ is violated. Again by replacing the three $\bar{y}_i$ values by their weighted least squares fitted values, we obtain $\hat{u}_i^{(2)}$ at the second iteration. Since all the constraints are satisfied, the algorithm terminates in two iterations. Note that $\{\hat{u}_i\}$ and $\{\hat{u}_i^{(2)}\}$ are very close to each other. We have also tried algorithm 3 by starting with the other two violated constraints. The algorithm terminates in three iterations in both cases and the estimates are very close to $\{\hat{u}_i\}$, further lending support to the simulation study of Dent et al. [3].

Acknowledgments

This research was supported by National Science Foundation Grant MCS 79-01846. The author is grateful to Wing Hung Wong for suggestions which greatly simplify the proof of the lemma, and to Ron Kenett and Sue Leurgans for suggestions on references.

References

[1] R.E. Barlow, D.J. Bartholomew, J.M. Bremner and H.D. Brunk, *Statistical Inference under Order Restriction* (John Wiley, New York, 1972).
[2] W. Dent, "A Note on Least Squares Fitting of Function Constrained to be Either Nonnegtive, Nondecreasing or Convex", *Management Science* 20 (1973) 130–132.
[3] W. Dent, T. Robertson and M. Johnson, "Concave Regression", Unpublished manuscript, University of Iowa (1977).
[4] D.L. Hanson and G. Pledger, "Consistency in Concave Regression", *Ann. Statistics* 4 (1976) 1038–1050.
[5] C. Hildretch, "Point Estimates of Ordinates of Concave Functions", *Journal of the American Statistical Association.* 49 (1954) 598–619.
[6] C.A. Holloway, "On the Estimation of Convex Functions", *Operations Research* 27 (1979) 401–407.
[7] R. Kenett, "Statistical Analysis of Enzyme Kinetic Data", Technical Report 580, Department of Statistics, University of Wisconsin, Madison (1979).
[8] J.B. Kruskal, "Nonmetric Multidimensional Scaling: A Numerical Method", *Psychometrika* 29 (1964) 115–129.
[9] A. Takayama, *Mathematical Economics* (The Dryden Press, Hinsdale, Ill., 1974).

CONCORDANT AND DISCORDANT MONOTONE CORRELATIONS AND THEIR EVALUATION BY NONLINEAR OPTIMIZATION

George KIMELDORF *

University of Texas at Dallas

Jerrold H. MAY and Allan R. SAMPSON [†]

University of Pittsburgh

Consider a two-way frequency table classifying all corporate bonds rated by both Moody's and Standard and Poor's, or the familiar father–son social mobility data matrix often used as a human resources Markov chain example. Both are examples of contingency tables whose rows and columns represent ordinal categorical variables, since such a natural ordering exists for bond ratings and for social class. In this paper we consider an approach to measuring and quantifying the degree of relationship between such row and column variables.

In the case of cardinal variables, correlation is a well-defined measure of association, and no scaling on the categories would be required. If there is no need to respect the order relationships involved, sup correlation, an eigenanalysis related scheme, could be used to compute a measure of association. Procedures for ordinal categorical data, such as Kendall's tau or Spearman's rho, impose numerical values on the categories.

We present some new procedures to measure the relationship between ordinal variables. Four new statistical measures of monotone relationship are derived from the concept of monotone correlation. The appropriateness of each of these measures depends on the specific situation giving rise to the data. These measures of association can be obtained only by using an iterative procedure. A nonlinear optimization algorithm is employed to evaluate these measures and to obtain the associated monotone scalings.

A computer program to implement the algorithm is developed, and is applied to several insightful examples to provide further understanding of the usefulness of these measures.

1. Introduction and statistical background

Measuring and understanding the basis for the association between two random variables, X and Y, is extremely important for the intelligent application of statistics, as well as for more insight into the underlying bivariate probabilistic structures. The focus of this paper is on the association between

---

Received October 7, 1980; revised July 31, 1981.

* The work of this author was supported by the National Science Foundation under Grant MCS-80-02152.

[†] The work of this author is sponsored by the Air Force Office of Scientific Research, Air Force Systems Command under Contract F49620-79-C-0161.

ordinal random variables, i.e. random variables where the observed values have a natural ordering without necessarily having naturally ascribed numerical values. For example, the values may arise from questionnaire responses based on the five-point scale: strongly disagree, disagree, no opinion, agree, strongly agree. In measuring the association between two ordinal variables using a measure based on assigning numerical values to the possible data values, it is natural to require that the resultant numerical measure of association does *not* depend on the actual numerical values but only the orderings. This property can be described as monotone scale invariance. When Pearson's correlation coefficient is used by assigning the values $1, \ldots, N$ to the scale levels, the resulting measure is not monotone scale invariant. For the five-point scale example, assigning 1 to strongly disagree, up to 5 to strongly agree and then computing the Pearson correlation would not provide a monotone invariant measure of association.

One monotone invariant scale measure is the *sup correlation* ρ' introduced by Gebelein [6], developed further by Sarmanov [16,17], Rényi [15], and Lancaster [12], and defined by $\rho'(X, Y) = \sup \rho(f(X), g(Y))$, where the supremum is taken over all Borel-measurable functions, f and g, such that $0 < \text{var } f(X) < \infty$ and $0 < \text{var } g(Y) < \infty$, where ρ is the Pearson correlation coefficient. For random variables (X, Y) jointly taking values on a finite rectangular lattice, there are computational methods for computing ρ' using eigenvalue routines (see Sarmanov and Lancaster). For continuous random variables, X and Y, the sup correlation is computable only in special cases where the joint p.d.f. admits a certain type of bivariate orthogonal expansion (see [12]).

An important dependence concept between two random variables is that of complete dependence, introduced by Lancaster [11]. A random variable Y is said to be *completely dependent* on a random variable X if there exists a function g such that

$$\text{prob}[Y = g(X)] = 1. \tag{1}$$

If Y is completely dependent on X and vice versa, then X and Y are said to be *mutually completely dependent*; in this case X and Y are perfectly predictable from each other. Observe that if X and Y are mutually completely dependent, then $\rho'(X, Y) = 1$.

Kimeldorf and Sampson [9] provided an example of random variables X and Y which were mutually completely dependent and yet were "almost" stochastically independent. To circumvent this difficulty, Kimeldorf and Sampson defined Y to be *monotone increasing (decreasing) dependent* on X if (1) holds for a monotone increasing (decreasing) function g. Furthermore, motivated by trying to measure the degree of monotone dependence, they defined the *monotone correlation* between random variables X and Y by

$$\rho^*(X, Y) = \sup \rho(f(X), g(Y)), \qquad (2)$$

where the supremum is taken over all monotone functions f and g for which $0 < \text{var } f(X) < \infty$ and $0 < \text{var } g(Y) < \infty$. The monotone correlation is a monotone scale-invariant measure of association and the maximizing functions (assuming they exist) for (2) are the "best" monotone scalings for cross linear predictability of X and Y. (Monotone scalings are order-preserving assignments of numerical values to ordinal data.) Kimeldorf and Sampson evalated the monotone correlation in only two special situations: (i) X and Y bivariate normal, in which case $\rho^* = |\rho|$; and (ii) X and Y independent, in which case $\rho^* = 0$.

The purposes of this paper are twofold. One is to derive new measures associated with the monotone correlation and to study their applicability. A second is to provide a computational procedure and computer program to evaluate the monotone correlation and these derived measures for the case when X and Y assume a finite number of values. The approach is to find an equivalent nonlinear program and then employ a modification of the optimization algorithm of May [13] to compute the maximizing values and the points at which they occur. In section 2 we introduce the concepts of concordancy, discordancy, and isoscaling for measuring the monotone association. The equivalent nonlinear programs are given in section 3. The specific algorithm and the computer program, which we call MONCOR, are described in section 4. A number of interesting applications and examples are considered in section 5. In section 6 we discuss how these methods might be used for scale reduction.

2. Concordancy, discordancy, and isoscaling

The concept of a monotone correlation can be refined by measuring separately the strength of the relationship between X and Y in a positive direction and the strength of the relationship in a negative direction, i.e. to measure separately the extent of concordancy and of discordancy between X and Y. These concepts are related to so-called measures of disagreement and measures of dissociation. If in (2) f and g are both restricted to be (nonstrictly) increasing (or equivalently both decreasing (nonstrictly)), the resulting measure is called the *concordant monotone correlation* (*CMC*). When f is restricted to be increasing and g decreasing (or equivalently f decreasing and g increasing), we find it convenient to examine -sup $\rho(f(X), g(Y))$, which in turn can be expressed as -sup $\rho(f(X), -g(Y))$, where both f and g are increasing. This leads naturally to defining the *discordant monotone correlation* (*DMC*) by inf $\rho(f(X), g(Y))$, where f and g are both restricted to be increasing.

The *DMC* and *CMC* have natural interpretations as measures of negative and positive association, respectively, for ordinal random variables. They also can be interpreted as providing bounds for the correlation between any arbitrary monotone scalings; specifically, for arbitrary increasing f and g,

$$DMC \leq \rho(f(X), g(Y)) \leq CMC. \tag{3}$$

Suppose it is desired to impose numeric monotone scalings for a pair of new tests; if the *CMC* and *DMC* are close, then by (3) it makes little difference which monotone scales are used. Also, if $CMC = DMC = 0$, then X and Y are independent random variables; however, it is possible for $DMC < CMC = 0$ and X and Y not to be independent. Furthermore, note that if X and Y are increasing monotone dependent then $CMC = 1$; and if X and Y are decreasing monotone dependent, then $DMC = -1$.

Sometimes the situation occurs when X and Y should have the same scaling. For example, suppose that X is a psychological test score pre-treatment and Y is the score post-treatment on the same test. This leads to another extension of the monotone correlation, which we refer to as isoscaling. If in (2) we restrict $f = g$, the resulting measure is called the *isoconcordant monotone correlation* (*ICMC*). Analogous to the *DMC* definition, the *isodiscordant monotone correlation* (*IDMC*) is given by inf $\rho(f(X), g(Y))$, where $f = g$. Obviously, isoscaling is not appropriate in practice when X and Y have essentially different ranges of values.

If X and Y are exchangeable ordinal random variables it might be conjectured, due to all the symmetries involved, that $ICMC = CMC$ (and $IDMC = DMC$). However, as is shown in section 5, surprisingly this is not the case.

The actual functions that maximize the correlations (assuming they exist) are generically called *monotone variables*; their specific interpretation depends upon which monotone correlation measure is used in their derivation. When measuring the monotone association using one of our monotone measures, we strongly advocate simultaneously examining the values of the corresponding monotone variables. Otherwise there can be potential misinterpretations. For instance, Hall [8, example 7] presents an example where the support of X and Y is three monotone disjunct pieces and the $CMC = 1$. In this example, corresponding monotone variables are $I_{(N,\infty)}(x)$ and $I_{(N,\infty)}(y)$, where $I_A(s)$ is the indicator function of the set A. Also, as we note in section 6 below, the monotone variables themselves may be quite useful in constructing scales for ordinal data.

3. Program formulation

The preceding extensions of the monotone correlation are applicable to all suitable pairs of random variables, continuous or discrete. We now focus on the case where X and Y jointly take on a finite number of values (a_i, b_j), $i = 1, \ldots, I$, $j = 1, \ldots, J$, and $\text{prob}(X = a_i, Y = b_j) = p_{ij}$. Then:

$$CMC = \max \left\{ \left[\sum_{i=1}^{I} \sum_{j=1}^{J} f(a_i) p_{ij} g(b_j) - \left(\sum_{i=1}^{I} f(a_i) p_{i \cdot} \right) \left(\sum_{j=1}^{J} g(b_j) p_{\cdot j} \right) \right] \right.$$

$$\times \left[\left(\sum_{i=1}^{I} f^2(a_i) p_{i \cdot} - \left(\sum_{i=1}^{I} f(a_i) p_{i \cdot} \right)^2 \right)^{1/2} \right.$$

$$\left. \left. \times \left(\sum_{j=1}^{J} g^2(b_j) p_{\cdot j} - \left(\sum_{j=1}^{J} g(b_j) p_{\cdot j} \right)^2 \right)^{1/2} \right]^{-1} \right\}, \tag{4}$$

subject to f and g being increasing functions for which the denominator in (4) is nonzero and where $p_{i \cdot} = \sum_{j=1}^{J} p_{ij}$ and $p_{\cdot j} = \sum_{i=1}^{I} p_{ij}$. Denote the values $f(a_i)$ by x_i, $i = 1, \ldots, I$, and $g(b_j)$ by y_j, $j = 1, \ldots, J$, so that (4) can be reformulated as

$$CMC = \max \frac{x'Py - (x'Pe)(y'P'e)}{\left(\Sigma x_i^2 p_{i \cdot} - (x'Pe)^2 \right)^{1/2} \left(\Sigma y_j^2 p_{\cdot j} - (y'P'e)^2 \right)^{1/2}} \tag{5}$$

subject to

$x_1 \leq \ldots \leq x_I,$

$y_1 \leq \ldots \leq y_J,$

$x \neq c_1 e, y \neq c_2 e,$

where $x = (x_1, \ldots, x_I)'$, $y = (y_1, \ldots, y_J)'$, $P = \{p_{ij}\}$ and $e = (1, \ldots, 1)'$. Thus, to compute the CMC all that is required is the matrix P of probabilities. For instance, the values $a_1, \ldots, a_5$ could be the five-point scale strongly disagree,..., strongly agree. The resultant monotone variable x would then provide a numerical scale assigning x_1 to strongly disagree,..., x_5 to strongly agree.

Analogous formulations of (5) can be given for $ICMC$, DMC, and $IDMC$. Again, the $ICMC$ and $IDMC$ are not defined when $I \neq J$.

When reporting the monotone variables, we standardize them without loss of generality so that in (5), for example, $x_1 = y_1 = 0$ and $x_I = y_J = 1$.

Until this point, the *CMC*, etc. have been defined as population quantities. For data from finite discrete distributions, the joint probabilities can be estimated from the data viewed in ordinal contingency table form. Then the *CMC* can be evaluated based upon the estimated probabilities. In this situation the *CMC* can either be viewed as an estimate of the "true" *CMC* or be viewed as a measure of monotone association for the ordinal contingency table.

4. Optimization approach and MONCOR description

The nonlinear programming problem (5) involves the optimization of a nonlinear fractional form subject to linear constraints. Note that if it were not for the monotone constraints, (5) would be an eigenvalue problem. The objective function in (5) is not pseudoconcave. To see this, consider the simple case of evaluating the

$$ICMC = \max(x'Px - (x'Pe)^2)/(\Sigma x_i^2 p_{i\cdot} - (x'Pe)^2)$$

for a symmetric probability matrix P. While both numerator and denominator are continuously differentiable on the feasible region, and $(\Sigma x_i^2 p_{i\cdot} - (x'Pe)^2)$ is a positive convex function of x, $(x'Px - x'Pe)^2$ would have to be non-negative and concave for pseudoconcavity (see Avriel [1]). This latter condition does not hold in general for symmetric P. Hence, in general, the *CMC*, and *ICMC*, *DMC* and *IDMC* will involve the optimization of a function with local optima. Although much work is presently being done in the area of global optimization (see, for example, Dixon and Szego [3,4]), we follow the standard procedure of using various starting points, computing the optima, and then choosing the best result based upon the different starting points.

Note that since correlation is unique in x and y only up to location and scale change, we could express (5) as

maximize $x'Py$

subject to

$$\sum_i x_i p_{i\cdot} = 0; \quad \sum_j y_j p_{\cdot j} = 0; \quad \sum_i x_i^2 p_{i\cdot} = 1; \quad \sum_j y_j^2 p_{\cdot j} = 1,$$

$$x_1 \leq x_2 \leq \ldots \leq x_I, \tag{6}$$

and

$$y_1 \leq y_2 \leq \ldots \leq y_J.$$

The formulation of (6), because of its nonlinear constraints, is not a desirable formulation since complexity in the objective function is much easier to deal with than complexity in the constraints. The constraints $x \neq c_1 e$ and $y \neq c_2 e$ in (5) are not computationally implementable in continuous variables. However, without loss of generality, we eliminate those constraints by fixing x_1 and y_1 at zero and x_I and y_J at one.

Specifically, the computation of the *CMC* (*DMC*) involves optimizing a nonconcave (nonconvex) function in $I + J - 4$ independent variables subject to monotonicity constraints. (The *ICMC* and *IDMC* involve $I - 2$ independent variables.) Since P is envisioned to be not much larger than 10×10, a modified Newton method was considered desirable because it should converge in a small number of iterations. *QRMNEW* (see [13]), an optimization method not requiring analytical derivatives, was employed because of its ease of adaptation and computational use.

QRMNEW is a hybrid local variations-modified Newton method, using orthogonal (*QR*) matrix factorization to derive a representation for the locally feasible region. It has been shown by May that starting from any initial point, *QRMNEW* converges to a point satisfying both first- and second-order necessary optimality conditions, so that any solution generated is at least a local optimum. Superlinear and order 2 convergence rates can be established under somewhat stronger conditions. Denote by $\{(x, y)^k\}$ the iterative sequence of points generated by the algorithm. In general, because of the lack of pseudo-concavity (pseudoconvexity) for the *CMC* and *ICMC* (*DMC* and *IDMC*), an iterate $(x, y)^k$ will usually be in a region not locally concave (convex). The

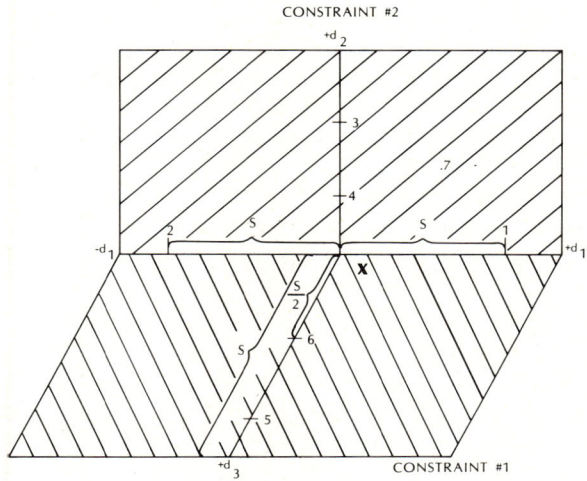

Fig. 1. An iteration of QRMNEW.

algorithm does have a rather sophisticated method for dealing with the indefinite projected matrix of second derivatives implied by the lack of local concavity (convexity).

While a complete mathematical description of *QRMNEW* is given by May, a general iteration is illustrated in fig. 1 to show the underlying logic. Given the current point x and a stepsize $s > 0$, the constraints, if any, that are satisfied exactly at x are used to generate, via orthogonal matrix factorization, a set of n coordinate directions. If u constraints are active, u directions lie in the manifold determined by those constraints, and the remaining $n - u$ directions are determined by computing a generalized inverse and are orthogonal to that manifold. (If no constraints are active, the standard Cartesian coordinate system is used.) The objective function is then evaluated at two points along each of these coordinate axes. For example, in fig. 1 two constraints are active in R^3. Three directions are generated: d_1, which lies in the manifold, so that movement away from x in either $+d_1$ or $-d_1$ is feasible; d_2, where going along $+d_2$ leads to infeasibility; and d_3, which is analogous to d_2. The function is evaluated at points 1 through 6, yielding second-order approximations to first and second partial directional derivatives along d_1, d_2, and d_3. Assume a maximum is being sought, e.g. computing the *CMC* or *ICMC*, and that the first derivatives along d_1, d_2, and d_3 are, respectively, positive, positive, and negative. Then the objective function cannot be increased by movement along d_3, so that it is dropped from consideration. The function is then evaluated at point 7, which is needed to approximate the second mixed partial directional derivative with respect to d_1 and d_2. A Newton-type search direction is computed and searched, and the algorithm moves to the best of the points found by Newton search procedure and points 1 through 7.

MONCOR is an interactive package designed to analyze probability matrices, *P*, of dimension less than or equal to 20×20. The user may input a single starting point for an optimization run, or allow the program to generate its own multiple starting points. In both cases the constraint set corresponding to the correlation measure requested is generated internally, and *QRMNEW* is used to compute an optimum. Additionally, two different strategies are employed in seeking an optimal solution. Numerical experience indicates that optimum values sometimes lie at monotone extreme points, i.e. points where all the *x* and *y* entries are either zero or one. This appears to be especially the case when computing the *DMC* or *IDMC* for a matrix with highly positive *CMC*, and vice versa. In fact, for certain cases the optima for all four monotone correlation measures might be achieved only at such points. Additionally, because nonoptimal monotone extreme points can be local optima (satisfying the Karush–Kuhn–Tucker second-order necessary optimality conditions (see Fiacco and McCormick [5]), *QRMNEW* starting from a random point might well be trapped by these local optima. Note that for an $I \times J$ matrix, there are only $(I-1)(J-1)$ monotone extreme points to consider for the *CMC* and

DMC $((n-1)$ for $ICMC$ and $IDMC$, assuming $I=J=n$). Hence, in order to avoid stopping at a local optimum when the global optimum is a monotone extreme point, $MONCOR$ evaluates the correlation of all monotone extreme points. Moreover, $MONCOR$ generates ten random monotone points, with coordinates selected on (0, 1), using the DEC random number generator (see [16]), and calls $QRMNEW$ to compute an optimum starting from each of them. The user may select to see only the final output, or an iteration-by-iteration output of the monotone correlations and monotone variables.

5. Applications

By means of the algorithm and the $MONCOR$ program, we now compute the CMC, etc. for several insightful examples. Let (X, Y) be a discrete bivariate random vector taking values in a 6×6 lattice: $\{a_1,\ldots,a_6\} \times \{b_1,\ldots,b_6\}$. Furthermore, suppose $\text{prob}(X = a_i) = 1/6$, for all i, and $\text{prob}(Y = b_j) = 1/6$, for all j; i.e. X and Y have uniform marginals. If X and Y are monotone increasing dependent then $\boldsymbol{P} = (1/6)\,\boldsymbol{I}$, where $\boldsymbol{P} = \{\text{prob}(X=i, Y=j)\}$, and $\boldsymbol{I}$ is the 6×6 identity matrix; if X and Y are monotone decreasing dependent, then $\boldsymbol{P} = (1/6)\,\boldsymbol{I}^*$, where $\boldsymbol{I}^* = \{\delta(i+j-7)\}$, and $\delta(x)$ is 1 if $x=0$, and is 0, otherwise. Now consider a one-parameter family of distributions indexed by θ, i.e. for a given θ, $\text{prob}(X=i, Y=j)$ is the (i,j)th element of $\boldsymbol{P}_\theta$, where

$$\boldsymbol{P}_\theta = \left(\frac{1+\theta}{2}\right)(1/6)\,\boldsymbol{I} + \left(\frac{1-\theta}{2}\right)(1/6)\,\boldsymbol{I}^*, \tag{7}$$

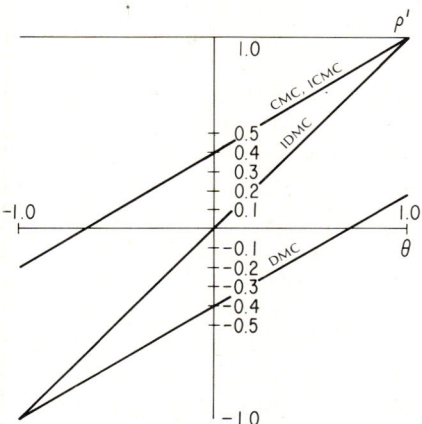

Fig. 2. CMC, DMC, IMC, $IDMC$ vs θ. $\boldsymbol{P}(\theta)$ given by (7).

where $-1 \leq \theta \leq 1$. Note that X and Y still have uniform marginal distributions for all θ. For $\theta = 1(-1)$, P_θ corresponds to the most monotone increasing (decreasing) dependent case; and intermediate values of θ describe varying degrees of mixtures of the two dependent extremes. In fig. 2 we graph the values of the *CMC, ICMC, DMC* and *IDMC* as functions of θ for P_θ given by (7). (Moreover, because the support of X and Y is two "disjunct" pieces (see Lancaster [12, p. 111]) it follows that the sup correlation ρ' is 1 for all θ in (7).)

From fig. 2 it can be observed that the *CMC, DMC, ICMC*, and *IDMC* are all linear functions of θ. Moreover, the $IDMC = \theta$. The *CMC* and *ICMC* coincide, and the *CMC* at θ is equal to the negative of the *DMC* evaluated at $-\theta$.

Now consider (X, Y) defined on a 3×3 lattice with

$$P = \begin{bmatrix} 0 & 1/4 & 0 \\ 1/4 & 0 & 1/4 \\ 0 & 1/4 & 0 \end{bmatrix}, \tag{8}$$

so that, for example, $\text{prob}(X = a_1, Y = b_2) = 1/4$. Note that P is a symmetric probability matrix, so that X and Y are exchangeable random variables. It follows in this case, by direct computation or by use of *MONCOR*, that the *ICMC* is 0, and the monotone variables for X and Y are $(0, 0.5, 1)'$. However, the *CMC* is $1/3$, and the monotone variables for X and Y, respectively, are either $(0, 1, 1)'$ and $(0, 0, 1)'$ or $(0, 0, 1)'$ and $(0, 1, 1)'$. Thus, (8) provides an example of exchangeable random variables where $ICMC \neq CMC$.

We now consider applying these monotone measures to an actual data example, taken from Bishop, Fienberg and Holland [2, p. 100], which in turn was adapted from Glass and Hall [7, p. 183]. These data are given in table 1.

Table 1
British mobility data (3,500 father–son data values).

Father's occupational status	Son's occupational status				
	S1	S2	S3	S4	S5
S1	50	45	8	18	8
S2	28	174	84	154	55
S3	11	78	110	223	96
S4	14	150	185	714	447
S5	3	42	72	320	411

Note: Status S1 is professional, and high administrative; status S2 is managerial, executive and higher grade supervisory; status S3 is lower grade supervisory; status S4 is skilled manual; and status S5 is semi-skilled and unskilled manual.

Table 2
ICMC, IDMC and monotone variables for British mobility data.

Measure	Value of measure	Monotone variable values				
ICMC	0.496	0	0.627	0.842	0.923	1.0
IDMC	0.242	0	0	0	0	1.0

Because the same categories are used to measure father's and son's occupational status, it is appropriate to use isoscaling. The ICMC, IDMC and the associated monotone variables were computed by the MONCOR program based on the empirical probability matrix specified by table 1. The values of the ICMC and IDMC as well as the monotone variables are presented in table 2.

The analogous version of (3) for isoscaling, namely $IDMC \leq \rho[f(X), f(Y)] \leq ICMC$, shows that regardless of the assignment of numerical values to the five ordinal categories, the resultant correlation is between 0.242 and 0.496.

6. Scaling

One important use of monotone variable theory is the ability to develop meaningful scales for ordinal variables. For example, suppose the five-point scale response to some question is elicited pre- and post- some experimental intervention. Through the use of the ICMC, we can provide a numerical scale for this five point response; this numerical scale has the property that among all possible such ordinal scalings, the post-response for this scaling is most linearly predictable from the pre-response. In table 2 the row corresponding to ICMC provides this scaling for the occupational status variable based on the British mobility data. Specifically, the numerical values for S1, S2, S3, S4, and S5 are 0, 0.627, 0.842, 0.923, and 1.0, respectively.

Often, the number of distinct values for the numerically scaled variables is substantially less than the number of values for the original ordinal variables. This reduction occurs when the optimizing f, g in (2) are not one-to-one functions. To illustrate this phenomenon, we consider the following example. A 10×10 matrix is generated where each entry is a randomly generated number on (0, 1), each generated independently of the other entires. In order to generate a "slightly" positive dependent distribution, the constant 2 was added to each diagonal term and the entire matrix scaled so as to add to one. The resultant matrix is given in table 3.

The CMC for the matrix in table 3 is 0.443, and the monotone variables for $a_1, \ldots, a_{10}$ and $b_1, \ldots, b_{10}$, are, respectively (0.000, 0.461, 0.461, 0.461, 0.872,

Table 3
Random 10×10 probability matrix.

X	Y									
	b_1	b_2	b_3	b_4	b_5	b_6	b_7	b_8	b_9	b_{10}
a_1	0.0331	0.0111	0.0092	0.0049	0.0016	0.0028	0.0009	0.0108	0.0096	0.0007
a_2	0.0101	0.0361	0.0057	0.0081	0.0133	0.0062	0.0121	0.0066	0.0003	0.0020
a_3	0.0102	0.0059	0.0347	0.0027	0.0055	0.0020	0.0104	0.0046	0.0069	0.0056
a_4	0.0144	0.0018	0.0065	0.0342	0.0006	0.0071	0.0055	0.0066	0.0084	0.0113
a_5	0.0006	0.0016	0.0087	0.0132	0.0435	0.0061	0.0100	0.0046	0.0044	0.0053
a_6	0.0022	0.0035	0.0151	0.0015	0.0056	0.0427	0.0062	0.0035	0.0089	0.0125
a_7	0.0002	0.0084	0.0026	0.0020	0.0005	0.0086	0.0387	0.0007	0.0034	0.0111
a_8	0.0084	0.0100	0.0079	0.0036	0.0100	0.0128	0.0044	0.0303	0.0121	0.0065
a_9	0.0028	0.0079	0.0141	0.0008	0.0133	0.0077	0.0064	0.0139	0.0402	0.0068
a_{10}	0.0009	0.0149	0.0042	0.0108	0.0022	0.0144	0.0130	0.0151	0.0146	0.0438

0.872, 0.872, 0.872, 0.873, 1.000)′ and (0.000, 0.537, 0.541, 0.541, 0.842, 0.842, 0.842, 0.842, 0.842, 1.000)′. Note that while the original variables each had ten separate values, there are only five distinct monotonely scaled values for X and five for Y. While this scale reduction phenomenon is based upon empirical observation, it is clear that it has great potential value in deriving simplified scales for large data sets.

7. Program availability

A description of the MONCOR program, and examples of its input and output procedures, is given in Kimeldorf, May and Sampson [10]. The FORTRAN program itself, and a user's manual, are available for distribution. For specific details contact Professor Jerrold May, Graduate School of Business, University of Pittsburgh, Pittsburgh, PA 15260, U.S.A.

References

[1] M. Avriel, *Nonlinear Programming: Analysis and Methods* (Prentice-Hall, Inc., Englewood Cliffs, NJ, 1976).
[2] Y.M.M. Bishop, S.E. Fienberg and P.W. Holland, *Discrete Multivariate Analysis: Theory and Practice* (MIT Press, Cambridge, 1975).
[3] L.C.W. Dixon and G.P. Szego, *Towards Global Optimisation* (North-Holland Publishing Company, New York, 1975).
[4] L.C.W. Dixon and G.P. Szego, *Towards Global Optimisation, 2* (North-Holland Publishing Company, New York, 1978).
[5] A. Fiacco and G. McCormick, *Nonlinear Programming: Sequential Unconstrained Minimization Techniques* (Wiley, New York, 1968).
[6] H. Gebelein, "Das Statistiche Problem der Korrelation als Variations und Eigenwert problem und sein Zusammenhang mit der Ausgleichungsrechnung", *Z. Angew Math. Mech* 21 (1941) 364–379.
[7] D.V. Glass and J.R. Hall, "Social Mobility in Britain: A Study of Intergeneration Changes in Status, in: D.V. Glass, ed., *Social Mobility in Britain* (Routledge and Kegan Paul Ltd., London, 1954).
[8] W.J. Hall, "On Characterizing Dependence in Joint Distributions", in: Bose et al., eds., *Essays in Probability and Statistics* (University of North Carolina Press, Chapel Hill, 1969).
[9] G. Kimeldorf and A.R. Sampson, "Monotone Dependence", *Annals of Statistics* 6 (1978) 895–903.
[10] G. Kimeldorf, J.H. May and A.R. Sampson, "MONCOR – A Program to Compute Concordant and Other Monotone Correlations, in: W.F. Eddy, ed., *Proceedings of Computer Science and Statistics: 13th Symposium on the Interface* (Springer-Verlag, New York, 1981).
[11] H.O. Lancaster, "Correlation and Complete Dependence of Random Variables", *Ann. Math. Statist.* 34 (1963) 1315–1321.
[12] H.O. Lancaster, *The Chi-Squared Distribution* (Wiley, New York, 1969).
[13] J.H. May, "Solving Nonlinear Programs Without Using Analytic Derivatives", *Operations Research* 27 (1979) 457–484.

[14] W.H. Payne, J.R. Rabung and T.P. Bogyo, "Coding the Lehmer Pseudo-Random Number Generator", *Communications of the ACM* 12 (1969) 85–86.
[15] A. Rényi, "On Measures of Dependence", *Acta. Math. Acad. Sci. Hungar.* 10 (1959) 441–451.
[16] O.V. Sarmanov, "The Maximal Correlation Coefficient (Symmetric Case)", *Dokl. Akad. Nauk. SSSR* 120 (1958a) 715–718 (English translation in *Sel. Transl. Math. Statist. Probability* 4, 271–275).
[17] O.V. Sarmanov, "The Maximal Correlation Coefficient (Non-Symmetric Case)", *Dokl. Akad. Nauk. SSSR* 121 (1958b) 52–55 (English translation in *Sel. Transl. Math. Statist. Probability* 4, 207–210).

PART II

MULTIVARIATE DATA ANALYSIS AND DESIGN OF EXPERIMENTS

INTRODUCTION TO CONTRIBUTIONS IN MULTIVARIATE DATA ANALYSIS AND DESIGN OF EXPERIMENTS

J.S. RUSTAGI and S.H. ZANAKIS

Multivariate statistical analysis deals with multiple responses [dependent variable $y = (y_1, y_2, \ldots, y_m)$] to several measurements taken simultaneously on some independent variable vector $x = (x_1, x_2, \ldots, x_k)$. That is, unlike Part I we are interested at the same time in more than one attribute (dependent variable y_i). The components x_i and x_j of x are assumed to be correlated. Many classical statistical procedures assume that the underlying model for x is that of multivariate normal with mean vector μ and covariance matrix Σ. Estimation of μ and Σ is one of the basic problems in multivariate data analysis. A large number of statistical techniques for the multivariate normal case are given in standard textbooks; for example, see Anderson [1], Morrison [5], and Srivastava and Khatri [8]. A journal devoted entirely to this subject, *Journal of Multivariate Analysis*, is published by the Academic Press.

When the assumption of normality is not feasible, nonparametric procedures are generally used. The field of nonparametric multivariate analysis is growing vigorously. A theoretical survey has been given by Puri and Sen [7].

One of the important areas of application in multivariate data analysis is that of *classification*. Given s mutually exclusive populations $\pi_1, \pi_2, \ldots, \pi_s$ and an observation x, the classification problem is to determine from which population the observation arose. The optimality criterion in such a case may be minimizing some measure of the error of misclassification. When the procedure reduces to a simple rule involving a function of x, it is called a *discriminant function*. Fisher [2] was the first to propose a rule of classification in terms of a linear discriminant function. Simple linear programming procedures for discriminant analysis were recently suggested by Freed and Glover [3]. Recent nonparametric formulations of such discriminant functions have emphasized the use of ranks, i.e. replacement of individual measurements with their ranked order value. The paper by Wang, "Derivation of a maximum rank sum statistic and application to discriminant analysis", addresses the classification problem in the nonparametric case. The suggested statistic is obtained in terms of "maximum rank sum". This criterion leads to linear and mixed-integer programming models for finding the optimal discriminant. Wang provides Monte-Carlo comparisons of this statistic with standard discriminants, e.g. Fisher's, in terms of errors of misclassification. In most cases the results are fairly comparable.

In the paper "Constrained multivariate analysis", Hausman also provides optimizing procedures in the context of discriminant analysis and principal component analysis. Various criteria of goodness of fit for statistical models can be proposed. In the context of discriminant analysis, Hausman poses an optimization problem for finding a linear function of observations with convenient *integer* weights so as to optimize some optimality criterion. The problem reduces to a constrained optimization problem, and a general branch-and-bound algorithm is provided to solve it (which may also be used in canonical correlation and multiple regression analysis). The same optimization approach is also applied by the author to the important multivariate technique of *principal components*. In order to reduce dimensions in multivariate problems, the technique of principal components is commonly used. Here one finds those linear functions of the components x which provide maximum variance under certain constraints. Usually the overall variability of the data can be explained by two or three principal components. These components help in the understanding of the structure of multivariate data and are also helpful in clustering techniques.

Clustering and classification (or discriminant analysis) both seek to assign objects to groups. However, the grouping rule in classification (discriminant analysis) is derived from a sample of objects whose group membership is known a priori, so that the misclassification error in the sample is minimized. In *clustering*, a set of individual sample elements is partitioned into various subsets (clusters), with boundaries not pre-defined, based on some measure of similarity or closeness of these points. Euclidean distance is commonly used as a measure of distance between points in multivariate contexts, but many clustering problems use other dissimilarity measures such as Mahalanobis distance and Hellinger distance. Clustering is regularly used in numerical taxonomy and marketing research, and has recently been applied to aid decision-makers in selecting the "most satisficing" solution to multiobjective optimization problems by grouping similar solutions together (Morse [6]). This may reduce greatly the number of optimal solutions one needs to consider. In the paper "An algorithm using Lagrangian relaxation and column generation for one-dimensional clustering problems" by Stanfel, a linear/dynamic programming approximation algorithm is suggested for the one-dimensional clustering problem formulated as an integer programming model. The optimality criterion is to choose the best "goodness of cluster" measure. Although the examples chosen are one- dimensional, the technique can be extended to more dimensions.

Estimation of a population total is an important problem in sampling from finite populations. When the population values are zeros and ones, the total reduces to the number of ones in the sample. Such a total arises in the context of auditing in accounting, carcinogenesis, air pollution, and other yes–no type problems. Plante and Sinha in their paper "Algorithmic improvements for obtaining the upper multinomial bound" have provided efficient algorithms for optimization problems which arise in the context of estimating the total absolute error in accounting records. Introducing a nesting concept in simplifying multinomial probabilities by binomial probabilities, the problem is formulated as a nonlinear optimization model. A generalized reduced gradient search procedure is provided by the authors.

A related problem in the context of data from finite sampling, censuses, survey, etc. is that there are many errors in record-keeping. With the exception of genuine outliers, which can be statistically treated, there are several other kinds of errors in the data which may lead to completely wrong conclusions. Garfinkel, Liepins and Kunnathur provide a timely survey on "Error localization for erroneous data". They examine various strategies for editing continuous, integer and categorical data. These strategies lead to standard optimization problems, such as set covering.

An important class of designs is concerned with factorial experiments. When the number of factors is large, the number of total experiments to be performed becomes impracticable. For example, ten factors, each at two levels, require $2^{10} = 1024$ treatment combinations for complete experimentation. Fractional factorials are quite commonly used in practice, especially in industrial research and development, to reduce the size of such an experimentation. For an introduction to fractional factorials and other designs, the reader is referred to Kempthorne [4]. Cost-optimal fractional factorials have recently been introduced, whose solutions require mathematical programming techniques. In the paper "Selection of cost-optimal frsactional factorials, $2^{m-r} 3^{n-s}$ Series" by Mount-Campbell and Neuhardt, an algorithm is provided to compare costs of various fractional factorials so as to choose the best one. Besides costs, other criteria such as estimator efficiency are also of interest, but they are not considered in this paper. An example from a material requirement planning environment is also discussed. The paper provides an important contribution to the design of experiments, especially in an industrial and business setting.

References

[1] T.W. Anderson, *An Introduction to Multivariate Statistical Analysis* (John Wiley and Sons, New York, 1958).
[2] R.A. Fisher, "The use of Multiple Observations in Taxonomic Problems", *Ann. Eugen.* 7 (1936) 179–188.
[3] N. Freed and F. Glover, "A Linear Programming Approach to the Discriminant Problem", *Decision Sciences* 12 (1981) 68–74.
[4] O. Kempthorne, *Design and Analysis of Experiments*, (Krieger Publishing Company, Huntington, New York, 1975).
[5] D.F. Morrison, *Multivariate Statistical Methods* (McGraw-Hill, New York, 1967).
[6] J.N. Morse, "Reducing the Size of the Nondominated Set: Pruning by Clustering", *Comp. & Oper. Res.* 7 (1980) 55–66.
[7] M. Puri and P. Sen, *Nonparametric Methods in Multivariate Analysis* (John Wiley and Sons, New York, 1971).
[8] M.S. Srivastava and C.G. Khatri, *An Introduction to Multivariate Analysis* (North-Holland, New York, 1979).

CONSTRAINED MULTIVARIATE ANALYSIS

Robert E. HAUSMAN, Jr.
Bell Laboratories

Fitting a statistical model to a set of data generally consists of maximizing some measure of the goodness of fit of the model, such as correlation or likelihood, over the parameter space of the model. We attempt in this paper to provide more easily interpretable parameter estimates by restricting the parameter space to multiples of a large but finite set of easily interpretable points. For instance, the coefficients of the variables in a linear discriminant analysis might be restricted to the set $\{1, 0, -1\}$, so that the discriminant function would be made up of simple sums and differences of the original variables. If the goodness of fit is not substantially degraded by this restriction, then these parameter estimates provide a reasonable alternative to those produced by the unrestricted analysis.

We develop a general branch-and-bound algorithm for performing this constrained optimization within the context of a variety of multivariate statistical techniques. In particular, we describe the application of the algorithm to principal components analysis and linear discriminant analysis, and we point out that it can also be applied to canonical correlation and multiple regression analyses among others. Finally, we provide an illustrative application of its use in principal components.

Our experience suggests that the goodness of fit is usually not substantially degraded by these constraints. In several constrained principal components analyses, we have never seen a degradation as large as 2%.

1. Introduction

Fitting a statistical model to a set of data generally consists of maximizing some measure of the goodness of fit of the model, such as correlation or likelihood, over the parameter space of the model. While this procedure yields "optimal" parameter estimates, we note that nearby points in the parameter space may fit almost as well. The degree to which this is true depends on the flatness of the objective function (the goodness-of-fit measure) near the optimum.

In multivariate analyses the parameters are often weights (coefficients) for the variables. In such cases we usually wish to compare these weights with each other in order to determine which are important. This task would be greatly simplified if the vector of weights were porportional to a vector containing only elements from some small set such as $\{1, 0, -1\}$ or $\{1, 0\}$. A vector of such weights would be relatively easy to interpret. We assume here that the variables

---

Received November 13, 1980; revised April 20, 1981.

are in comparable units; otherwise, the use of standardized variables should be considered.

Consider, for instance, a test designed to separate qualified and unqualified job applicants. A standard linear discriminant analysis might be used to determine the optimal weights for each question for use in determining a composite score. If, however, these weights are restricted to the set {1,0}, then they may be interpreted as telling us which questions should be included in the test and which should not. This will result in a potentially shorter test (questions given zero weight need not be asked) and a much simpler scoring algorithm (total the number of correct answers).

In this paper we examine the problem of maximizing the goodness of fit over only the set of points in the parameter space which are considered to have relatively easy interpretations. In integer programming a similar type of constrained optimization is performed. One common technique which is used is a branch-and-bound procedure [2]. We develop a similar branch-and-bound procedure for solving constrained multivariate analysis problems.

We begin, in section 2, by discussing the branch-and-bound algorithm. The branch-and-bound tree is described and the subproblems at each node of the tree are defined. A general method for solving these subproblems is then given in section 3. In section 4 we describe the application of the procedure to principal components analysis and linear discriminant analysis. We also mention that the procedure has been applied before to canonical correlation and can be applied to multiple regression. As will be discussed, it is sometimes useful in these analyses to require that the solution parameter vector be orthogonal to some specified set of vectors. A technique for incorporating constraints of this type into the maximization process is covered in an appendix.

We present an example of the use of constrained principal components analysis in section 5. Section 6 provides a summary and describes some possible directions for future work.

2. The branch-and-bound algorithm

Suppose we have a set of data, X, and a model which we will try to fit to the data. The data are in whatever form is appropriate for the analysis. Typically, X is an $n \times p$ matrix representing n observations on each of p variables. If the parameters of the model are coefficients of the variables, then θ is a p-dimensional vector of parameters. We also have a goodness-of-fit criterion, $f(X; \theta)$, which is a function of the data and the parameter vector. Additionally, we may have some constraints on θ of the form

$$g(X; \theta) = 0. \tag{1}$$

The standard analysis would consist of obtaining the solution to the problem

$$\max_{\boldsymbol{\theta}} f(\boldsymbol{X}; \boldsymbol{\theta})$$

s.t. (2)

$$g(\boldsymbol{X}; \boldsymbol{\theta}) = \boldsymbol{0}.$$

As indicated above, $\boldsymbol{\theta}$ often represents a set of coefficients or weights for the variables. Comparing these weights might be greatly simplified if the optimal $\boldsymbol{\theta}$ satisfied the constraint $\boldsymbol{\theta} = \alpha \boldsymbol{t}$ for some real number α and some vector $\boldsymbol{t} = (t_1, \ldots, t_p)$ with elements t_k all chosen from some small set $\mathcal{S}$. Note that α allows the magnitude of the weights to vary while holding constant the relative importance of the individual variables as reflected by the weights. In many analyses, such as principal components and canonical correlation, α is a normalizing constant which has no effect on the goodness of fit of the model. In symbols, the "discreteness constraint" can be written

$$t_k \in \mathcal{S}, \quad k = 1, \ldots, p. \tag{3}$$

In the examples in this paper, we take

$$\mathcal{S} = \{1, 0, -1\}. \tag{4}$$

While the analysis applies equally well to any other finite set of values, we have found that this set has yielded almost as good a fit as an unconstrained analysis in the constrained analyses which we have performed. We have therefore seen no reason to expand this set for our own applications.

One approach to obtaining a parameter vector of form $\alpha \boldsymbol{t}$, where $\boldsymbol{t}$ satisfies (3), is to apply some rounding rule to the vector $\boldsymbol{\theta}$ which solves (2). This approach, however, may lead to a vector $\boldsymbol{\theta}$ which has a substantially lower goodness of fit than the optimal $\boldsymbol{\theta}$ and which may not even be feasible.

In this paper we develop a simple procedure which provides the solution to the problem (2) under the restriction that $\boldsymbol{\theta}$ be of the above form. If the value of the goodness-of-fit criterion for the resultant solution is not too different from that of the unrestricted optimal solution, then the restricted estimate of $\boldsymbol{\theta}$ is a reasonable alternative to the unrestricted estimate.

Note that if the goodness of fit is not substantially decreased by this restriction, then the objective function must be rather flat (in at least one dimension) near the optimum. Thus, there may be many restricted solutions with roughly equal goodness of fit. In other words, the constrained solution represents one view of the model subspace, but there are probably many other easily interpretable views which are equally useful.

If every element of the p-element vector t is a member of S, we shall say that $t \in S^{(p)}$. The problem which we wish to solve, then, is:

$$\max_{t,\alpha} f(X; \alpha t)$$

s.t.

$$g(X; \alpha t) = 0 \tag{5}$$

and

$$t \in S^{(p)}.$$

The most straightforward approach is to try all of the $|S|^p$ possible choices for t, where $|S|$ is the number of elements in S. Unfortunately, this usually becomes much too expensive computationally, even for small to moderate values of p and $|S|$. Heuristic search procedures can also be used, but they do not guarantee that the optimum will be found. In this paper we propose a branch-and-bound procedure, which in most cases will be much more efficient than complete enumeration, but will still guarantee that the optimal solution to (5) will be found.

Consider the tree shown in fig. 1. This is the tree for a problem for which

$$p = 3$$

and

$$S = \{1, 0, -1\}.$$

The depth of the tree is $p + 1$, that is, beyond the root there is one level for each element of t. Except for the leaf nodes (13 through 39 in fig. 1), the degree of each node is $|S|$, which in this case is three. Thus, each of the children of a node at level k corresponds to one of the three possible values for t_k.

Each node in this tree corresponds to a subproblem of the form:

$$\max_{t,\alpha} f(X; \alpha t)$$

s.t.

$$g(X; \alpha t) = 0 \tag{6}$$

and

{all the constraints listed above that node}.

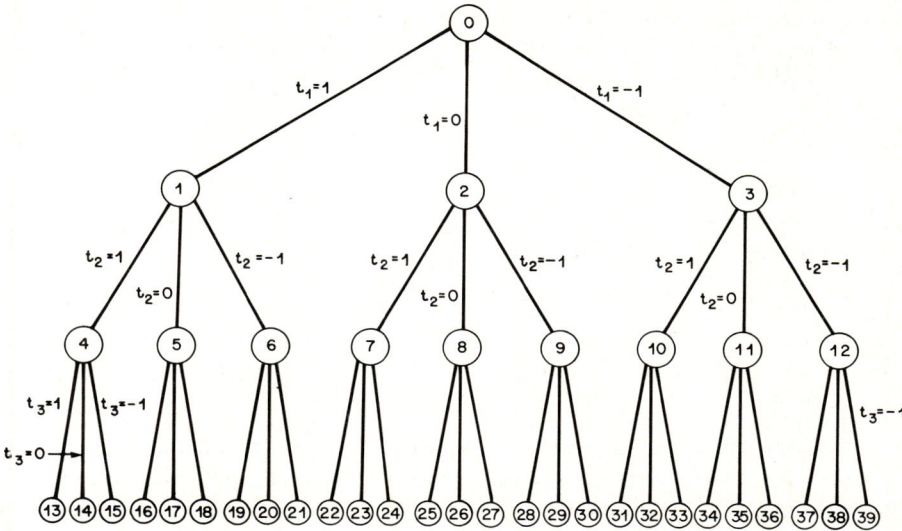

Fig. 1. A branch-and-bound tree.

For example, the subproblem at node seven is

$$\max_{t,\alpha} f(X; \alpha t)$$

s.t.

$$g(X; \alpha t) = 0 \tag{7}$$

and

$t_1 = 0, t_2 = 1$.

Note that t is not required to belong to $\mathbb{S}^{(p)}$ unless we are examining a leaf node. Furthermore, these leaf nodes correspond to all possible solutions of (5).

Let the maximal value of $f(X; \alpha t)$ at node i be f_i^*. Then if i is an ancestor of j (that is, there is a path in the three leading downward from i to j),

$$f_i^* \geq f_j^*. \tag{8}$$

As usual in branch-and-bound algorithms it is this relationship that allows us to limit the number of nodes which we must evaluate and still find the optimal solution. Suppose we have evaluated the leaf node l and the corresponding value is f_l^*. suppose, further, that node i has value $f_i^* < f_l^*$. Then we know that leaf l is better than any node, and specifically any leaf, which is a descendant of node i (that is, below node i in the tree). Thus, none of those descendants needs to be evaluated.

The branch-and-bound algorithm is as follows.

1. $\mathcal{C}$ is a set of nodes, initially empty.
2. Let $i = 0$.
3. Let $\mathcal{B}(i)$ be the set of children of i.
4. Evaluate the nodes in $\mathcal{B}(i)$.
5. Let $\mathcal{C} = \mathcal{C} \cup \mathcal{B}(i)$.
6. Let i be the node in $\mathcal{C}$ having the greatest value f_i^*.
7. If node i is a leaf, it is the solution: stop; otherwise go to 8.
8. Delete node i from $\mathcal{C}$.
9. Go to 3.

We work our way down the tree, visiting nodes with high values first. We stop when we have found a leaf node with a value which we can guarantee is higher than that of any other leaf node.

$\mathcal{C}$ is the set of "live" nodes. These are nodes which have been evaluated, but whose children, if any, have not. At all times, each leaf node either is alive or has a live node as an ancestor.

Each time through the loop, we select the live node whose value is greatest. That node is then labeled node i. The value of node i is an upper bound on the optimal $f(X; \alpha t)$ in (5). If node i is a leaf, then it is also feasible and hence optimal. If it is not a leaf, then it dies and its children become alive.

This algorithm ensures that we evaluate only those nodes whose parents have values at least as great as that of the optimal leaf node.

3. Solving the subproblems

We must still develop a general method of determining the solution of each of the subproblems. At the $(q+1)$th level of the tree, where $1 \leq q \leq p$, each node is associated with a subproblem of the form

$$\max_{t,\alpha} f(X; \alpha t)$$

s.t.

$$g(X; \alpha t) = 0 \tag{9}$$

and

$$t_k = t_{k0}, \quad k = 1, \ldots, q,$$

where the numbers t_{k0} are specified members of S. In the remainder of this section we show how this subproblem may be rewritten in a form similar to that of the original unconstrained problem. Because of this similarity, we are able to solve the subproblems for many multivariate analyses by means of the techniques available for the solution of the corresponding unconstrained problems.

We first note that the last set of constraints in (9) can be replaced by

$$\alpha t = T\psi, \tag{10}$$

where

$$T = \begin{bmatrix} t_{10} & \vdots & \\ \vdots & \vdots & 0 \\ t_{q0} & \vdots & \\ \cdots & \cdots & \cdots \\ 0 & & I \end{bmatrix} \tag{11}$$

and I is the $(p-q) \times (p-q)$ identity matrix. The first element of ψ is α and the other $(p-q)$ elements are the last $(p-q)$ elements of αt.

We can now use (10) to rewrite the subproblem (9) as:

$$\max_{\psi} f(X; T\psi)$$

s.t. $\tag{12}$

$g(X; T\psi) = 0$.

Note the similarity between this statement of the subproblem and the statement of the unconstrained problem given in (2). In the next section we illustrate, through two applications, how this similarity allows us to use the methodology for the unconstrained problem to obtain solutions to the constrained subproblems.

4. Applications

The algorithm described above can be applied to a number of different techniques of statistical analysis. Among them are:
   (1) principal components analysis;
   (2) linear discriminant analysis;
   (3) canonical correlation analysis; and
   (4) multiple regression.

We have defined constrained versions of each of these techniques and formulated algorithms for performing these constrained analyses based on the general branch-and-bound algorithm described above. Two of these algorithms, those for performing constrained principal components analysis and constrained canonical correlation, have been implemented and tested. Results on a number of real data sets indicate that the imposition of the discreteness constraint is accompanied by very little degradation of the goodness of fit. In earlier efforts we examined and compared three alternative algorithms for performing constrained canonical correlation and examined the response surface of the objective function for standard canonical correlation.

In this section we discuss the adaptation of the general algorithm to constrained principal components analysis and constrained linear discriminant analysis.

4.1. Constrained principal components analysis

Principal components analysis is used to study the structure of a $p \times p$ covariance (or correlation) matrix R calculated from a set of measurements on a p-dimensional vector z of random variables. The primary objective of principal components analysis is to calculate a set of derived random variables, called the principal components, which are linear combinations of the original variables with certain properties to be discussed below.

The first principal component is given by

$$y_1 = a_1' z, \tag{13}$$

where a_1 is the parameter vector. Most texts determine a_1 by maximizing the variance of y_1 subject to the constraint that $a_1' a_1 = 1$. We prefer an alternative formulation which leads to the same parameter estimates of a_1 in the unconstrained problem, but not necessarily in the constrained problem.

In this alternative formulation the variance in z accounted for by y_1 is defined as the difference between the sum of the variances of the elements of z and the sum of the variances of the residuals from the p regressions of each element of z on y_1. This variance accounted for can be shown to be

$$f(R; a_1) = \frac{a_1' R^2 a_1}{a_1' R a_1}. \tag{14}$$

The maximum value of this function of a_1, f^*, is given by the largest characteristic root of R^2 with respect to R and the maximum is attained when a_1 is the corresponding characteristic vector, a_1^*. That is, the optimal solution is characterized by a_1^* and f^*, where

$$(R^2 - f^* R) a_1^* = 0. \tag{15}$$

A second principal component can then be defined as that linear combination of z which, taken together with the first component, accounts for as much variation in z as possible. A total of p principal components may be defined in this manner.

We now consider the first constrained principal component (CPC), which is that linear combination of z,

$$\hat{y}_1 = \hat{a}'_1 z, \tag{16}$$

that explains as much variance as possible, subject to the constraint that $\hat{a}_1 = \alpha t$ for some $t \in S^{(p)}$. In order to determine the vector $\hat{a}_1$, we use the branch-and-bound algorithm.

Referring back to (12), we note that the subproblem can be formed by substituting $T\psi$ for a_1 in the unconstrained problem (14). Thus, the subproblem at each node can be written as:

$$\max_{\psi} \left\{ f(R; T\psi) = \frac{\psi' T' R^2 T \psi}{\psi' T' R T \psi} \right\}. \tag{17}$$

The value of this node is just the largest characteristic root of $T'R^2T$ with respect to $T'RT$.

The second CPC is that linear combination of z,

$$\hat{y}_2 = \hat{a}'_2 z, \tag{18}$$

which, taken together with the first CPC, explains as much variation as possible. The coefficient vector $\hat{a}_2$ may be found by substituting the partial covariance matrix of z given $\hat{y}_1$ for R and then repeating the procedure used to find the first CPC.

One characteristic of unconstrained principal components analysis is that the coefficient vector, a_1, which determines the first principal component, is orthogonal to that which determines the second principal component. In a constrained principal components analysis this is not necessarily true. The analyst, may, however, desire to force the coefficient vector of the second CPC to be orthogonal to that of the first. The appendix discusses the advantages of requiring orthogonality and describes a technique which enforces it.

A commonly used measure for determining the goodness of fit in principal components analysis is the cumulative percent of variance in z explained by the principal components. Obviously, we cannot increase this measure by adding the discreteness constraint. In studies by the author of several data sets, however, the use of CPCs rather than the principal components from a standard analysis never resulted in a decrease of more than 2% in the value of this measure.

4.2. Constrained linear discriminant analysis

Linear discriminant analysis is used to help decide to which of two or more populations a particular multivariate observation belongs. In this section we consider only the two-population case.

Suppose that from each of two populations we have collected a set of observations on a p-dimensional vector z of random variables. Let S be the pooled estimate of the within-population covariance matrix and $\bar{z}_1$ and $\bar{z}_2$ the usual estimates of the two-population means. Along any single dimension in the p-dimensional variable space, we have a within-population standard deviation and a between-populations difference in means. The objective of linear discriminant analysis is to identify the dimension which maximizes the ratio of the latter to the former. This is the dimension along which the populations are most distinct.

The linear discriminant function represents this dimension as a linear combination of the elements of z,

$$y = a'z. \tag{19}$$

In order to determine the vector a, we write the problem described above as:

$$\max_{a} \{ f(\bar{z}_1, \bar{z}_2, S; a) = a'\bar{z}_1 - a'\bar{z}_2 \}$$

s.t. (20)

$$a'Sa = 1.$$

Using a standard Lagrangian multiplier approach, we find that the maximum value

$$f^* = \left[(\bar{z}_1 - \bar{z}_2)' S^{-1} (\bar{z}_1 - \bar{z}_2) \right]^{1/2} \tag{21}$$

is attained when

$$a^* = \lambda S^{-1}(\bar{z}_1 - \bar{z}_2), \tag{22}$$

where λ is simply a scale factor chosen to satisfy the constraint in (20).

When we perform a constrained linear discriminant analysis, the subproblem at each node can be written as

$$\max_{\psi} \psi'T'\bar{z}_1 - \psi'T'\bar{z}_2$$

s.t. (23)

$$\psi'T'ST\psi = 1,$$

where T depends on the node being evaluated. Thus, by analogy with the solution of (20), the value of the node is

$$f_i^* = \left[(\bar{z}_1 - \bar{z}_2)'T(T'ST)^{-1}T'(\bar{z}_1 - \bar{z}_2)\right]^{1/2}. \tag{24}$$

Notice that, as in (12), we obtained the subproblem given in (23) by merely substituting $T\psi$ for a in the unconstrained problem (20).

5. An example of constrained principal components

In this section we describe a constrained principal components analysis and compare the results to those obtained from a standard principal components analysis of the same data.

A large number of recent high-school graduates were given tests in several subject areas to determine their readiness to do college-level work. The results of these tests in terms of the percentage of students passing each test were reported by school district in a local newspaper [1]. These data were used to form a correlation matrix for the proportion of students passing the various tests. Then both standard and constrained principal component analyses were performed on this correlation matrix in an effort to determine the relationships among these tests.

Table 1 lists the coefficients obtained for the first two components using the standard and constrained analyses. In both cases the first component is obviously an overall level; that is, some school districts tended to do better than other school districts on all tests.

The second component addresses the differences between the tests. Specifically, the second component from the standard analysis seems to be a measure of the quantitative/qualitative tilt of each school district's graduates. The

Table 1
Principal components example. Coefficients of the first two components.

Test	First component		Second component	
	Standard	Constrained	Standard	Constrained
Sentence structure	0.43	1	−0.28	0
Logical relationships	0.44	1	0.12	0
Essay	0.32	1	−0.66	1
Composition	0.46	1	−0.18	1
Computation	0.39	1	0.53	−1
Algebra	0.40	1	0.40	−1

Table 2
Principal components example. Goodness of fit.

Component	Cumulative percent of variance explained	
	Standard	Constrained
1	74.16	74.11
2	91.03	90.77
3	96.17	95.99
4	98.98	98.91
5	99.92	
6	100.00	

second constrained component measures a similar quantity, but more specifically. The essay and composition tests examine the students' writing abilities, while the computation and algebra tests measure their high-school mathematics abilities. Also, note that the second constrained component could not have been obtained from the second standard component by merely rounding the coefficients.

Looking at the two sets of coefficients, one might wonder how much information we lose by using the constrained rather than the standard components. To answer this question, we examine the cumulative percent of variance explained by the first two components of each of the two analyses. These measures of the goodness of fit are reported in table 2. The difference in the explanatory power of the two sets of components is only 0.26%.

Note that only four components could be extracted using the constrained analysis. This is because we imposed an additional constraint, as discussed in the appendix, that the coefficient vectors of different components be orthogonal to each other. The only member of $S^{(p)}$ which is orthogonal to the first four coefficient vectors is the null vector.

Finally, we are interested in the efficiency of the branch-and-bound algorithm. For each component the complete tree has 1,093 nodes of which 729 are

Table 3
Principal components example. Nodes evaluated.

Component	Nodes
1	16
2	35
3	13
4	14

leaf nodes. Table 3 lists the number of node evaluations which were required to determine each of the components. Even the most difficult component required the solution of only 35 subproblems.

6. Summary

We have developed a methodology for obtaining easily interpretable parameter estimates for many multivariate statistical models. The methodology is based on optimizing a goodness-of-fit criterion over a restricted parameter space. The algorithm employed to perform this optimization makes use of a branch-and-bound tree. We have shown how this algorithm can be employed for principal components analysis and linear discriminant analysis.

This methodology will be useful in situations in which the goodness-of-fit function is relatively flat near its optimum. In the analyses we have tried, the measure of the goodness of fit has been reduced very little by restricting the parameter space. However, this will not be the case for all types of analyses or even for all data sets for a particular type of analysis. Much work still remains to be done to determine when we can add the discreteness constraint without greatly affecting the optimal goodness of fit.

Finally, the branch and bound tree, as described, branches first on the first parameter, then on the second, and so on down to the leaves. While the tree must have $p + 1$ levels, the branching need not occur in this order. An optimal order might be defined as that which allows us to evaluate the smallest number of nodes. It is hoped that further work will shed light on the decision of which parameter to branch on next.

Appendix: enforcing orthogonality in the algorithm

As noted in section 4, we may wish to require the parameter vector θ to be orthogonal to some set of vectors. For example, in constrained principal components we may wish the coefficient vector which defines the kth component to be orthogonal to those defining the first $k - 1$ components. This orthogonality is an implicit property of regular principal components, but not of CPCs.

There are three major reasons why we might wish to enforce orthogonality in constrained principal components. First, after the first few CPCs have been extracted, there are many constrained components which are equivalent in terms of the proportion of variance that they explain. Requiring that the coefficient vector of the component be orthogonal to those of previous components represents a useful means of choosing among these otherwise equivalent components. Secondly, enforcing orthogonality results in a large amount of de facto tree pruning. This means that many nodes need not be evaluated and the solution is achieved much more quickly. Finally, the first CPC is often an overall average of all the variables. In this case, orthogonality ensures that all subsequent CPCs will be contrasts of the original variables.

Other constrained analyses may also benefit from an orthogonality constraint, especially because of the computational savings associated with the tree pruning. In this appendix we develop a general method for solving the subproblems subject to such a constraint.

The constraint may be stated as

$$A\theta = 0, \tag{A1}$$

where the rows of A are the vectors to which θ must be orthogonal. After substituting $T\psi$ for θ, we can write each subproblem as:

$$\max_{\psi} f(X; T\psi)$$

s.t.

$$g(X; T\psi) = 0 \tag{A2}$$

and

$$AT\psi = 0.$$

Let AT be $s \times t$ and have rank r. If $r = t$, then ψ must be the null vector, which corresponds to a goodness of fit of zero. This goodness of fit is the node's value.

If $r < t$, then define $(AT)^*$ as a $t \times (t-r)$ matrix of full column rank $(t-r)$ such that

$$(AT)(AT)^* = 0. \tag{A3}$$

Then, the statements

$$AT\psi = 0 \tag{A4}$$

and there exists an η such that

$$\psi = (AT)^*\eta \tag{A5}$$

both define $(t-r)$-dimensional linear subspaces for ψ. Furthermore

$$\psi = (AT)^*\eta \tag{A6}$$

implies

$$AT\psi = AT(AT)^*\eta$$
$$= 0, \tag{A7}$$

so that they both define the same subspace. Hence, we can rewrite (A2) as

$$\max_{\eta} f(X; T(AT)^*\eta)$$

s.t. (A8)

$$g(X; T(AT)^*\eta) = 0.$$

This is in the same form as (12) and so the same method employed to solve it may also be used here.

References

[1] L. Cohen, "Basic Skills Test Scores Down for Third Straight Year", *Asbury Park Press* (22 March, 1981) C8.
[2] R.S. Garfinkel and G.L. Nemhauser, *Integer Programming* (John Wiley & Sons Ltd., New York, 1972).

DERIVATION OF A MAXIMUM RANK SUM STATISTIC AND APPLICATION TO DISCRIMINANT ANALYSIS *

Chiang WANG

The University of Arizona

This paper presents a linear discriminant function based on the development of a nonparametric statistic. The statistic is defined as a maximum rank sum statistic which maximizes the sum of the ranks associated with the observations from one sample, based on the distances of all observations from two samples to a hyperplane. The optimal hyperplane which gives the maximum rank sum statistic is considered as a discriminant function for the two-population discriminant analysis. Mixed-integer and linear programming formulations are then derived for obtaining this type of nonparametric statistic. An efficient algorithm for computing this statistic is developed for the special case in which both samples are two-dimensional. Monte Carlo studies are conducted to evaluate the performance of the discriminant function derived from the maximum rank sum statistic in comparison with some statistical discriminant functions. The results show that the new discriminant function is competitive in various noncontaminated situations and performs better in some contaminated situations.

1. Introduction

This paper presents formulations and methods of obtaining a nonparametric statistic and discusses an example usage of this statistic for constructing a two-population discriminant function. The statistic to be considered may be defined through the solution of the following problem. Let x_i, $i = 1,\ldots,n_1$, y_j, $j = 1,\ldots,n_2$, be the m-dimensional samples from continuous populations I and II, respectively. The problem is to determine an m-dimensional column vector $c = (c_1,\ldots,c_m)'$, $-\infty < c_k < \infty$, $k = 1,\ldots,m$, and not all c_k's equal zero, such that the statistic

$$S = \sum_{i=1}^{n_1} \sum_{j=1}^{n_2} \Psi\big[(y_j - x_i)c\big] \tag{1}$$

is maximized, where

$$\Psi(t) = \begin{cases} 1, & \text{if } t \geq 0, \\ 0, & \text{if } t < 0. \end{cases}$$

Received October 2, 1980; revised April 5, 1981.

  * This research was partially supported by a National Science Foundation Grant CPE-8006757 and by a National Institute of Health Grant GM22271-02.

The problem is equivalent to finding an m-dimensional hyperplane with coefficient vector c such that after ranking in ascending order the distances for all x's and y's to the hyperplane, the sum of the ranks associated with the y's is maximized.

The equivalence of these two problems can be illustrated as follows. Let $d_i = x_i c$, $i = 1,\ldots,n_1$, and $e_j = y_j c$, $j = 1,\ldots,n_2$, be the distances of x_i and y_j to a hyperplane with coefficient vector c. Let the x's and y's be ranked based on the combined values of d_i and e_j, ranking in ascending order. When $(y_j - x_i)c = e_j - d_i \geq 0$, y_j has a higher or equal rank than x_i. The value of $\sum_{i=1}^{n_1} \Psi(y_j - x_i) c$ then gives the number of times y_j has a higher or equal rank than x's, $j = 1,\ldots,n_2$. Also, S then represents the total number of times the ranks of the y's are higher or equal to the ranks of the x's. Let R be the sum of the ranks of $y_1,\ldots,y_{n_2}$ among $x_1,\ldots,x_{n_1}$ and $y_1,\ldots,y_{n_2}$. The relationship between S and R is given by $S = R - n_2(n_2 + 1)/2$. Therefore, the vector c which gives the maximum value of S, denoted as S^*, also gives the maximum value of R. The statistic S^* is thus defined as the maximum rank sum statistic.

The motivation of deriving this nonparametric statistic is to use the corresponding hyperplane as a linear discriminant function for the two-population discriminant analysis problem. The problem can be defined as follows. Let x_i, $i = 1,\ldots,n_1$, and y_j, $j = 1,\ldots,n_2$, be the m-dimensional training samples from populations I and II, respectively. A linear decision rule with coefficient vector c and a constant λ is to be derived to classify a new observation z into one of the two populations: if $L(z) = zc + \lambda \geq 0$, classify z into population I; otherwise, classify z into population II. The objective is to find a decision rule such that some measure of the probability of misclassification is minimized. For a thorough discussion of discriminant analysis, see the books by Anderson [1], Goldstein and Dillon [3], and Lachenbruch [4].

Before the hyperplane obtained from the maximum rank sum statistic is to be considered as a discriminant function, it should be noted that the value of λ is undetermined from the statistic. Therefore, the hyperplane cannot be used directly as a regular discriminant function. But this problem can be solved if we assume that the hyperplane passes through the origin (i.e., $\lambda = 0$) and use a ranking procedure to assign a new observation z as follows.

Step 1. Compute $d_i = x_i c$, $i = 1,\ldots,n_1$ and $e_j = y_j c$, $j = 1,\ldots,n_2$, where d_i and e_j are the distances of x_i and y_j to the hyperplane with coefficient vector c and passing through the origin. For a new observation z, compute $h = zc$, the distance of z to the hyperplane.

Step 2. Let γ_x be the rank of h among h and d_i, $i = 1,\ldots,n_1$, ranking in descending order. Let γ_y be the rank of h among h and e_j, $j = 1,\ldots,n_2$, ranking in ascending order.

Step 3. Compute $p_x = \gamma_x/(n_1 + 1)$ and $p_y = \gamma_y/(n_2 + 1)$. The ranking procedure is then of the form:

(i) if $p_x > p_y$, assign z to population I;

(ii) if $p_x < p_y$, assign z to population II; and
(iii) if $p_x = p_y$, use a nonranking procedure to assign z.

In the above ranking procedure $p_x(p_y)$ represents the probability of having an observation at least as extreme as z in the direction of the y's (x's). Therefore this procedure has the objective of minimizing the total probability of misclassification. When $p_x = p_y$, any nonranking procedure (e.g. Fisher's linear discriminant function [1]), can be used to assign z.

The rank discriminant function obtained from the maximum rank sum statistic also has the following properties.

(1) If the two training samples are linearly separable, this rank discriminant function will classify all the observations correctly. This is because the largest possible rank sum will be obtained if there exists a hyperplane such that, relative to the hyperplane, every observation in one sample has a higher rank than each of the observations in the other sample. This is one of the desirable properties that every discriminant function should have.

(2) The relative positions among the observations contribute significantly to the determination of this discriminant function in addition to the magnitude of the data. This is due to the characteristics of both the maximum rank sum statistic and the rank procedure. The existence of outliers in the samples will have less effect on this discriminant function than on some parametric discriminant functions. This property is also useful if the samples are contaminated. This discriminant function should be more robust against contamination than those which do not have this property.

In section 2 some mathematical formulations and their solution procedures are suggested for the determination of the maximum rank sum statistic. Monte Carlo studies are reported in section 3 for the performance of the rank discriminant function from this statistic and several other discriminant functions. A summary is presented in section 4.

2. Derivation of the maximum rank sum statistic

The problem (1) discussed in section 1 may be formulated as a mixed-integer programming problem as follows:

$$\text{minimize } T = \sum_{i=1}^{n_1} \sum_{j=1}^{n_2} P_{ij}$$

subject to

$$(y_j - x_i) c + K P_{ij} \geq 0, \quad i = 1, \ldots, n_1, \quad j = 1, \ldots, n_2, \tag{2}$$

$$\sum_{k=1}^{m} c_k^2 > 0,$$

where K is an arbitrarily large positive constant, $c = (c_1, \ldots, c_m)'$ are continuous variables unrestricted in sign, and

$$P_{ij} = \begin{cases} 1, & \text{if } (y_j - x_i) c < 0, \\ 0, & \text{otherwise}. \end{cases}$$

In other words, P_{ij} equals 1 if y_j has a smaller rank than x_i relative to the hyperplane with coefficients c. Therefore the vector c which minimizes T will also give the maximum rank sum statistic S^*. To avoid the difficulty of assigning the value of K for a given set of samples and nonlinearity associated with the last constraint, the problem may be reformulated as follows:

$$\text{minimize } T = \sum_{i=1}^{n_1} \sum_{j=1}^{n_2} P_{ij}$$

subject to

$$\sum_{k=1}^{m} (y_{jk} - x_{ik}) c_k + K P_{ij} \geq 0, \qquad i = 1, \ldots, n_1, \quad j = 1, \ldots, n_2,$$

$$-1 + 2 D_k \leq c_k \leq 1 - 2 E_k, \qquad k = 1, \ldots, m, \tag{3}$$

$$\sum_{k=1}^{m} D_k + \sum_{k=1}^{m} E_k = 1,$$

where P_{ij} and c are the same as defined in model (2), D_k and E_k are also zero–one variables to restrict one c_k equal to $+1$ or -1, and the other c's within $[-1, +1]$; k has a lower bound:

$$K^* = m \max_{i,j,k} (|y_{jk} - x_{ik}|).$$

Note that formulation (3) is very similar to the mixed-integer program given in [5] except that there are more constraints and zero–one variables in (3). Formulation (3) can, therefore, be solved by either a branch-and-bound algorithm directly or with the use of Benders' decomposition [2]. But because of the large number of zero–one variables involved in (3), the use of those solution procedures is quite difficult except when the sample size is very small. An efficient algorithm for the two-characteristic problem ($m = 2$), however, can be developed as follows.

Let $s_{(1)} \leq \ldots \leq s(N)$, $N = n_1 n_2$, be the ordered slopes of the lines each of which passes through a combination of one x and one y. Assume that a rank sum statistic S is computed from a line with slope s such that $s_{(i)} < s < s_{(i+1)}$, $i \in \{1, \ldots, N-1\}$. If the position of the line is changed, the value of S will be the same as long as the slope of the line is still between $s_{(i)}$ and $s_{(i+1)}$.

Therefore, in this case there are at most $(N+1)$ situations which may give different values of the rank sum statistic. In addition, since the intercept of a line has no effect on the value of the statistic, without loss of generality we may assume that the line to be determined passes through the origin. The complete algorithm for the two-characteristic problem is now presented below.

Step 1. Initialize $l=0$ and $S^*=0$. Compute the slopes

$$s_t = (y_{j2} - x_{i2})/(y_{j1} - x_{i1}),$$

$i = 1, \ldots, n_1$, $j = 1, \ldots, n_2$, where $t = 1, \ldots, n_1 n_2 = N$ denotes an index for each combination of i and j.

Step 2. Order the slopes s_t, $t = 1, \ldots, N$, such that $s_{(1)} < \ldots < s(N)$.

Step 3. Let $ss_{(1)} = s_{(1)} - \delta$, $ss_{(t)} = (s_{(t)} + s_{(t+1)})/2$, $t = 2, \ldots, N$, and $ss(N+1) = s_{(N)} + \delta$, where δ is a small positive constant.

Step 4. Let $l = l+1$. Compute the distances d_i and e_j of each x_i and y_j to the line with slope $ss_{(l)}$. The distances are proportional to the Eucledian distance from the points to the line.

Step 5. Rank the combined distances $d_1, \ldots, d_{n_1}$ and $e_1, \ldots, e_{n_2}$ in ascending order and compute the sum of the ranks of the y's, S, from the ranked distances. If S is greater than the current maximum rank sum S^*, let $S^* = S$. Go to step 4 if $l < N+1$; otherwise, terminate.

To compute the distance d of a point $z = (z_1, z_2)$ to a line with slope s in step 4, the following simple formula may be used:

$$d = -z_1 s + z_2.$$

A FORTRAN program has been developed for the above algorithm. For simplicity only the first optimal slope is recorded if multiple solutions exist. Some limited computational experience indicates that this algorithm can solve problems with almost any sample size in a reasonable amount of computing time.

To obtain a statistic which has similar properties as the maximum rank sum statistic S^* but is computationally feasible for problems with more than two characteristics ($m > 2$), a new formulation has to be developed. One of the alternatives which may provide a reasonable statistic is as follows. For any hyperplane with coefficient vector c, let $t_{ij} = (x_i - y_j) c$ be the difference of the distances from x_i and y_j to the hyperplane. Note that y_j will have a smaller rank than x_i if and only if $t_{ij} > 0$. To maximize the rank sum of the y's, a penalty t_{ij} may be assigned to the x_i y_j pair if $t_{ij} > 0$. Therefore, a statistic may be developed by determining a hyperplane such that the total penalty is minimized. The problem may be formulated as follows:

$$\text{minimize } T_1 = \sum_{i=1}^{n_1} \sum_{j=1}^{n_2} t_{ij}$$

subject to

$$\sum_{k=1}^{m} (x_{ik} - y_{jk}) c_k - t_{ij} \leq 0, \qquad i=1,\ldots,n_1, \quad j=1,\ldots,n_2, \qquad (4)$$

$$\sum_{k=1}^{m} c_k^2 = 1,$$

$t_{ij} \geq 0, \qquad c_k$ unrestricted in sign.

The equality constraint in (4) is required to ensure that the trivial solution $c_k = 0$, $t_{ij} = 0$, for all i, j, and k, is infeasible. Note that although statistics T_1 and T are not the same, T_1 also has the property that the y's will tend to have larger ranks than the x's.

To avoid the difficulty of solving formulation (4) due to the existence of a nonlinear constraint, a small-value "threshold" $\delta(<0)$ may be used to replace each zero on the right-hand side of the first $(n_1 + n_2)$ constraints. In this way the trivial null solution will be infeasible and the formulation becomes a linear program.

The new formulation may be written as follows:

$$\text{minimize } T_2 = \sum_{i=1}^{n_1} \sum_{j=1}^{n_2} t_{ij}$$

subject to

$$\sum_{k=1}^{m} (x_{ij} - y_{jk}) c_k - t_{ij} \leq \delta, \qquad i=1,\ldots,n_1, \quad j=1,\ldots,n_2, \qquad (5)$$

$t_{ij} \geq 0, \qquad c_k$ unrestricted in sign.

Another statistic which might provide a good approximation to T can be obtained from the following formulation:

$$\text{maximize } T_3 = \sum_{k=1}^{m} \sum_{i=1}^{n_1} \sum_{j=1}^{n_2} (y_{jk} - x_{ik}) c_k = \sum_{k=1}^{m} \alpha_k c_k = \alpha'c$$

subject to

$$\sum_{k=1}^{m} c_k^2 = 1, \qquad (6)$$

where α and c are m-dimensional column vectors.

Table 1
Values of skewness and kurtosis of the five distributions used in the Monte Carlo study.

Distribution	Skewness	Kurtosis
1	0.0	1.8
2	0.0	3.0
3	0.0	9.0
4	0.5	8.8
5	1.6	8.8

In formulation (6) the statistic T_3 is obtained by maximizing the sum of differences of the distances from y_j and x_i to the optimal hyperplane with coefficients c. This problem may be solved quite easily by the method of Lagrange multipliers. The optimal solution of c_k is given by $c_k^* = \alpha_k / \|\alpha\|$. The difference between formulations (5) and (6) is that in (5) only the positive values of $(x_{ik} - y_{jk}) c_k$ are considered in the objective function, whereas in (6) both positive and negative values are considered.

To evaluate how well formulations (5) and (6) approximate formulation (3), a Monte Carlo study was conducted to compare the sum of the ranks associated with the y's obtained from those formulations. Random samples of sizes 24 or 30 were generated for both x's and y's from five bivariate distributions in the lambda family [6,7]. These distributions are characterized by their skewness and kurtosis and their values are given in table 1. The means for the x and y samples were at (0, 0) and (1, 1), respectively. Both samples had identity covariance matrices. For a given sample size and type of distribution, a pair of random samples for the x's and y's were generated and the rank sums of the y's were computed from the solutions of formulations (3), (5), and (6). The process was repeated 100 times and the average rank sum from each formulation was computed. The ratios of the average rank sums among those

Table 2
Ratios of average rank sums from formulations (3), (5) and (6) for a sample size of 30.

Distribution	(3)/(5)	(3)/(6)	(5)/(6)
1	0.87	0.78	1.11
2	0.86	0.83	1.03
3	0.94	0.95	0.99
4	0.87	0.88	0.99
5	0.87	0.87	1.00

three formulations for a sample size of 30 are shown in table 2. The result for a sample size of 24 is not given because it is very similar to the result given in table 2.

The result in table 2 shows that except for distribution 1 both formulations (5) and (6) approximate (3) quite well. In addition, the differences between (5) and (6) are negligible. Therefore, formulation (6) may be a better approximation to (3) since its optimal solution can be obtained more easily than formulation (5).

3. Monte Carlo studies

This section reports the results of Monte Carlo studies for evaluating the performance of the rank discriminant function obtained from the maximum rank sum statistic, denoted as RMRS, in comparison with some statistical discriminant functions. The maximum rank sum statistic was computed using the algorithm for the bivariate data presented in section 2. The statistical discriminant functions included in the studies are the Fisher's linear discriminant function (LDF) [1] and the LDF with Huber-type robust estimates of means and coveriances (LDF–Huber) [8].

The Monte Carlo studies include several noncontaminated and contaminated cases which are characterized by their respective bivariate distributions. In the noncontaminated cases the components of the bivariate random samples were generated from one of the five lambda distributions specified in table 1. In all those five cases the two populations had variances one and correlation zero. The mean of the first population, μ_1, was always at the origin, and the mean of the second population, μ_2, was at either (0.7071, 0.7071) or (1.0, 1.0) so that the Mahalanobis distance Δ^2 equals 1 and 2, respectively.

In the contaminated cases, both the contaminating and contaminated distributions for each population were specified as normals and were assumed to have the same mean. The means for populations 1 and 2 were the same as the means in the noncontaminated cases. All the contaminated distributions had identity covariance matrices. The contaminating distributions had correlation zero and standard deviations 5, 10, or 20, representing mild, moderate, and severe contamination, respectively. The percentage of contamination was assumed to be 20% of the total sample size for both populations.

The Monte Carlo studies proceeded as follows: First, training samples of sample size either 16 or 30 from each population were generated from a given set of populations. Various discriminant functions were derived from those training samples. Two new sets of samples, each of size 60, were then generated from the same populations and were classified by the various sample discriminant functions. The percentage of misclassification was recorded for each function. The process was repeated 100 times to compute the average percent

Table 3

Average percentages of misclassification for the noncontaminated cases with $\Delta^2 = 1$ and sample size of 16.

Procedure	Distribution type	1	2	3	4	5
LDF		35.9	33.0	11.8	29.3	30.7
LDF–Huber		36.1	33.0	11.3	29.1	29.9
RMRS		35.5	32.8	13.3	29.6	30.2

Table 4

Average percentages of misclassification for the noncontaminated cases with $\Delta^2 = 1$ and sample size of 30.

Procedure	Distribution type	1	2	3	4	5
LDF		33.1	31.3	12.2	28.4	29.7
LDF–Huber		33.3	31.4	11.4	28.1	29.4
RMRS		33.4	31.7	12.8	28.8	29.7

Table 5

Average percentages of misclassification for the noncontaminated cases with $\Delta^2 = 2$ and sample size of 16.

Procedure	Distribution type	1	2	3	4	5
LDF		28.6	25.2	7.0	21.1	22.9
LDF–Huber		28.9	25.0	6.6	21.8	22.0
RMRS		28.9	25.1	7.9	22.9	22.4

Table 6

Average percentages of misclassification for the noncontaminated cases with $\Delta^2 = 2$ and sample size of 30.

Procedure	Distribution type	1	2	3	4	5
LDF		26.5	24.5	6.9	21.3	22.4
LDF–Huber		26.6	24.7	6.5	20.8	21.6
RMRS		26.9	24.7	8.0	21.5	22.4

Table 7
Average percentages of misclassification for the contaminated cases with $\Delta^2 = 1$ and sample size of 16.

Procedure	Contamination level	Mild	Moderate	Severe
LDF		40.8	43.7	51.0
LDF–Huber		38.5	43.1	44.7
RMRS		40.9	40.5	40.5

Table 8
Average percentages of misclassification for the contaminated cases with $\Delta^2 = 1$ and sample size of 30.

Procedure	Contamination level	Mild	Moderate	Severe
LDF		38.6	45.0	49.4
LDF–Huber		37.4	39.1	41.2
RMRS		40.2	40.8	41.1

Table 9
Average percentages of misclassification for the contaminated cases with $\Delta^2 = 2$ and sample size of 16.

Procedure	Contamination level	Mild	Moderate	Severe
LDF		33.9	40.2	49.7
LDF–Huber		32.4	39.0	41.7
RMRS		35.9	36.5	36.0

Table 10
Average percentages of misclassification for the contaminated cases with $\Delta^2 = 2$ and sample size of 30.

Procedure	Contamination level	Mild	Moderate	Severe
LDF		33.5	39.7	46.7
LDF–Huber		33.3	35.2	36.7
RMRS		35.4	35.6	36.2

of misclassification for each discriminant function. The whole process was then repeated with another set of populations. Since all the distributions in the contaminated cases were normals, the bivariate random variates were generated directly from the GGNQF routine in IMSL.

The average percentages of misclassification for the non-contaminated cases are given in tables 3–6. The standard errors of the averages are all less than 0.01. As shown in the tables, the three procedures performed equally well in all situations. The differences between them are not statistically significant in any of the cases.

The results for the contaminated cases are given in tables 7–10. As expected, the LDF-Huber procedure is more robust against contamination than the regular LDF procedure. The RMRS procedure performed better than the LDF–Huber procedure for moderate and severe contamination with a sample size of 16, and the two procedures did equally well with a sample size of 30. On the other hand, the LDF–Huber procedure performed better than the RMRS procedure for mild contamination. The extent of overlapping between the two populations (i.e. $\Delta^2 = 1$ vs. $\Delta^2 = 2$) does not have a significant effect on the difference of the two procedures.

4. Conclusions

A maximum rank sum statistic was defined and the methods of obtaining this nonparametric statistic were proposed. Some other nonparametric statistics, similar to this statistic but computationally more efficient, were also discussed and evaluated. This maximum rank sum statistic was then used to construct a linear discriminant function with the rank procedure. Results from a Monte Carlo study showed that the rank classification rule derived from this statistic was quite competitive for various noncontaminated situations and was robust against moderate and severe contamination.

If a quadratic or higher-order classification rule is to be derived from the maximum rank sum statistic, the solution will be considerably more difficult to obtain even for the two-variable case. It will be more practical if either the statistic T_1^* or T_2^* defined in section 2 is used to provide an approximate solution.

References

[1] T.W. Anderson, *Introduction to Multivariate Statistical Analysis* (John Wiley & Sons, New York, 1958) ch. 6.
[2] J.F. Benders, "Partitioning Procedures for Solving Mixed-Variables Programming Problems", *Numerische Mathematik* 4 (1962) 238–252.

[3] M. Goldstein and W.R. Dillion, *Discrete Discriminant Analysis* (John Wiley & Sons, New York, 1978).
[4] P.A. Lachenbruch, *Discriminant Analysis* (Hafner Press, New York, 1975).
[5] J.M. Liittschwager and C. Wang, "Integer Programming Solution of a Classification Problem", *Management Science* 24 (1978) 1515–1525.
[6] J.S. Ramberg and B.W. Schmeiser, "An Approximate Method for Generating Symmetric Random Variables", *Communications of the ACM* 15 (1972) 987–990.
[7] J.S. Ramberg and B.W. Schmeiser, "An Approximate Method for Generating Asymmetric Random Variables", *Communications of the ACM* 17 (1974) 78–82.
[8] R.H. Randles, J.D. Broffitt, J.S. Ramberg and R.V. Hogg, "Generalized Linear and Quadratic Discriminant Functions Using Robust Estimates", *Journal of the American Statistical Association* 73 (1978) 564–568.

AN ALGORITHM USING LAGRANGIAN RELAXATION AND COLUMN GENERATION FOR ONE-DIMENSIONAL CLUSTERING PROBLEMS

Larry E. STANFEL
Clarkson College

Clustering problems arise in a variety of both statistical and deterministic forms, and have been found to arise naturally in management science and other areas of study. Heuristic approaches are commonly employed, since finding a best clustering typically demands the solution of an integer programming problem, and the latter may possess nonlinearities as well. In the present paper the clustering problem has a nonlinear objective. By means of a suitable constraint, the problem is linearized but leads to the creation of a family of problems which must be solved in order to obtain a solution to the given problem. A solution method is devised which utilizes linear programming to solve the integer linear programs and which guarantees exact solutions for the case of collinear objects, unless there are duality gaps.

Since the number of LP columns may be so large as to prohibit explicit storage, an efficient dynamic programming algorithm is utilized to create columns for LP basis insertion. Lagrangian relaxation provides the collection of solutions necessitated by the linearized objective function. In the event of duality gaps, a simple approximation scheme bounds the unknown objective function values. Noncollinear data must be mapped into one dimension and some tactics for that transformation are mentioned and referenced. The computer implementation is described and solved example problems are included. For larger problems computational improvements are suggested.

1. Introduction

Clustering problems involve partitioning a finite set of objects into subsets (clusters, groups) so as to optimize some function which measures the homogeneity of objects in subsets and perhaps the heterogeneity of different subsets. In other cases an object may belong to more than a single cluster (an example is information retrieval, where topical clusters may overlap), but such problems are not considered here. In many other cases the requirement of optimality may be relaxed and the procedure employed may terminate when some "a priori" criteria for acceptability are met, such as in [3].

When an optimal partition is sought there are two common variations:
(1) the number of clusters is fixed in advance; and

Received October 15, 1980; revised March 19 and June 25, 1981.

(2) the determination of the number of clusters is a part of the optimization problem.

The objective function considered here does not constrain the number of clusters.

Several books [2,10,15,32] survey the many problem varieties and solution approaches and contain valuable bibliographies. Related methodologies include "taxonomy", "numerical taxonomy", "aggregation", and "classification". We exclude hierarchical problems in this paper.

Clustering problems belong to a general class of combinational problems, many of which have been formulated as integer programming problems; an early example of such a formulation can be found in [37]. Optimization need not lead to an explicit IP, as illustrated by [27], where a rather specialized objective function is minimized using an iterative scheme. The computational requirements of that method are not reported. Since clustering requires the partition of a set, then there should exist a relationship to set partitioning problems, SPP, [13]. If the clustering objective function may be expressed as a sum of contributions from the subsets belonging to the partition, then the two problems are equivalent. This occurs, for example, in [26], although the inability to solve SPPs efficiently [19] is a disappointment. An interesting relationship to location problems may be found in [8]. A two-phase approach is utilized in [23], where the first generates a collection of candidate clusters and the second pursues an optimal clustering solution, perhaps with the aid of SPP algorithms. We, incidentally, shall exploit the SPP view and realize some easing of the computational burden.

Branch-and-bound algorithms have been offered ([21], for example), as well as a dynamic programming formulation [18] which was applicable to arbitrary sets of points, but which proved dimensionally infeasible for even small numbers of objects. We shall utilize DP to good advantage, but in a supporting role.

A number of heuristics for solving the optimization problems have been developed. Experience with one of these is reported in [34].

The spectrum of problems in which clustering has found application is very broad: disease incidences, patterns, and the inference of causes [5]; the classification of archaeological artifacts [9]; organization of human behavior [1,36]; clustering index terms in an information retrieval system [14]; marketing research [7,28]; creation and search of large data files [29]; inference of ground features from air-borne or satellite sensors [24,4]; and various aggregation problems in economics, where near-optimal clusterings of great quantities of data are necessary [12].

A principal difficulty in clustering problems is the selection of an objective function. In some instances an observer may have intuitive notions about what constitutes a good solution if he is able to view a representation of the objects. Certainly, in fig. 1 the solid lines appear to provide a better solution than the

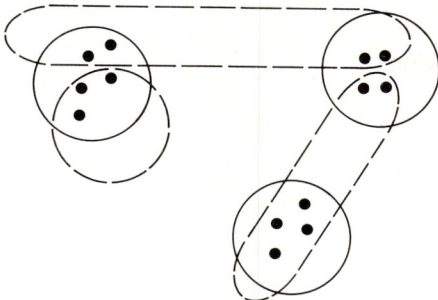

Fig. 1. Alternative partitions.

dashed lines. The former appear to mark an optimal solution. Yet, intuition falters for the arrangement in fig. 2. It is a lack of structure in fig. 2 that makes attractive the objective function approach: when subjective criteria fail we rely upon objective criteria. The challenge is to put into objective form the criteria in which we have explicit confidence in the easy circumstances.

Unlike problems involving dollars or a tangible quantity, there is no obvious choice; there are many alternatives available, each with its own advantages and disadvantages. A typical procedure then is to select a function which seems to embody the intuitive requirements that one can at least roughly formulate. We expect the solutions obtained to be intuitively satisfying, although with rather unstructured data a best solution and a mediocre solution may not appear vastly different.

Thus, it is possible to anticipate the existence of a great many objective functions. It is not the intent to include a catalog of these, but among the references cited several will be found applied, and [10] for example lists several.

It seems obvious that a recognizably good partition has two properties:

(i) the objects contained in a single cluster are similar to the other objects in it; and

(ii) a cluster should be dissimilar to other clusters.

It is conventional to assume that a distance function or metric is defined upon the set of objects and to measure similarity inversely with the distance

Fig. 2. Objects difficult to cluster.

between two objects. The way to extend that notion to the partition itself, however, is not apparent. It would seem that the global measure must involve an overall measure of within-cluster similarity and an overall measure of between-cluster dissimilarity if (i) and (ii) are to be preserved. Within-subset similarity might be measured by the diameter of the set, the average distance to a center of gravity, or the average of the distances within the cluster.

Analogously, one might measure the difference between two subsets by the set-theoretic distance between them [16], the average of the inter-cluster distances, etc.

The objective function used here is the average within-group distance (wgd) minus the average between-group distance (bgd), i.e.

$$f = \overline{W} - \overline{B}.$$

For the six objects and partition shown in fig. 3, we compute

$$f = \frac{d_{23} + d_{24} + d_{34} + d_{56}}{4}$$

$$- \frac{d_{12} + d_{13} + d_{14} + d_{15} + d_{16} + d_{25} + d_{26} + d_{35} + d_{36} + d_{45} + d_{46}}{11},$$

where d_{ij} = distance between objects i and j.

In the case where a denominator of one of f's two terms is zero, we may either neglect that term or prohibit solutions consisting of one cluster or n clusters, n being the number of objects given.

The origin of this objective function is somewhat uncertain. The paper [22] suggested average between-group distance as a measure of subset dissimilarity. The use of average similarity to measure cluster homogeneity may be traced to [33], which sought to avoid the extremes of clustering caused by complete

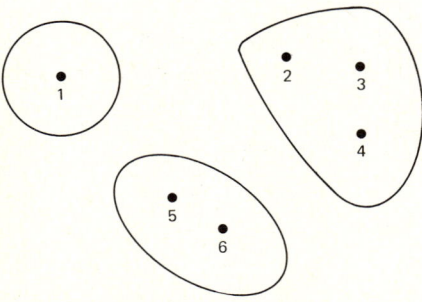

Fig. 3. A sample partition.

linkage and single linkage criteria. The difference of the two terms appeared in an unpublished dissertation [25].

The function is motivated by the previous items (i) and (ii) and one hopes that by minimizing it he achieves those two conditions. On the empirical side the function seems, in practice, to preserve clusters that seem natural to an observer in cases where the dimensionality allows viewing and where distance is measured in a natural way; Euclidean or rectangular, for example.

The dissertation [6] addressed the problem of the cluster-preserving properties of this function. A particular test with $n = 36$ objects and Euclidean distances proceeded as follows. Given an optimal partition, one cluster acted as a fixed reference and the remaining clusters were moved uniformly toward the center of that reference set. The clustering problem was solved at several stops in this process to ascertain whether the original partition had been preserved. The sequence of diagrams, figs. 4–7, portrays the results. The scale of distance is unimportant, since our f has the property that if distances are multiplied by a non-negative scalar C, then the function value is multiplied by C. Consequently, scaling the distances does not change the optimal partition. The drawings shown are as informative as a matrix of distances would be. In the author's opinion the invariance of optimal solutions under scalar multiplication of distances is a desirable property.

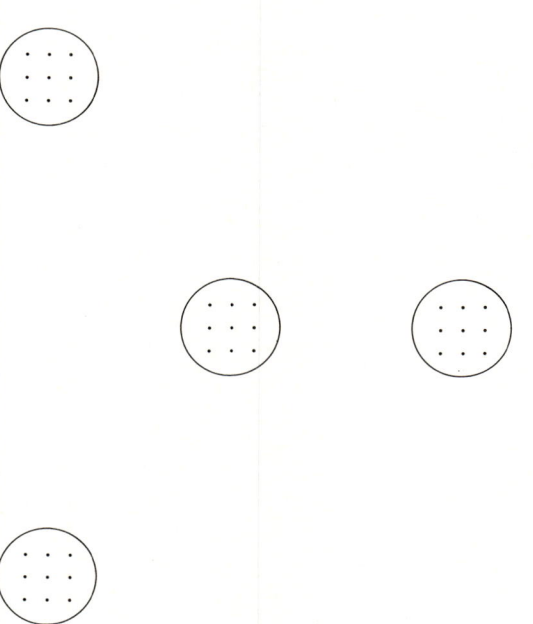

Fig. 4. An optimal solution with f, stage 1.

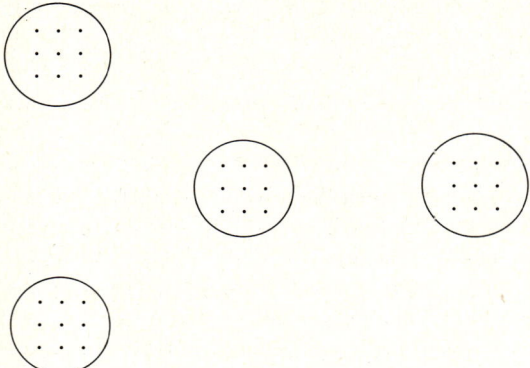

Fig. 5. An optimal solution with f, stage 2.

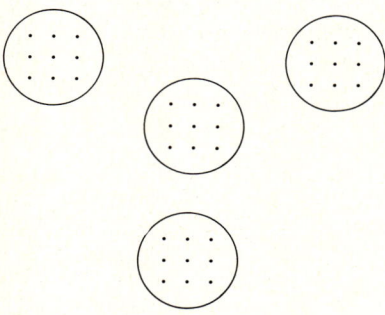

Fig. 6. An optimal solution with f, stage 3.

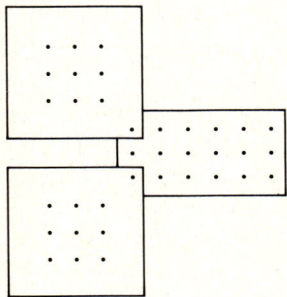

Fig. 7. An optimal solution with f, stage 4.

It is difficult to compare different objective functions applied to a set of test problems if we are not assured of obtaining optimal results in each instance. The present paper offers the possibility of optimal solutions for a class of problems as well as the ability to measure nearness to optimality in the event such a solution is not achieved.

The question of what objective function is appropriate in a given situation is difficult to answer. If the clustering problem is to satisfy a criterion having physical significance, there is no hesitation – we formulate the physical aspects mathematically and are satisfied. A good example is in [20].

As emphasized, however, criteria tend to be much less tangible. It would not be surprising if an objective function were to include parameters then to allow a user to create his own specific functions from a family of them. An obvious way to accomplish this is to attribute differing weights to the two terms in our objective function. Our use of a Lagrange multiplier gives exactly that capability.

2. Problem formulation and solution

Let

$$X_{ij} = \begin{cases} 1, & \text{if } i \text{ and } j \text{ belong to the same cluster,} \\ 0, & \text{otherwise,} \end{cases}$$

$$f = \frac{\sum\sum_{i<j} d_{ij} X_{ij}}{\sum\sum_{i<j} X_{ij}} - \frac{\sum\sum_{i<j} d_{ij}(1 - X_{ij})}{\sum\sum_{i<j}(1 - X_{ij})}. \tag{1}$$

In (1) the first term is $\overline{W}$, the second, $\overline{B}$.

The constraints on the minimization problem are of the form

$$X_{ij} + X_{jk} + X_{ik} \neq 2, \quad i < j < k,$$

which guarantee that if i and j are together and j and k are together, then i and k must be together.

By using additional variables and constraints, the problem can be written as an integer programming problem with linear constraints and a nonlinear objective function. The final formulation is not given here because it is computationally intractable; the most direct formulation is not fruitful. We address, instead, some problem simplifications.

Suppose, however, that the number of wgd's is fixed equal to K. The right-hand side of (1) may then be written:

$$\frac{\sum\sum_{i<j} d_{ij} X_{ij}}{K} - \frac{\sum\sum_{i<j} d_{ij}(1-X_{ij})}{\binom{n}{2} - K}.$$

Simplifying, we obtain:

$$\frac{\left[\binom{n}{2} - K\right] \sum\sum_{i<j} d_{ij} X_{ij} - K \sum\sum_{i<j} d_{ij}(1-X_{ij})}{K\left[\binom{n}{2} - K\right]}.$$

Finally, then,

$$f = K_1 \sum\sum_{i<j} d_{ij} X_{ij} + K_2; \qquad K_1, K_2 \text{ constants}; \quad K_1 \geq 0.$$

If we could solve such a subproblem for each possible value of K, we would have solved the given problem.

Suppose, next, that the objects to be clustered lie on a line and number them consecutively. If they do not lie on a line we must map them into one. The topic of dimension reduction is addressed later.

With the collinear points numbered we have available a convenient representation of clusters. The cluster $(1, 3\,4, 6)$, for example, is represented by the column vector $(1011010\ldots0)^T$. Let

n_j = number of objects in cluster j,
w_j = $\binom{n_j}{2}$ = number of wgd's in cluster j,
d_j = sum of wgd's in cluster j,
X_j = 1, if cluster j is selected,
    = 0, otherwise,
$\mathbf{1}$ = a column of n 1's
A = the 0–1 matrix whose columns represent every possible cluster.

The problem with K wgd's becomes:

$$\text{minimize} \sum_j d_j X_j$$

subject to

$$AX = \mathbf{1}, \tag{2}$$

$$\sum_j w_j X_j = K,$$

$$X_j = 0, 1, \quad \text{all } j.$$

In this way we have arrived at a SPP with one additional constraint.

It may be shown that our function f has the property that an optimal partition of collinear points will yield only clusters consisting of contiguous points; or, in other words, of consecutively numbered points. For example, the cluster $(1, 3, 4, 6)$ would not occur. Thus, it is apparent that if we have numbered a set of points in an arbitrary space or mapped them into a line, the contiguous point limitation could cause an optimal solution to be overlooked. We could solve a problem exactly, but still fail to find the best solution to the given problem. For structureless data it would be advisable to accomplish several mappings, solve the corresponding problem for each, and take as our solution the best of those obtained.

The assumption of only consecutively-numbered cluster elements causes each of A's columns to have its 1's in consecutive elements. As shown in [13] such a matrix is unimodular, so that if we neglected the constraint involving K, the integer programming problem (2) could be solved with the simplex method.

But solving the clustering problem involves solving a problem (2) for each feasible value of K permitted by the n objects. Solutions for a range of K are required, so the odious constraint may be eliminated and our collection of problems may be generated by a Lagrangian relaxation upon that constraint.

The problem becomes

$$\text{minimize } \sum_j (d_j + \lambda w_j) X_j$$

subject to

$$AX = 1.$$

(3)

The paper [35] pursued the solution of (3) via IBM's MPSX, generating and storing explicitly the matrix A and making use of that package's parametric programming capabilities to modify λ.

Because A has $\binom{n+1}{2}$[1] columns an alternative to the generation of that matrix is required. We use a column generating procedure based on the solution of a dynamic programming problem.

Suppose at some stage of one of the LP problems B is a basis, C_B a cost vector, and a_j a nonbasic column of A. For some values of $r \leq s$, a_j will have the form

$$a_j^T = (0, 0, \ldots, 1, 1, \ldots, 1, 0, \ldots, 0),$$

---

[1] Actually, the column of all 1's is eliminated from A since (1) is undefined if the number of bgd's is zero. One avoids having (1) undefined by virtue of the number of wgd's being zero, by deleting a unit vector corresponding to one of a pair of closest points, since those two will be grouped if any points are. If these two extreme solutions are considered admissible, they are examined separately.

where the first 1 is in the rth component, the last in the sth.

Pricing the column demands the calculation of the quantity $C_j - Z_j$, where

$$Z_j = C_B^T B^{-1} a_j = \pi^T a_j = \sum_{i=r}^{s} \pi_i$$

and

$$C_j = \sum_{m=r+1}^{s} \sum_{l=r}^{m-1} d_{ml}, \quad \text{if } r < s$$
$$= 0, \quad \text{if } r = s.$$

The double summation gives the total within-group distances of the cluster corresponding to a_j. For example, if $n = 5$, and consecutive points are unit distance apart, the vector $(0, 1, 1, 1, 1)$ gives $C_j = 1 + 2 + 3 + 1 + 2 + 1$ and Z_j = sum of the last *four* components of π.

It is the form of $C_j - Z_j$ that is suggestive of separability, which in turn motivates a DP approach. When we consider applying the present methodology to other clustering objective functions we recognize two requirements: first, that a linearity be achievable from the relaxation, and secondly that there be separability for the DP application. Thus, we are not limited to the function f employed here, but the function selected must have these properties.

We envision the construction of a column, $(Y_1, Y_2, \ldots, Y_n)^T$, in stages, as shown in fig. 8. Each $Y_i = 0$ or 1, 1's constrained to be consecutive, and R_i is defined so that $\sum_{i=1}^{n} R_i = C_j - Z_j$.

The state variable s_i is the string compiled up to stage i. While the number of such binary patterns is large, our restricted column structure causes the number of possible values of s_i to be quite small.

Table 1 illustrates the totality of states, possible values of Y_i, and corresponding R_i.

Table 1 is described by reference to its rows and columns. The states in row 1, column 1–potentially vast in number for large i – need be neither stored nor treated. A state in (1, 1) is reached when the decision $Y_i = 0$ is made when the process occupies a state in (3, 1). The program realizes that the string has been terminated and either stores the vector (actually the integer pair marking the beginning and end of the 1 subsequence) or neglects it, depending upon its $C_j - Z_j$ value, which could not change once a zero has followed 1's. At stage i, then, we have but i states to consider.

The states in (2, 1) and (3, 1) are uniquely represented by the integer describing the number of consecutive 1's preceding through stage $i - 1$, so no actual vectors need be stored. In fact, as will be seen, no representation of state is stored – the integer is an index to a position in an array where the information is stored.

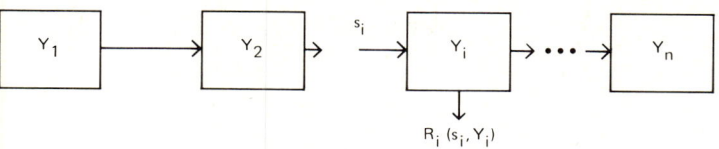

Fig. 8. Column construction in stages.

The return in (2, 3) corresponds to a single point subset in which there is zero within group distances. The Z_j contribution still applies.

The return in (3, 3) reflects the definition of R_i which insures separability: we add the distances d_{1i} as wgd's since the ith point has joined the $(i-1)$th, $(i-2)$th,..., $(i-w)$th points in a subset; $w\lambda$ is added because the addition of a wth point contributes $w\lambda$ to the objective function in (3) (λ is nonpositive throughout the iterations); and $-\pi_i$, of course, is the contribution from $-C_B^T B^{-1} a_j$. Thus, with R_i so defined, $\Sigma_i R_i = C_j - Z_j$. The sums $\Sigma_{l \in L} d_{li}$ are not recomputed each time they are referenced. (The DP routine will be executed a number of times for each value of λ, whereas λ must vary over a specified range.) Rather, as the inter-point distances are computed from the coordinates read in at the beginning of the program, the sums are computed recursively, stored in a two-dimensional array, and retrieved immediately given i and w.

As mentioned, an array index serves to identify state. The correspondence is as in table 2, at stage j.

A further simplification is that optimal decisions may be stored *cumulatively*. For example, suppose at stage j that for a state 0...1 1 ($j-1$ characters in the string), $d_j^* = $ optimal number of following 1's. Then at stage $j-1$, for the state consisting of the first $(j-2)$ characters in the above string, 0...1, if

Table 1
Possibilities for state and decision variables and stage returns.

s_i	Y_i	R_i
Contains 1's but $Y_{i-1}=0$	Must=0	0
All 0's	0 or 1	$-\pi_i Y_i$
0...0 0 1		$[\Sigma_{l \in L} d_{li} + w\lambda - \pi_i] Y_i$
0...0 1 1	0 or 1	$\|L\| = w$
0...1 1 1		
⋮		
1...1 1 1		

Table 2
Condensing the representation of state.

State	Index
0 0 ... 0 1	1
0 0 ... 1 1	2
⋮	⋮
1 1 ... 1 1	$j-1$
0 0 ... 0 0	j

$Y^*_{j-1} = 1$, we store as an optimal decision $d^*_j + 1$. If $Y^*_j = 0$, the state is terminated, as previously described.

The above scheme of storage and reference allows an enormous saving in storage for, regardless of n, the DP manipulations require just two tables, and not one per stage. The reason is revealed in the example of the previous paragraph. Consider the state corresponding to the ith position of the list at stage j, i.e. the last i positions are unity. If the optimal choice is $Y^*_j = 1$, then at stage $j-1$ the corresponding state is $(i+1)$. Therefore, the optimal return and decision are conveniently stored for the next iteration. $4n$ words of storage are sufficient, then, for all the DP work ($2n$ for the "next" stage and $2n$ for the "present" stage).

As mentioned, where a state is terminated, its objective function value is examined. To locate a best column, we compare any terminated state column to the best incumbent and either store or reject it. At the final stage (stage 2, owing to the structure of state variables) comparisons are made between the two tabulated values for $f_2(s_2)$ and the best of those states terminated to choose a best column. Other selection criteria are possible, as mentioned below.

Before describing some sample results and computation times, let us mention briefly the overall construction and flow of the program.

Essentially there are only two subprograms, one to accomplish the revised simplex technique and the other for the dynamic programming problems.

Aside from details too minute to warrant mention, the dynamic programming processing has been described. Let us, then, explicate the manner in which these two subprograms cooperate. First, the LP corresponding to a fixed λ is considered.

Given an LP basis, the first task is to price the (unavailable) columns from the SPP. The DP routine is entered and to this point one of two user-elected alternatives pursued: finding a column with most negative unit price or finding a column with a negative unit price (that is, one that offers some improvement). In practice, no overall time differentials have resulted from this choice — the *sequence* of columns entered have been diferent, although the sets entered have been identical in the two cases. (It should be pointed out that, owing to the complexity of f, unless the data are contrived or of especially simple structure, the chances of alternative optima are negligible.)

It is in the DP (pricing) phase that the greatest potential for increased efficiency lies. Few LP programs undertake a full pricing each time a column is to be entered. What should be done here is to derive from one DP pass a *list* of favorable columns and enter these, successively, so long as any remain favorable. There is, of course, a multitude of clever pricing strategies.

It should be kept in mind that each time pricing is required, an n stage DP with up to n states per stage is solved. Though we have economized greatly on the DP computational load, this will be seen to represent substantial processing once the numbers of LP iterations are exhibited. A line is printed for each entering-departing vector pair and the values of all the basic variables are printed for each optimal solution. The variables are conveniently notated X_{rs} where $r=$ first point in the cluster and $s=$ the number of points in the cluster.

Once an optimal LP is found, the next λ value is taken, causing C_B to change. The new $C_B^T B^{-1}$ is used to price the *optimal solution* for the previous λ value, a presumably good starting point for the new problem. As a rule, it is just that.

The paper [35] observed a near unimodality in $f^*(\lambda)$. By this is meant both that the number of values of λ violating the property were few and that the sizes of the bumps were relatively small. Furthermore, theoretical support was found for that behavior in some simple cases where the set of distances could be characterized overall.

A justifiable pursuit, then, is to program the quest for an optimal solution via an intelligent search over λ. Given unimodality, the journey to a best λ could be greatly accelerated. Given a function that is almost unimodal, considerable improvements should still be attainable. To date the generation of the next λ value has been most elementary, $\lambda_{i+1} = \lambda_i - 1$. This was for the purpose of gaining knowledge as to the typical sensitivity of solutions to λ values, rates of change of objective function values, etc. The experiments to be discussed below have been conducted with the multiplier varying in unit steps over a pre-established range. It should be clear that since the w_j are positive and we are minimizing, positive λ values would generate the same solution as $\lambda = 0$.

3. Results

Several trial problems are described and their solutions and computation times summarized. The machine was an IBM 370-155, the program was written in FORTRAN, and aWATFIV was the compiler. The identity matrix was used as a first basis, but negative λ prevents that solution from being optimal.

Example 1. For the 20 collinear points shown in fig. 9, an optimal solution has the subsets drawn there and table 3 illustrates the sequence of $f^*(\lambda)$ values. Values of λ omitted from table 3 indicate ranges of unchanged solutions. Points not listed in table 3 entries are points of single element subsets. The 20 problems, $-20 \leq \lambda \leq -1$, were solved in 0.17 minutes. The same example was presented in [35], and solved on the identical computer by MPSX for $-19 \leq \lambda \leq 0$, with λ varying in steps of length 0.5. The time required was 1.5 minutes. Thus, the original trial accomplished about double the work while requiring nine times the computation time.

The coordinates of the points in fig. 9 are 0, 1, 5, 7, 10, 15, 17, 18, 19, 20, 25, 27, 30, 32, 34, 35, 40, 45, 47, and 55, respectively.

Notice that the sequence of f^* values is unimodal in this case and that significant gaps in K resulted.

In [11] the geometry of gaps is analyzed in the space where the optimal function value is plotted against all possible right-hand sides of the constraint(s). Each vector λ corresponds to a hyperplane tangent to that surface, so the absence of a tangent hyperplane at a point implies nonconvexity of the surface. The hyperplanes corresponding to points in the vicinity of the gap serve to provide *lower bounds* on the inaccessible solution.

Since we have only one relaxed constraint, our hyperplanes are lines, and fig. 10 illustrates the scheme of the foregoing paragraph, although the drawing is not intended to portray our particular problem. Knowledge of the shape of the surface we illustrate is generally unavailable, and we emphasize that convex objective functions may lead to surfaces that are nonconvex. Fig. 10 shows f^*, not f.

Whatever the unknown $f^*(K_0)$ it satisfies

$$f^*(K_0) \geq \max[l_1(K_0), l_2(K_0), l_3(K_0)],$$

since l_1, l_2, l_3 are tangent to the surface. Consequently, although we may realize gaps, it is simple to discover if they *potentially* contain optima, since the l's are easily determined by the λ's which provided the respective K's.

The above approximation technique was employed for all the gaps in example 1 and $\lambda = -5$ does indeed yield an optimal solution; that is, the clustering problem has been solved exactly.

Example 2. The 50 collinear points, along with the best solution discovered, are shown in fig. 11 and the function values, etc. appear in table 4.

Fig. 9. Test data and optimal partition.

Table 3
Sequence of solutions for example 1.

$-\lambda$	K	Solution	f^*
1	0	all single element subsets	0
2	5	(1, 2) (8, 9, 10) (15, 16)	−17.66
3	13	(1, 2) (3, 4) (7–10) (11, 12) (14–16) (18, 19)	−17.85
4	20	(1, 2) (3, 4) (6–10) (11, 12) (13–16) (18, 19)	−17.88
5	22	(1, 2) (3–5) (6–10) (11, 12) (13–16) (18, 19)	−17.93 (optimal)
7	38	(1–5) (6–10) (11–16) (17–19)	−17.64
11	43	(1–5) (6–11) (12–17) (18–20)	−17.35
13	52	(1–5) (6–12) (13–19)	−16.70
14	71	(1–5) (6–16) (17–20)	−16.40
16	81	(1–10) (11–19)	−16.23
20	94	(1–12) (13–20)	−15.53

The forty problems, $-40 \leq \lambda \leq -1$, were solved in 1.43 minutes. The values obtained, it will be noted in table 4, are not unimodal, but the best solution found persists *for more than one-third the λ range*.

The coordinates of the points in example 2 are, from left to right, 0, 1, 3, 4, 7, 9, 10, 11, 12, 13, 14, 16, 17, 18, 19, 20, 21, 22, 26, 27, 28, 29, 30, 31, 32, 40, 41, 43, 45, 46, 49, 51, 53, 54, 55, 56, 58, 60, 70, 73, 76, 78, 79, 80, 86, 87, 88, 95, 96, 99.

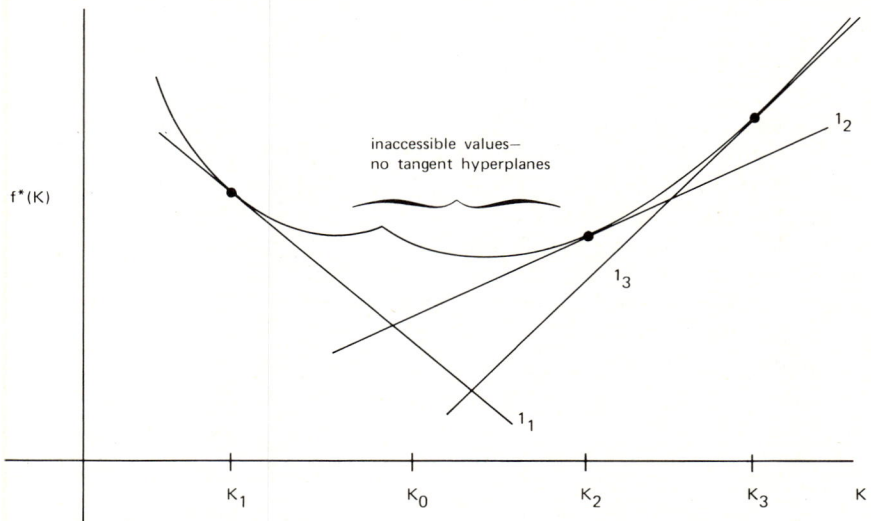

Fig. 10. Hyperplane approximations.

Fig. 11. Test data and optimal partition.

Example 3. The second 50-point example is useful for noticing solution sensitivities, but of less value for time comparisons, since it was solved in three separate runs. Better solutions seemed to appear just beyond the ends of specified λ intervals, a condition which argues for a search over λ as opposed to an enumeration.

The time therefore includes three compilations, plus two problems where the number of LP iterations is inordinate, namely for the first λ values of the second and third runs.

Table 4
Sequence of solutions for example 2.

−λ	K	Solution	f*
2	32	(1,2)(3,4)(6–8)(9–11)(13–15)(16–18)(20–22)(23–25) (26,27)(29,30)(34–36)(42–44)(44–47)(48,49)	−33.07
3	50	(1,2)(3,4)(5,6)(7–11)(12–14)(15–18)(19–21)(22–25) (26,27)(28–30)(31,32)(33–36)(37,38)(42–44) (45–47)(48,49)	−33.21
4	93	(1–4)(6–11)(12–18)(19–25)(26–28)(29–31)(32–36) (37,38)(39,40)(41–44)(45–47)(48–50)	−33.58
5	108	(1–4)(5–11)(12–18)(19–25)(26–30)(31,32)(33–38) (39,40)(41–44)(45–47)(48–50)	−33.70
6	113	(1–4)(5–11)(12–18)(19–25)(26–30)(32–38)(39–40) (41–44)(45–47)(48–50)	−33.717
7	123	(1–4)(5–11)(12–18)(19–25)(26–30)(31–38)(40–44) (45–47)(48–50)	−33.724
8	168	(1–5)(6–18)(19–25)(26–30)(31–38)(39–44)(45–47) (48–50)	−33.76
10	177	(1–5)(6–18)(19–25)(26–30)(31–38)(39–44)(45–50)	−33.74
11	186	(1–4)(5–18)(19–25)(26–30)(31–38)(39–44)(45–50)	−33.719
12	226	(1–4)(5–18)(19–25)(26–38)(39–44)(45–50)	−33.64
13	263	(1–11)(12–25)(26–38)(39–47)(48–50)	−33.84
14	291	(1–18)(19–25)(26–38)(39–47)(48–50)	−33.99
16	333	(1–4)(5–25)(26–38)(39–47)(48–50)	−34.20
18	444	(1–25)(26–38)(39–50)	−35.58
32	625	(1–30)(31–50)	−34.28
34	636	(1–31)(32–50	−34.31
36	744	(1–37)(38–50)	−35.00
37	769	(1–38)(39–50)	−35.43
40	Same as λ = −37.		

The times were

$-20 \leq \lambda \leq -1$, 1.26 min,
$-35 \leq \lambda \leq -21$, 0.66 min,
$-56 \leq \lambda \leq -36$, 0.76 min.

It may seem unusual at first inspection that the first 20 iterations of example 3 required nearly the same time as the first 40 iterations of example 2. Clearly, solutions are much more sensitive to λ values in the early going, where the percentage-wise weight attributable to unit increase in that parameter is much greater. For large values of λ, solutions do not change radically, and therefore the LP's for higher $|\lambda|$ require fewer iterations and consequently fewer dynamic programming executions.

To return to example 2, the numbers of iterations to optimality for the successive λ values were 0, 23, 14, 21, 7, 7, 6, 13, 8, 8, 6, 6, 7, 3, 3, 5, 2, 3, 2, 2, 2, 2, 1, 2, 3, 2, 3, 4, 3, 5, 7, 5, 5, 5, 4, 4, 2, 2, 2, 2, so that nearly 79% of the effort is expended in the first half of the iterations.

For the first 20 problems of example 3 the numbers of iterations were 0, 22, 19, 10, 5, 12, 19, 10, 3, 6, 8, 17, 15, 8, 11, 10, 11, 5, 7, 17. The total here is 215 which is 68 more than for the first 20 iterations of example 2. The final 20 problems of example 2 required 65 iterations.

As a set of data, example 3 took the 20 points of example 1 along with thirty additional points whose coordinates were 56, 57, 60, 62, 64, 65, 66, 67, 70, 71, 72, 73, 74, 75, 76, 77, 78, 79, 80, 81, 85, 86, 87, 88, 89, 92, 93, 94, 95, 96.

The best clustering discovered occurred for $\lambda = -36$, with $K = 664$, $f^* = -38.86$, and just two clusters, (1–17) and (18–50). f^* is not unimodal over the λ values selected, whereas the best solution persists over eleven consecutive values of λ, or nearly 20% of the range. Furthermore, solutions within 0.09 or about 0.2% of the best value are provided by another nine consecutive λ values. It would be difficult for a search routine to avoid an excellent solution.

Example 4. A final 50-point example was solved for $-40 \leq \lambda \leq -1$ in 2.19 minutes. The point coordinates were 0, 1, 2, 3,7, 9, 10, 11, 12, 13, 14, 17, 18, 19, 20, 23, 26, 27, 28, 29, 30, 31, 32, 33, 34, 35, 36, 40, 42, 43, 44, 45, 50, 51, 53, 54, 56, 57, 58, 60, 65, 70, 75, 80, 82, 83, 84, 90, 91, 98. The best solution occurred first at $\lambda = -35$ and persisted through $\lambda = -40$. It gave $K = 825$, $f^* = -32.17$, and two clusters, (1–40) and (41–50).

It will be noticed that our examples have been chosen to be somewhat unstructured; that is, optimal solutions are not conspicuous. In this way we promote more unusual f^* behavior and we gain an appreciation of the variety of situations that may arise. Were they more structured, we would observe sharper distinctions in function values. Example 4, it will be noted, may quite possibly have been terminated prematurely, since the best value occurred at the end of the region of experimentation (more evidence of the necessity for a variable step length, adaptive search over λ).

4. Accommodating multi-dimensional data

The development and the examples included in this paper are for clusterings of collinear point sets. The methodology should *not* be construed as limited to cases where the data exist in that form, however. The problem of mapping higher dimensional data into a smaller dimension, with tolerable distortion in inter-point relationships, has been studied. Mapping into a line, however, introduces another level of approximation. It becomes desirable to solve *several* problems, one for each of several different mappings when the data are unstructured. Any problem requiring clustering relative to a single attribute is, of its nature, one-dimensional, and thus some problems do not require transformation.

The interested reader is referred to several sources for dimension reduction strategies. In [31] a procedure is found to minimize total inter-point distance distortions after the linearization of the data. Unfortunately, a requirement is the solution of a traveling salesman problem over the given points. As a compromise, a subset of the data is mapped into one dimension according to subjective considerations, after which a transformation is derived which approximates the mapping of the sample. The remainder of the data are then mapped into the line by means of that transformation.

The existence of efficient methods for gaining good approximation to traveling salesman problems, not mentioned in [31], could eliminate the necessity of the transformation and allow the entire data set to be mapped at once.

The group method of data handling (GMDH) [17] has been suggested as a general method for modeling systems and can also be used to reduce data dimensionality to unity. The method works as follows: a data vector (one object) is presented as an input and a collection of second-degree polynomials operate upon these components to produce a collection of intermediate outputs. Selection criteria are invoked to determine which intermediate outputs are significant and only these are input to the next stage, where they are operated upon by another collection of second-degree polynomials. Such stages continue until a single output (a one-dimensional representation) survives. A mapping of quite high degree can be synthesized solely in terms of second-degree polynomials. A subset of data for which the results are thought to be known is used for training; that is, to specify the coefficients of all the second degree polynomials according to a minimum mean-square error criterion. The topics of training and selection require greater attention than is affordable here.

One feels that dimensionality is less a problem if the given data are geometrically visible. For example, if they may be exhibited in two or three dimensions, we feel more confident in mapping them into one dimension. The approach in [30] is to reduce any data set to a two-dimensional representation.

The given set of vectors is partitioned by means of a grid of hyperrectangles in that space. The location of the rectangle and the number of objects contained in it become the two-dimensional coordinates, and a final sequence of iterations seeks to preserve in a plane the original inter-point distances.

5. Summary

We have treated the class of clustering problems involving one-dimensional sets of data and a particular nonlinear objective function. The only obstacle to an exact solution is the existence of duality gaps, and in that case a very simple approximation scheme shows whether a better solution possibly was missed.

A subsidiary advantage of the Lagrange multiplier is in allowing differing weights to be attributed to distances within and between clusters, respectively. The user, being presented with a range of solutions changing with that weight, is capable of exercising a degree of judgment in choosing a solution. Since there is often no overwhelming, correct choice for an objective function, that ability may be valuable.

Potential computational economies were noted, but a final point merits particular attention. In tracing the sequence of solutions in a particular problem it is seen that a significant fraction of the LP iterations effect no change in the objective function value. For example 2, 27% of the iterations produced no change, and in example 4 that fraction is 22%. (Owing to the partition of the work in example 3, the fraction is not reliable.) Recalling that a DP is solved for each of these iterations, as much as three-fourths of our effort is unproductive. A remedy was suggested in the context of pricing schemes. With a list of basic candidates selected for the $C_j - Z_j$ value, we may compute the value of the corresponding entering variables and reject those providing no improvement.

To conclude, the technique appears to be a viable one and without radical modification should accommodate problems with several hundred points.

References

[1] I. Adler and D. Kafry, "Capturing and Clustering Judges' Policies", *Organizational Behavior and Human Performance* 25, 3 (1980) 384–394.
[2] M.R. Anderberg, *Cluster Analysis for Applications* (Academic Press, New York, 1973).
[3] R.E. Bonner, "On Some Clustering Techniques", *IBM Journal of R&D* 22 (1964) 22–32.
[4] J. Bryant, "On the Clustering of Multidimensional Pictorial Data", *Proceedings of the LACIE Symposium*, Lyndon B. Johnson Space Center, Houston (July 1979) pp. 647–659.
[5] F. Burbank, "A Sequential Space-Time Cluster Analysis of Cancer Mortality in the United States: Etiologic Implications", *American Journal of Epidemiology* 95 (1972).
[6] S. Chrandrasekharan, "An Implicit Enumeration Algorithm for Clustering Problems", Ph.D. Dissertation, University of Texas, Arlington, Texas (1975).

[7] J.M. Chaffray and G. Lilien, "A New Approach to Industrial Market Segmentation", *Sloan Management Review* 19, 3 (1978) 17–29.
[8] L. Cooper, "*N*-Dimensional Location Models: An Application to Cluster Analysis", *Journal of Regional Science* 18, 1 (1973) 41–54.
[9] G. Cowgill, "Archaeological Applications of Factor, Cluster, and Proximity Analysis", *American Antiquity* 33 (1968) 367–375.
[10] B.S. Duran and P. Odell, *Cluster Analysis, A Survey* (Springer-Verlag, Berlin, 1974).
[11] H. Everett, "Generalized Lagrange Multiplier Method for Solving Problems of Optimal Allocation of Resources", *Operations Research* 11 (1963) 399–417.
[12] W.D. Fisher, *Clustering and Aggregation in Economics* (Johns Hopkins Press, Baltimore, 1969).
[13] R. Garfinkel and G. Nemhauser, *Integer Programming* (John Wiley and Sons, New York, 1972).
[14] C. Gotlieb and S. Kumar, "Semantic Clustering of Index Terms", *Journal of the ACM* 15 (1968) 493–513.
[15] J.A. Hartigan, *Clustering Algorithms* (John Wiley and Sons, New York, 1975).
[16] E. Hewitt and K. Stromberg, *Real and Abstract Analysis* (Springer-Verlag, Berlin, 1965).
[17] A.G. Ivakhnenko, "Polynomial Theory of Complex Systems", *IEEE Transactions on Systems, Man, and Cybernetics* 1 (1971) 364–378.
[18] R. Jensen, "A Dynamic Programming Algorithm for Cluster Analysis", *Operations Research* 12 (1969) 1034–1057.
[19] R.M. Karp, "Reducibility Among Combinatorial Problems", in: R. Miller and J. Thatcher, eds., *Complexity of Computer Computations* (Plenum Publishing Co., New York, 1972) pp. 85–103.
[20] B. Kernighan, "Optimal Sequential Partitions of Graphs", *Journal of the ACM* 18 (1971) 32–40.
[21] W.L.G. Koontz, P. Narenda and K. Fukunaga, "A Branch and Bound Clustering Algorithm", *IEEE Transactions on Computers* (Sept. 1975) 908–915.
[22] G. Lance and W. Williams, "A Generalized Sorting Strategy for Computer Classifications", *Nature* 212 (1966) 218.
[23] L. Lefkovitch, "Conditional Clustering", *Biometrics* 36 (1980) 43–58.
[24] R. Lennington and M. Rassbach, "CLASSY – An Adaptive Maximum Likelihood Clustering Algorithm", *Proceedings of the LACIE Symposium*, Lyndon B. Johnson Space Center, Houston (July 1979) pp. 671–689.
[25] M. Padron, *An Axiomatic Basis and Computational Methods for Optimal Clustering* (University of Florida, Gainesville, 1969).
[26] M.R. Rao, "Cluster Analysis and Mathematical Programming", *Journal of the American Statistical Association* 66 (1971) 622–626.
[27] V.A. Rao and R. Umesh, "An Optimization Clustering Algorithm", *Proceedings of the Fourth International Joint Conference on Pattern Recognition*, Kyoto, Japan (IEEE, 1979) pp. 296–300.
[28] D.H. Robertson and D. Bellenger, "Identifying Bank Market Segments", *Journal of Bank Research* 7, 4 (1977) 276–283.
[29] G. Salton and A. Wong, "Generation and Search of Clustered Files", *ACM Transactions on Database Systems* 3 (1978).
[30] B. Schachter, "A Nonlinear Mapping Algorithm for Large Data Sets", *Computer Graphics and Image Processing* 8 (1978) 271–276.
[31] D.B. Simons et al., "An Approach to Nonlinear Mapping for Pattern Recognition", in: I. Tendam, ed., *Machine Processing of Remotely-Sensed Data* (IEEE, New York, 1979) pp. 323–330.
[32] P. Sneath and R. Sokal, *Numerical Taxonomy* (W.H. Freeman and Co., San Francisco, 1973).
[33] R. Sokal and C. Michener, "A Statistical Method for Evaluating Systematic Relationships", *University of Kansas Scientific Bulletin* 38 (1958) 1409–1438.

[34] L.E. Stanfel, "Experiments with a Very Efficient Heuristic for Clustering Problems", *Information Systems* 4 (1979) 285–292.
[35] L.E. Stanfel, "A Lagrangian Treatment of Certain Nonlinear Clustering Problems", *European Journal of Operations Research*, to appear.
[36] A. Van de Vew et al., "Coordination Patterns within an Interorganizational Network", *Human Relations* 32, 1 (1979) 19–36.
[37] H. Vinod, "Integer Programming and the Theory of Groups", *Journal of the American Statistica Association* 64 (1969) 506–519.

ALGORITHMIC IMPROVEMENTS FOR OBTAINING THE UPPER MULTINOMIAL BOUND

Robert PLANTE *
Purdue University

and

Prabhakant SINHA
Rutgers University

The multinomial bound is a nonparametric bound for a finite population total where most of the population elements have a value of zero and the remaining elements have positive values. Until now the computer processing time required to obtain bounds for samples containing more than 20 nonzero elements has been prohibitively high. In this paper the applicability of the multinomial bound is extended to samples containing up to 30 nonzero elements and the computer processing time of smaller problems is reduced by more than 99% of that previously required. These improvements involve the nesting of probability terms, the development of a customized nonlinear search algorithm, and the design of a tree data structure. The procedures presented in this research are developed in terms of a specific problem encountered in accounting, the determination of confidence bounds on the total overstatement, and understatement errors in accounting populations. These procedures have general applicability to a class of threshold problems as well, such as the determination of confidence bounds on the level of carcinogens in animals and in air pollutants.

1. Introduction

To avoid difficulties commonly encountered when large-sample confidence intervals are developed for estimating the total overstatement error in an accounting population that contains few such errors, Fienberg, Neter and Leitch [2] developed a nonparametric procedure based on the multinomial distribution that provides an upper bound for the total overstatement error.[1] They showed that the upper multinomial bound (UMB) was much less conservative than its chief competitor in auditing, the Stringer bound (Stringer [14] and Leslie, Teitlebaum and Anderson [10]). However, when there are more

---

Received September 15, 1980; revised March 10 and July 18, 1981.

* The authors acknowledge the helpful comments received from the referees.

[1] Other applications, particularly in the physical sciences, have been recognized. The National Academy of Sciences is currently organizing a committee to study these types of problems further.

than eight overstatement errors in a probability sample selected from the population, the practical implementation of the Fienberg et al. [2] approach is limited due to the prohibitive amount of computer time and storage required. Through the use of clustering,[2] which attains approximate (more conservative) bounds, this limitation was to a large extent removed by Leitch, Neter, Plante and Sinha [8] for up to 20 errors. However, the computer time required for more than 20 errors remained prohibitively high. It is important to handle samples containing up to at least 30 errors since large-sample theory can permit the use of classical estimators, such as ratio and difference estimators, in obtaining a confidence bound for the population total when there are more than 30 errors.

This paper extends the practical applicability of the multinomial bound to samples containing up to 30 errors through procedures designed to markedly reduce the time required to obtain the UMB. To maintain consistency with past and current research in this area the development of these procedures will be presented entirely in terms of a specific problem, the determination of confidence bounds on the total overstatement errors in accounting populations. However, there is no loss of generality in the application of these procedures to other problems for which a confidence bound on a finite population total is desired, where most of the population elements have a value of zero, and the remaining elements have positive values.

2. The upper multinomial bound model

Conceptually the UMB on total overstatement errors in an accounting population is obtained by:

(1) Expressing the total overstatement error, D, in terms of the population proportion, p_i, of each possible error amount, and the total book value, Y (in dollar units) as follows:

$$D = (Y/100) \sum_{i=0}^{M} i p_i, \tag{1}$$

where $M = $ the maximum possible overstatement error.

(2) Selecting a probability sample of n dollar units from the accounting population where the sample frequencies for each possible error amount follow the multinomial probability distribution (see Neter, Leitch and Fienberg [12] for an illustration).

---

[2] This procedure is similar to variable aggregation methods suggested by Fisher [4] and Geoffrion [5].

(3) Establishing a 1-α multidimensional confidence region for the multinomial parameters as follows:

$$\sum_S \frac{n!}{z_0! z_1! \ldots z_M!} \prod_{i=0}^{M} p_i^{z_i} = \alpha, \qquad (2)$$

where S represents the set of all outcomes $Z = (z_0, z_1, \ldots, z_M)$ which are "as extreme as or less extreme than" the observed sample outcome.

(5) Finding the maximum value of (1) over the confidence region established by (2), which is the upper bound.

Formally, the model used by Fienberg et al. [2] to determine the UMB is

$$\text{UMB} = \text{maximum:} \ (Y/100) \sum_{i=0}^{M} i p_i \qquad (3)$$

subject to

$$\sum_{i=0}^{M} p_i = 1, \qquad (4a)$$

$$\sum_S \frac{n!}{z_0! z_1! \ldots z_M!} \prod_{i=0}^{M} p_i^{z_i} \geq \alpha, \qquad (4b)$$

$$p_i \geq 0 \text{ for all } i. \qquad (4c)$$

Fienberg et al. [2] proposed the use of a "step-down" S set that defines the "less extreme" vectors of outcomes, Z, used in constraint (4b). The step-down S set consists of outcomes for which the following criteria are met:

Criterion (A): The total number of errors does not exceed the observed number of errors.

Criterion (B): Any individual "less extreme" error cannot exceed the corresponding observed error.

These criteria are operationalized by rank-ordering (high to low) the observed errors and not allowing the cumulative frequency of an error in an S set outcome to be more than the observed cumulative frequency. Mathematically these criteria may be expressed as:

$$\sum_{i=j}^{K} z_{e_i} \leq K - j + 1, \quad j = 1, \ldots, K, \qquad (5)$$

where, K = the number of errors; and e_i = the ith rank-ordered error such that $e_K \geq e_{K-1} \geq \ldots \geq e_1$.

Table 1
Results from a simple random sample of size n.

Sample results	
Observed frequency	Error (cents)
$n-2$	0
1	10
1	20
Total $\bar{n}$	$\overline{30}$

As an example, suppose a simple random sample of size n is selected with the observed errors given in table 1. Under criterion (A) no multinomial probability term in the step-down S set formed from the result of this sample can have more than two errors, and under criterion (B) the sum of the errors for each multinomial probability term cannot exceed 30 cents, i.e. an outcome of two 20-cent errors is not allowed in this step-down S set. Furthermore, the cumulative frequency of errors cannot exceed one for the 20 cent outcome or two for the 10 cent outcome. For the step-down S set it has been shown by Fienberg et al. [2] that the multinomial parameters corresponding to the observed errors and the maximum possible error, M, are the only parameters that take on nonzero values when the maximization problem is solved. As a result, the outcomes for each term of the step-down S set in the previous example can be represented by the first three columns of table 2.

As indicated by columns (4) and (5) of table 2, the two criteria which establish the formation of the step-down S set are both satisfied and the parameters corresponding to any other possible outcome satisfying these criteria are dominated in the maximum solution by the parameters p_0, p_{10}, p_{20}, and p_M, where p_M is the multinomial parameter corresponding to the maxi-

Table 2
Step-down S set for two errors (sample size $= n$).

Step-down S set			Total nonzero outcomes criterion (1)	Total error criterion (2)
Zero (1)	Ten (2)	Twenty (3)	(4)	(5)
$n-2$	1	1	2	30
$n-2$	2	0	2	20
$n-1$	0	1	1	20
$n-1$	1	0	1	10
n	0	0	0	0

mum possible overstatement error. Fienberg et al. [2] implemented this model and were able to obtain confidence bounds for samples containing up to eight errors. For samples with more than eight errors, the number of terms in the step-down S set imposes a computational burden that is untenable even in today's computer environment. To partially remove this limitation Leitch et al. [8] proposed the use of clustering. This approach yields good approximate bounds, and reduces the number of terms in the step-down S set. Since the computer processing time (CPU) required to obtain the UMB is due in large part to the amount of time spent evaluating constraint (4b) (for ten errors there are 58,786 terms in (4b)), achieving a reduction in the number of probability terms in the constraint correspondingly reduces the CPU time.

Three procedures are described in this paper which further reduce the CPU time required to obtain the UMB by reducing the amount of time spent computing (4b). In brief, these procedures are (a) the nesting of probability terms to represent the probability terms of (4b) compactly, (b) the development of a customized nonlinear optimization algorithm which reduces the frequency of computing (4b), and (c) the development of a tree data structure to curtail the frequency of duplicate multiplications.

Computational results are reported for each procedure separately as well as for the cumulative impact from a combined implementation of the procedures for various cases which appear in an appendix. These cases are based on the distributions observed by Johnson, Leitch and Neter [7] for overstatement errors in accounting populations.

3. The nesting of probability terms

The nesting of probability terms focuses on a compact representation of constraint (4b). A simple example will serve to demonstrate the "nesting" concept. Consider the following summation of multinomial probability terms:

$$\frac{n!}{(n-2)!1!1!} p_0^{n-2} p_1^1 p_2^1 + \frac{n!}{(n-2)!2!0!} p_0^{n-2} p_1^2 p_2^0 + \frac{n!}{(n-2)!0!2!} p_0^{n-2} p_1^0 p_2^2.$$

This summation of three terms is clearly equivalent to the following single term:

$$\frac{n!}{(n-2)!2!} p_0^{n-2} (p_1 + p_2)^2.$$

Thus, the amount of information contained in three multinomial probability terms is condensed into one binomial term and the portion of this term, denoted $(p_1 + p_2)^2$, is designated "nested".

Table 3
S set outcomes for two sample errors when only criterion (A) for the step-down S set is imposed.

S set Zero (1)	Nesting of 10 and 20 cent errors (2)	Outcome no. (3)
$n-2$	2	1
$n-1$	1	2
n	0	3

The procedure for nesting probability terms involves two steps. First, the criteria on the formation of the step-down S set are relaxed by only restricting the S set to outcomes for which the total number of errors does not exceed the observed number of errors, this corresponds to criterion (A). Secondly, the excess terms, those terms resulting from the S set relaxation that violate criterion (B), are substracted out.

3.1. Step 1 of the nesting procedure

The probability terms resulting from the S set relaxation are binomial. Further, the number of terms resulting from the S set relaxation equals the number of sample errors plus one, i.e. for 10 errors the number of outcomes in the S set is 11. Mathematically the S set relaxation can be expressed as follows:

$$\sum_{i=0}^{K} \binom{n}{i} p_0^{n-i} P^i, \tag{6}$$

where $P =$ the sum of the parameters corresponding to the sample errors.

As an example, consider the case for the observed sample errors given in table 1. Table 3 contains the S set outcomes based on these observations that represent the probability terms resulting from the use of nesting. For instance, the first outcome of table 3 corresponds to the following binomial probability term:

$$\frac{n!}{(n-2)!2!} p_0^{n-2} (p_{10} + p_{20})^2. \tag{7}$$

3.2. Step 2 of the nesting procedure

The added criterion imposed by the step-down S set, criterion (B), is enforced by enumerating those outcomes in (6) that violate this criterion and subtracting their contributions from the probability sum (6). For the example in table 3 there is only one outcome to eliminate, namely that outcome corresponding to two 20 cent errors, and its associated probability term is:

$$\frac{n!}{(n-2)!2!} p_0^{(n-2)} p_{20}^2. \tag{8}$$

Constraint (4b) is therefore formulated by subtracting this term from the probabilities associated with the S set relaxation as follows:

$$\frac{n!}{(n-2)!2!} p_0^{(n-2)} (p_{10}+p_{20})^2 + \frac{n!}{(n-1)!1!} p_0^{(n-1)} (p_{10}+p_{20}) + p_0^n$$

$$- \frac{n!}{(n-2)!2!} p_0^{(n-2)} p_{20}^2. \tag{9}$$

Thus, for two errors constraint (4b) now requires fewer S set terms (compare to table 2). Although the number of probability terms to subtract out can become large, it is often possible to nest many of these terms which further reduces the number of terms required to compute constraint (4b). For example, consider the sample results shown in table 4. The number of errors in this sample is five and thus there are six binomial terms in the S set that result from the relaxation of criterion (B). Table 5 shows the excess terms resulting from this relaxation that are subtracted from the sum of the six binomial terms to enforce criterion (B). Many of these subtraction terms have been formulated by nesting probability terms. For example, $p_5^2(p_3+p_2)p_1^2 p_0^{n-5}$ (the first term in table 5) is a nested term which represents two unnested subtraction terms, $p_5^2 p_3 p_1^2 p_0^{n-5}$ and $p_5^2 p_2 p_1^2 p_0^{n-5}$.

An algorithm has been developed and encoded in *FORTRAN* IV to identify and nest where possible the terms to subtract out. For the sample results given in table 4, table 5 shows the complete set of subtraction terms resulting from the algorithm's implementation. To facilitate the computation of these terms the algorithm also generates an array index which is used to locate the value of each unique combination of a parameter and its associated exponent. These indices are shown in table 5; i.e. 22 corresponds to the 22nd element of an

Table 4
Sample observations (sample size $=n$).

Error	Observed frequency
0	$n-5$
1	1
2	1
3	1
4	1
5	1

Table 5
Subtraction terms resulting from the sample results given in table 4.

Array index	Corresponding probability terms (w/o coefficients)
22 14 18 65	$p_5^2(p_3+p_2)p_1^2 p_0^{n-5}$
22 34 65	$p_5^2 p_1^3 p_0^{n-5}$
22 4 3 2 65	$p_5^2 p_3 p_2 p_1 p_0^{n-5}$
22 19 2 65	$p_5^2 p_2^2 p_1 p_0^{n-5}$
39 18 65	$(p_5+p_4)^3 p_1^2 p_0^{n-5}$
39 3 2 65	$(p_5+p_4)^3 p_2 p_1 p_0^{n-5}$
56 2 65	$(p_5+p_4+p_3)^4 p_1 p_0^{n-5}$
73 65	$(p_5+p_4+p_3+p_2)^5 p_0^{n-5}$
22 29 49	$p_5^2(p_2+p_1)^2 p_0^{n-4}$
22 4 13 49	$p_5^2 p_3(p_2+p_1)p_0^{n-4}$
39 13 49	$(p_5+p_4)^3(p_2+p_1)p_0^{n-4}$
56 49	$(p_5+p_4+p_3)^4 p_0^{n-4}$
22 12 33	$p_5^2(p_3+p_2+p_1)p_0^{n-3}$
39 33	$(p_5+p_4)^3 p_0^{n-3}$
22 17	$p_5^2 p_0^{n-2}$

array which has the value p_5^2, 14 corresponds to (p_3+p_2), 18 corresponds to p_1^2, and 65 corresponds to p_0^{n-5}, such that the product of the 22nd, 14th, 18th and 65th elements corresponds to the term $p_5^2(p_3+p_2)p_1^2 p_0^{n-5}$. A detailed description of the algorithm used to generate the nested subtraction terms is presented in Plante [13].

3.3. Computational results

The use of nested probability terms allows for the representation of the step-down S set by fewer terms. The reduction in size of the step-down S set achieved through the nesting of probability terms is exhibited in table 6 for cases selected from the Appendix. For example, 58,786 outcomes were previously required in the step-down S set for 10 distinct sample errors. Through the nesting of probability terms this step-down S set can now be represented by ll binomial terms of the S set and 1,407 terms to subtract out, a reduction of 97.59%. Unlike the clustering approach of Leitch et al. [9] the use of nested probability terms provides an exact representation of the step-down S set based on the original sample observations and thus yields the same multinomial bound as would be obtained without nesting.

For a given problem the nesting of probability terms reduces (i) the time required to compute the confidence region, (ii) the time required to solve the maximization problem, and (iii) the amount of computer storage required to retain the S set.

With the use of nesting and clustering the UMB can be obtained for up to

Table 6
A comparison of the number of outcomes required to represent the step-down S set for nested and unnested probability terms.[a]

Case number	Numbers of errors	Numbers of clusters	Number of probability terms		Percent reduction
			Unnested	Nested	
1	6	6	429	39	90.91
2	8	8	4,862	186	96.17
3	10	10	58,786	1,418	97.59
3	10	6	2,167	248	88.56
4	15	6	5,911	847	85.67
5	20	6	16,368	3,157	80.71
6	25	6	29,131	6,382	78.09
7	15	6	6,489	433	93.33
8	20	6	21,348	1,749	91.81
9	25	6	59,696	3,086	94.83
10	15	6	11,473	1,582	86.21
11	20	6	26,116	4,541	82.61
12	25	6	128,948	14,547	88.72
13	15	6	22,888	1,502	93.44
14	20	6	80,909	4,303	94.68
15	25	6	255,631	14,061	94.50
16	30	6	824,439	29,911	96.37

[a] A reduction in the number of probability terms produces a corresponding reduction in the CPU time required to obtain the UMB, e.g. previously the CPU time for six and eight sample errors was 108.8 and 3,510 seconds, respectively. By nesting terms these times have been reduced to 8.6 and 34.9 seconds, reductions of 92.1 and 99.0%, respectively.

25 errors occurring in a sample. As indicated by Leitch et al. [9], at least six clusters should be used to obtain a good approximation of the bound when the number of errors is greater than 20. Unfortunately even when clustering and nesting are used jointly, the size of the step-down S set remains too large when the number of sample errors exceeds 25. Consequently, there exists a gap between 25 and 30 errors for which the computation of the UMB (or a close approximation thereof) becomes infeasible. In the next section we present an approach that is designed to close this gap.

4. Customized nonlinear optimization procedure

The use of clustered sample errors and nested probability terms are approaches to facilitate the computation and efficient representation of the confidence region for the multinomial parameters. The development of a customized nonlinear optimization procedure focuses on the overall efficiency, in terms of convergence rate and computational time, of the optimization algorithm employed.

Currently, the nonlinear optimization algorithm used to obtain the UMB is the generalized gradient search (GGS) procedure, developed by K.E. Cross and W.L. Kephart, Union Carbide Corporation, Oak Ridge, Tennessee. This is a projection method procedure that, although very robust, is slow when compared to other search algorithms (Himmelblau [6]). This procedure requires that the search be conducted only within the feasible region. When an infeasible solution is obtained the algorithm iterates back to feasibility prior to continuing the search. These iterations are performed in directions which are formed from the Jacobian of the active constraints.

When the active constraints are highly nonlinear a large number of iterations towards feasibility are often required. The frequency of evaluating the probability terms that represent the confidence region for the UMB is of concern since the bulk of an algorithm's computation time is directly attributable to the number of times this constraint must be computed, as can be seen in a comparison of search and total constraint evaluation times presented in table 7. For a customized optimization algorithm to be useful in reducing convergence time, it must either reduce the number of times the constraints are evaluated in the feasibility search, or reduce the number of objective search iterations required to prove optimality, or both. To achieve these ends algorithms within the projection method class were tried and resulted in the development of an optimization procedure for obtaining multinomial bounds which has proven to be very succesful.

The generalized reduced gradient (GRG) algorithm, a nonlinear search procedure developed by Wolfe [15] and extended to include nonlinear constraints by Abadie and Carpentier [1], reduces the parameter space over which the feasibility search is conducted.

To obtain the UMB, the feasibility search of the GRG is conducted over a

Table 7
Comparison of search and total constraint evaluation times.

Case number	Number of clusters	Search time	Constraint time	Percent of search time used to evaluate (4b)
1	6	8.64	6.17	71.4
2	8	34.87	31.62	90.7
3	10	386.13	380.25	98.5
4	6	174.80	168.91	96.6
7	6	51.84	48.51	93.6
10	6	131.64	129.12	98.1
13	6	112.98	110.62	97.9

two-dimensional space (number of constraints) whereas the GGS requires a $K+1$-dimensional (number of errors) search. Thus, for the GRG algorithm the number of constraint computations is reduced relative to the GGS provided the convergence rate of the GRG algorithm is at least that of the GGS.

Luenberger [11] gives some indication that the convergence rates are similar and further states that the difference (in measured time) between the two procedures is due to the dimensions of the parameter space required for the feasibility search.

In order to implement the GRG procedure the problem must be modeled such that the constraints are binding. For the multinomial bound, Leitch et al. [8] have shown that constraint (4b) is binding in the optimal solution. Thus, the model is reformulated directly without requiring the addition of slack or artificial variables. Furthermore, an initial feasible solution is required for the GRG algorithm. This is usually accomplished through a phase I–phase II procedure. Here, however, the phase I can be made trivial since an initial feasible solution is analytically available by setting the initial value of the multinomial parameters as follows:

$$p_0 = (\alpha)^{1/n},$$

$$p_i = 0, \quad i = 1, 2, \ldots, M-1,$$

$$p_M = 1 - p_0.$$

The GRG method was modified to provide a problem-specific procedure for obtaining the UMB. The procedure will hereafter be referred to as the reduced gradient (RG). Plante [13] presents the algorithm designed to implement the RG procedure.[3]

4.1. Computational results

The implementation of the reduced gradient algorithm greatly reduces the time for obtaining the UMB. This is readily apparent from the results tabulated in table 8. For cases of 6, 8, 10, and 15 sample errors selected from the appendix, the mean reduction in CPU was an unexpected 89.95%. The convergence rates of the compared algorithms are in most cases markedly different since the RG procedure requires on the average 74.87% fewer search iterations than the GGS for the test problems used. Thus, efficiency has been achieved through both a reduction in the number of constraint evaluations

---

[3] The RG algorithm has been augmented by search acceleration procedures similar to those employed in the GGS algorithm.

Table 8
Comparison of GGS and RG algorithms for obtaining the upper multinomial bound (constraint tolerance = 10^{-8}; objective tolerance = 10^{-15}; nesting of terms used for both GGS and RG procedures).

Case number	Number of clusters	Multinomial bound (1000's)	Iterations		Run time (seconds)[a]	
			GGS	RG	GGS	RG
1	6	38.3	282	76	8.6	0.8
2	8	43.5	288	62	34.9	2.5
3	10	48.4	400	44	386.1	13.9
4	6	36.7	529	133	174.8	16.1
5[b]	6	41.5	n.r.[c]	76	n.r.	54.0
6[b]	6	46.9	n.r.	68	n.r.	70.2
7	6	67.6	286	175	51.8	18.6
8	6	72.1	n.r.	172	n.r.	82.3
9[b]	6	90.5	n.r.	117	n.r.	164.0
10	6	71.8	205	45	131.6	23.2
11[b]	6	91.1	n.r.	113	n.r.	246.0
12[b]	6	113.0	n.r.	62	n.r.	400.0
13	6	116.0	183	41	113.0	17.8
14[b]	6	146.0	n.r.	34	n.r.	60.8
15[b]	6	176.7	n.r.	33	n.r	217.8
16[b]	6	206.1	n.r.	37	n.r.	540.7

[a] This represents the total central processing unit (CPU) time. IBM run times were converted to equivalent CDC time (IBM time = 4 * CDC time).
[b] These problems were run in double precision arithmetic under FORTRAN optimizer level H on an IBM 370-158. All other problems were run in single precision arithmetic under FORTRAN optimizer level 2 on a CDC 70/74.
[c] Not run due to high projected cost.

required by the feasibility search and a reduction in the number of objective search iterations. Furthermore, the results in table 8 for cases of 20, 25, and 30 sample errors suggest that the development of the customized nonlinear optimization procedure has effectively reduced the CPU time required to solve problems for samples containing up to 30 errors.[4] Even so, the CPU time remains relatively high for obtaining the UMB when there are from 25–30 errors occurring in a sample.

In the next section a procedure is presented that further reduces the computational time required to obtain upper multinomial bounds, should this be desired.

[4] As part of a simulation analysis of the upper multinomial bound, Leitch et al. [9] obtained the multinomial bound for 490 cases where the number of errors ranged from 16 to 25 (25% of these cases were for 20 errors or more). The average CPU time for the 490 cases was 57.34 seconds/case.

5. Tree data structure for upper multinomial bound

As previously noted from table 7, the bulk of time used by the search procedure to obtain the upper multinomial bound is attributable to the amount of time used in evaluating constraint (4b). The purpose of this section is to describe a tree data structure that efficiently represents the probability terms of constraint (4b) and thereby reduces the computational time required to evaluate this constraint. This increase in efficiency is due to the reduction in the number of multiplicative operations that are performed for each term of the step-down S set through the reduction of duplicate operations. For example, if the product $p_1^3 p_2^5$ is used more than once in the probability sum (4b), then efficiency is achieved if this product were computed once and the resulting value retrieved whenever it is required.

5.1. Tree structure

To facilitate an understanding of the concepts used in the tree data structure development, a small example will be used. Consider the multinomial probability terms, without coefficients, in table 9. Upon inspection of table 9 it is clear that there are duplicate operations, i.e. the product $p_0^{10} p_1^2$ appears twice. Note, for future reference, that a total of eight multiplications is required to compute these terms. A representation of the multinomial probability terms in table 9, which reduces the number of duplicate operations, is the task at hand.

Consider the tree structure presented in fig. 1. This structure has the following properties.

(1) The number in the left-hand side of each bracket corresponds to a tree node number.

(2) A node without a predecessor is designated a root node, i.e. node 1 is a root node.

(3) A node without a successor is designated a terminal node, i.e. nodes 4, 5, 6 and 7 are terminal nodes.

(4) The value of a root node is the value of the multinomial parameter term associated with the node, i.e. node 1 has a value corresponding to p_0^{10}.

Table 9
Multinomial probability term example.

Multinomial probability terms without coefficients
$p_0^{10} p_1^2 p_2^1$
$p_0^{10} p_1^2 p_2^2$
$p_0^{10} p_1^1 p_2^2$
$p_0^{10} p_1^1 p_2^1$

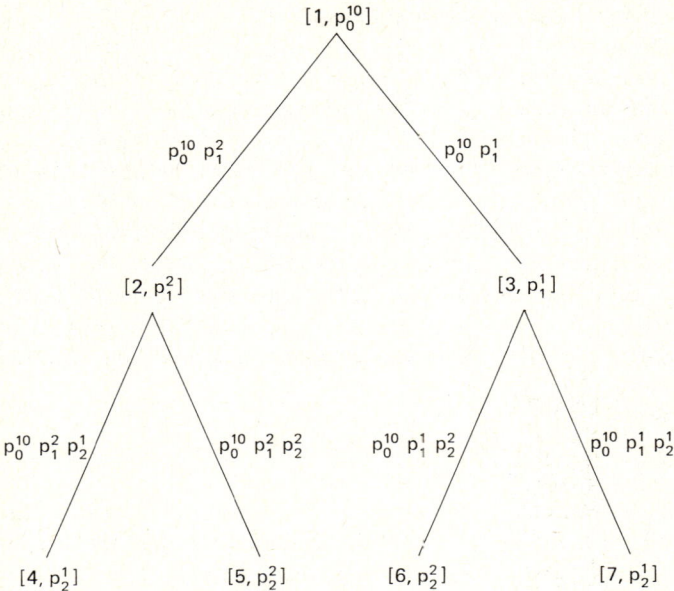

Fig. 1. Tree structure.

(5) Each arc of the tree structure represents a multiplication of the term associated with a node and the value of the node's predecessor. The value of a node, excluding root nodes, corresponds to this product, i.e. node 2 has a value corresponding to $p_0^{10}p_1^2$. These values appear next to the arcs on fig. 1.

(6) The multinomial parameter term associated with a node has at least as high a frequency of occurrence as those associated with any of the node's successors, i.e. p_0^{10}, associated with node 1, occurs four times in table 9 and each term of node 1's successors occurs only twice.

As a result each terminal node corresponds to the value of a multinomial probability term, without coefficients. For example, the value of node 4 corresponds to $p_0^{10}p_1^2p_2^1$, node 5 corresponds to $p_0^{10}p_1^2p_2^2$, etc. Furthermore, the number of multiplications required to compute these terms has been reduced from eight to six which, as a matter of design, happens to equal the number of arcs in the tree structure.

Plante [13] presents the algorithm designed to implement the tree structure for constraint (4b). An example of the tree structure resulting from this algorithm's implementation is shown in fig. 2, which contains the tree structure used to represent the probability terms given in table 5. For ease in presentation, the tree structure is decomposed into subtrees, such that each subtree has a root node corresponding to a unique exponent of the parameter p_0.

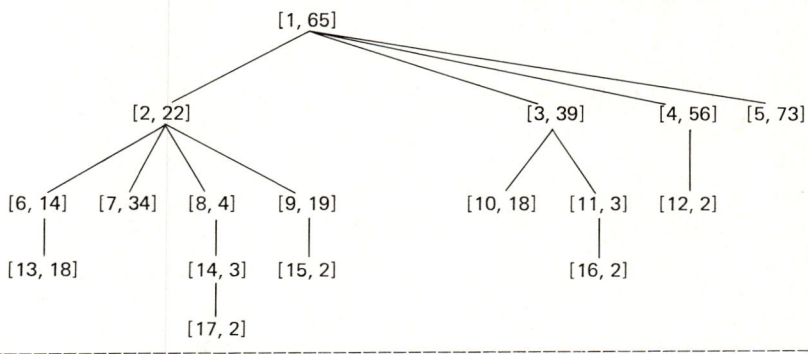
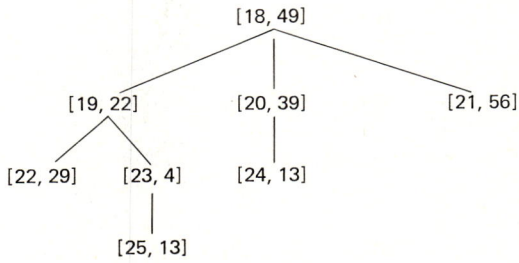
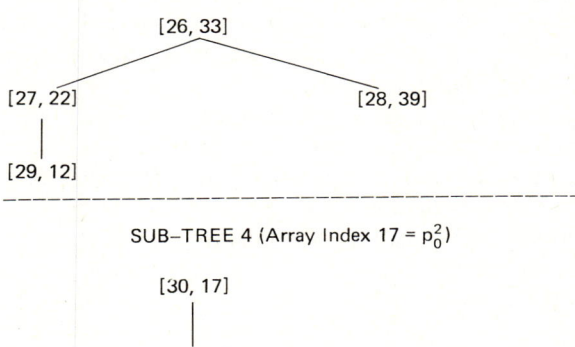

Fig. 2. Tree structure for each subtree generated to represent the terms in table 5 ($[i, j]$ = ith node, jth array index).

Table 10
Comparison between the time required with a tree structure and without a tree structure to obtain the upper multinomial bound (number of clusters = 6).

Case number	Number of errors	Previous time (table 9) (seconds) [a]	Time with tree added (seconds) [a]
1	6	0.8	0.5
2	8	2.1	0.9
4	15	16.1	14.5
7	15	18.6	15.7
10	15	23.2	20.3
12 [b]	25	400.0	343.5
13	15	17.8	16.1
15 [b]	25	217.8	190.8
16 [b]	30	540.7	448.8

[a] This represents the total central processing unit (CPU) time.
[b] These cases were run on an IBM 370-178.

Furthermore, the multinomial parameter term associated with a node is represented by its corresponding array index (previously described). Thus, the value of the first term in table 5 is the value of the 13th node in subtree 1, namely the product (arcs) of the array elements 65, 22, 14 and 18 or equivalently $p_0^{n-5} p_5^2 (p_3 + p_2) p_1^2$.

5.2. Computational results

The impact of reducing duplicate multiplications is illustrated by comparing the CPU time required to obtain the multinomial bound when duplications are not accounted for and when they are. These comparisons appear in table 10 for cases selected from the Appendix.

Table 10 contains the resulting reduction in time achieved by incorporating the tree structure into the upper multinomial bound algorithm. From these results it is evident that the incorporation of the tree structure has reduced the CPU time for the cases studied. Indeed, the average percent reduction in CPU time to obtain the UMB is 21.4% for these cases.

6. Discussion

Three algorithmic improvements have been presented which substantially reduce the computational time and thereby the cost for obtaining upper multinomial bounds.

The implementation of these procedures greatly reduces the computer processing time required to obtain the upper multinomial bound. Individually each procedure has reduced the computational time as follows.

(1) Nesting of probability terms reduces the number of terms in the step-down S set and thereby the CPU time by approximately 90% of that prior to this research.

(2) The reduced gradient algorithm (RG) reduces the CPU time by approximately 90% of that when the nesting of probability terms is employed.

(3) The tree data structure reduces the CPU time by approximately 20% of that when the RG algorithm and the nesting of probability terms are employed.

The combined use of these procedures has decreased the CPU time to obtain the upper multinomial bound by approximately 99% of that prior to this research. Achieving such a large reduction in computation time permits the determination of reasonably tight upper multinomial bounds for up to 30 errors occurring in a sample.

Appendix: Study cases used for the algorithmic development of the upper multinomial bound

Sample results for 16 random samples.

Case number	Number of nonzero sample errors	Nonzero sample errors (cents)
1	6	5,12,20,28,41,67
2	8	4,10,15,21,27,36,48,74
3	10	4,8,12,16,21,26,33,41,53,78
4	15	1,1,1,1,1,2,3,4,6,8,11,14,19,27,46
5	20	1,1,1,1,1,1,2,2,3,4,5,6,8,10,12,15,18,24,32,50
6	25	1,1,1,1,1,1,1,1,2,2,3,4,5,6,7,8,9,11,13,15,18,22,27,34,54
7	15	1,1,1,1,2,3,5,7,9,12,17,25,43,100,100
8	20	1,1,1,1,1,2,2,3,4,5,7,8,11,13,17,22,30,48,100,100
9	25	1,1,1,1,1,1,1,2,3,3,4,5,6,8,9,11,13,16,20,25,33,52,100,100,100
10	15	3,6,9,12,14,17,20,24,27,32,37,43,51,63,87
11	20	2,5,7,9,11,13,15,18,20,22,25,28,31,35,39,44,50,57,69,94
12	25	2,4,6,8,9,11,13,14,16,18,20,22,24,26,28,30,33,37,40,44,49,55,63,74,98
13	15	3,10,17,23,30,37,43,50,57,63,70,77,83,90,97
14	20	2,8,12,18,22,28,32,38,42,48,52,58,62,68,72,78,82,88,92,98
15	25	2,6,10,14,18,22,26,30,34,38,42,46,50,54,58,62,66,70,74,78,82,86,90,94,98
16	30	3,6,10,13,16,19,23,26,29,32,35,39,42,45,48,52,55,58,61,65,68,71,74,77,81,84,87,90,94,97

References

[1] J. Abadie and J. Carpentier, "Generalization of the Wolfe Reduced Gradient Method to the Case of Nonlinear Constraints", in R. Fletcher, ed., *Optimization* (Academic Press, New York, 1969).
[2] Stephen E. Fienberg, John Neter and Robert A. Leitch, "Estimating the Total Overstatement Error in Accounting Populations", *Journal of the American Statistical Association* 72, 358 (1977) 295–302.
[3] Walter D. Fisher, "On Grouping for Maximum Homogeneity", *Journal of the American Statistical Association* 53 (1958) 789–798.
[4] W.D. Fisher, *Clustering and Aggregation in Economics* (Johns Hopkins University Press, Baltimore, 1969).
[5] A. Geoffrion, "Customer Aggregation in Distribution Modeling", Working Paper 259, Management Science Center, UCLA (1967).
[6] David M. Himmelblau, *Applied Nonlinear Programming* (McGraw-Hill, New York, 1972).
[7] Johnny R. Johnson, Robert A. Leitch and John Neter, "Characteristics of Error in Accounts Receivable and Inventory Audits", *The Accounting Review* 56, 2 (1981) 279–293.
[8] Robert A. Leitch, John Neter, Robert D. Plante and Prabhakant Sinha, "Implementation of Upper Multinomial Bound Using Clustering", *Journal of the American Statistical Association* 76, 375 (1981) 230–234.
[9] Robert A. Leitch, John Neter, Robert D. Plante and Prabhakant Sinha, "Modified multinomial Bounds for Larger Numbers of Errors in Audits", *The Accounting Review* (April 1982), forthcoming.
[10] Leslie, Teitlebaum and Anderson, *Dollar Unit Sampling* (Copp Clark Pitman, Toronto, 1979).
[11] David G. Luenberger, *Introduction to Linear and Non-Linear Programming* (Addison-Wesley, Reading, Mass., 1979).
[12] John Neter, Robert A. Leitch and Stephen E. Fienberg, "Dollar Unit Sampling: Multinomial Bounds for Total Overstatement and Understatement Errors", *The Accounting Review* 53 (1978) 77–93.
[13] Robert D. Plante, "Algorithmic Improvements and Extensions for Obtaining the Multinomial Bound on Total Overstatement and Understatement Errors", unpublished Doctoral Dissertation, University of Georgia (1980).
[14] Kenneth W. Stringer, "Practical Aspects of Statistical Sampling in Auditing", in: *Proceedings of the Business and Economic Statistics Section* (American Statistical Association, 1963) pp. 405–411.
[15] P. Wolfe, *Recent Advances in Mathematical Programming* (McGraw-Hill, New York, 1963).

ERROR LOCALIZATION FOR ERRONEOUS DATA: A SURVEY

G.E. LIEPINS
Oak Ridge National Laboratory

R.S. GARFINKEL *
The University of Tennessee

and

A.S. KUNNATHUR [†]
University of Wisconsin, Milwaukee

Erroneous data is prevalent in large data sets. In many cases the user must revise the erroneous data so as to make it consistent with prespecified edits. The objective is to change the data as little as possible in the process of making it acceptable.

Various models have been proposed for the solution of this problem. To a large extent models depend on the form of the data (continuous, discrete, categorical) and on the form of the edits (linear, general). The most prevalent model in the literature requires that the fewest entries (fields) be changed in the process of making the record acceptable. It is denoted by minimum (weighted) fields to impute (MWFI) where the weights indicate levels of confidence in the reliability of the different fields.

It is shown that, subject to some assumptions, MFWI can be justified as a model of finding a "most likely" record. It is also shown that for categorical data MWFI is in the class of notoriously difficult *NP-complete* problems.

Finally, algorithms for MWFI are described for the continuous data-linear edit case as well as for the categorical data-general edit case. Preliminary computational experience with new algorithms for the latter case based on set covering is very promising.

1. Introduction

The fact that erroneous information is prevalent in large data sets is stressed by Terry [16] and by Pritzker et al. [13]. In these papers and others it is often estimated that the error rate on data gathered from various sources is typically between 1 and 10%, though at times it is significantly higher. Thus, the

---

Received August 4, 1980; revised April 26 and July 20, 1981.

* Research supported in part by Oak Ridge National Laboratory under a grant from the Department of Energy.

[†] Research supported in part by a Bureau of Census grant to the University of Tennessee.

magnitude of incorrect data can be staggering in, for instance, a population and housing census which processes tens or even hundreds of millions of records.

To make the ideas in the next paragraphs more precise, we introduce the following notation. A *record* x is an answer to a questionnaire containing a number of questions. Thus, if there are n questions, x is an n-dimensional vector. The components of the vector x are termed *fields*, so that the jth field refers to the jth question. An *edit* $E \subset R^n$ simply specifies a set of possible values for x which causes the record to be suspect. Thus, if $x \in E$ we say that x "fails" the edit E. In terms of the previous paragraph, the data were deemed "erroneous" because they failed certain edits.

Options for handling erroneous data

Given that a record has failed one or more edits, there remains the question of what to do with it. The options include:
(a) ask the respondent to revise the record;
(b) keep the record;
(c) discard the record; or
(d) revise the record.

The first option is the most desirable if it makes sense economically. For obvious reasons, often that is not the case.

In the case where statistical tests indicate that the record is an outlier, option (b) may be reasonable if one also notes the existence of the suspicious record and keeps track of it for some future statistical analysis. In the case where the record is known to be in error, it is clearly more difficult to justify this option.

Option (c) is attractive on the surface, but two arguments may render it unacceptable. In some cases individual records are irreplaceable and in other cases too large a proportion of the data would be discarded. (The latter is the case for at least one data base dealing with energy use in commercial buildings.) Another objection to this option is that discarding of erroneous records may bias the distribution of the data.

If, for any of the reasons given above, options (a)–(c) are unacceptable, then erroneous records must be changed. That is, new values must be *imputed* to some of the fields. There are many possible ways to change erroneous records. New values can be imputed directly to a subset of the fields, or a set of fields can be identified in one stage and imputation reserved for a second stage. These options are referred to as one- and two-stage imputation, and either can be accomplished within an optimization framework. Both options are discussed in the following sections.

2. Objectives and factors important to the error localization and imputation process

There are a number of factors which affect either error localization or imputation. In this section we discuss the following: the form of the data, the type of edits, the form of the edits, the error model, and the overall objective.

2.1. Form of the data and data records

Here we distinguish between continuous, discrete, and categorical data. By continuous data we mean data which can theoretically accept any value in some given interval, say $[l, u]$. For instance, the kwh of electricity used by an industrial concern in a given month can be considered continuous data.

By discrete data we mean that an acceptable response in a field is any one of a discrete set of entries. Family size would be an example of discrete data. Categorical data are a special case of discrete data. For categorical data, the coded values are merely identification codes. Relationship to head of household is one such example.

A data set is defined to be continuous, discrete, or categorical according to whether the data in each of the fields are continuous, discrete, or categorical, respectively. Data sets which have discrete data in some fields and continuous data in others are called mixed.

The distinction between noncategorical and categorical data is important because of the meaningfulness or lack thereof (respectively) of the usual Euclidean norm. For example, in the case of categorical data, it is plausible to postulate that any errors are independent of the apparent magnitude of error.

2.2. Type of the edits

We distinguish between two types of edits, namely deterministic and stochastic. If $\bar{x}$ is the event that x is not a valid record, then the edit E is deterministic if

$$P(\bar{x} | x \in E) = 1,$$

and stochastic otherwise. Clearly, the different types of edits lead to different decision processes. For instance, if a respondent indicates an age of eight and claims parentage of 23 children, we would all agree that the probability that the record is in error is one. On the other hand, the event "parentage of 23 children" may not be grounds for rejection of the field (or record), since the event has certainly occurred in the past. It would, however, have a low enough probability to be considered an outlier and to warrant the record being "flagged" for further investigation.

2.3. Form of the edits

With respect to the form of the edits, we distinguish first between algebraic and general edits. An algebraic edit, usually associated with continuous or with noncategorical discrete data, is of the form

$$x \in E \Leftrightarrow f(x) \neq 0, \tag{1}$$

where f is some algebraic function. For instance, if the first $n-1$ fields correspond to coal shipments to various regions, and if the nth field is total coal shipment, an edit would be:

$$x \in E \Leftrightarrow \sum_{j=1}^{n-1} x_j - x_n \neq 0. \tag{2}$$

By a *general edit*, usually associated with categorical data, we simply mean an edit which cannot (naturally) be put into the form (1). (We use the term "naturally" because it is always possible, with some manipulation and possibly expansion of the number of fields to make any edit algebraic (see section 4.3.) An example of a general edit might be:

$$x \in E \Leftrightarrow (\text{age} \leq 10 \text{ and marital status} \neq \text{single}).$$

Within the class of algebraic edits we may wish to distinguish further between linear and nonlinear edits. A *linear* edit (an example is (2)) is of the form

$$x \in E \Leftrightarrow \sum_{j=1}^{n} a_j x_j - b \neq 0, \tag{3}$$

where the a_j's and b are given. An algebraic edit is *nonlinear* if f is nonlinear.

2.4. The error model

By an error model we mean a representation of the data process

$$x = y + \epsilon,$$

where x is the observed data (record), y is the true data, and ϵ is the error.

The significance of the error model is that, in a theoretically unified approach to data editing, the error model should determine the form of the error localization and imputation insofar as significantly different error models would be expected to lead to different results for error localization and imputation.

2.5. The objective

Since the true data are unknown it seems reasonable to impute values which result in a record that passes the edits and is as "close" as possible to the observed record. This vague objective leads to more concrete models proposed in the following sections.

3. Optimization and underlying error models

Freund and Hartley [4] addressed a problem with continuous data and m linear edits represented by

$$\sum_{j=1}^{n} a_{ij} x_j - b_i \neq 0, \quad i = 1, \ldots, m.$$

Loosely speaking, their objective was to change the data as little as possible while coming as close as possible to satisfying the edits. They modeled this objective as:

$$\text{minimize } \sum_{j=1}^{n} c_j (x_j - y_j)^2 + \sum_{i=1}^{m} u_i \left(b_i - \sum_{j=1}^{n} a_{ij} y_j \right)^2, \tag{4}$$

where the non-negative weights c_j and u_i are supplied by the user and are taken to be measures of the respective importance of not changing the data and of satisfying the edits. The value y^0 which minimizes (4) is found by setting the derivative of (4) to zero, and is taken to be the imputed record. (Note that it is not required, as it is in subsequent approaches, that the resulting record satisfy all of the edits.)

Fellegi and Holt [3] developed an approach for problems with categorical data and general edits. Their objective was to determine the *minimum number of fields* which would have to be changed in order to produce a record which *satisfied all edits*.

We will denote the *minimum (weighted) number of fields to impute* model by MWFI. Before describing the Fellegi–Holt approach to MWFI, as well as the related approaches of Sande and McKeown, we discuss the linkage between the error model and the resulting mathematical program. Initial work on this problem is due to Naus et al. [12] whose approach is to try to approximate the probabilities of errors in the various fields, given that a specific set of edits has been failed by the record.

Letting $\bar{A}_j$ be the event that field j is in error, they approximate

$$P(\bar{A}_j | x \in E), \tag{5}$$

with

$$P\left(\bar{A}_j \middle| \bigcup_{k\alpha E} \bar{A}_k\right) \qquad (6)$$

where $k\alpha E$ means that field k "enters" (can contribute to the failure of) edit E. (In terms of a linear edit, field k enters the edit $\Sigma a_j x_j - b \neq 0$ if and only if $a_k \neq 0$.) The conditional probability (6) is then determined by Bayes rule, based on the assumed error model. Some intuitive justification is given in [12] for approximating (5) by (6), but it is also shown that they are very different entities. Finally, in an appendix, conditions are given under which the approximation is valid.

We conclude this section with an analysis of the error model conditions which render MWFI an appropriate optimization model.

Let $E(x)$ be the set of edits failed by record x and let J^0 denote the set of all subsets of the set of fields $\{1,\ldots,n\} = N$. Further suppose that the underlying objective can be stated: "find the set of fields which is most likely in error". Equivalently, this can be written as the problem of finding the set $K^* \in J^0$ which maximizes over all $K \in J^0$:

$$P\left\{(\bar{A}_j, j \in K) \cap (A_j, j \in N-K) \middle| x \in \bigcap_{E \in E(x)} E\right\}, \qquad (7)$$

where A_j is the event that field j is not in error. Now, assume that we are able to replace $x \in \bigcap_{E \in E(x)} E$ in (7) by $\bigcup_{k\alpha E(x)} \bar{A}_k$ as in (5) and (6), where $k\alpha E(x)$ means that $k\alpha E$ for some $E \in E(x)$. Then (7) becomes

$$\underset{K \in J^0}{\text{maximize}} P\left\{(\bar{A}_j, j \in K) \cap (A_j, j \in N-K) \middle| \bigcup_{k\alpha E(x)} \bar{A}_k\right\}. \qquad (8)$$

Furthermore, assume that for the error model the $P(\bar{A}_j)$'s are independent of each other and also independent of the magnitude of error. Then, with some additional restrictions on the error process (Naus et al. [12], Liepins [8]), (8) can be written as:

$$\max \prod_{j \in K} P(\bar{A}_j) \prod_{j \in N-K} (1 - P(\bar{A}_j))$$

subject to

$$x + t \in S, \qquad (9)$$

$$t_j = 0, \quad j \in N - K,$$

where t is a proposed correction to x and S is the set of acceptable records.

We would like to show that (9) can be transformed into MWFI. First note that the objective function of (9) can be written as:

$$\prod_{j \in N} (1 - P(\bar{A}_j)) \prod_{j \in K} \frac{P(\bar{A}_j)}{(1 - P(\bar{A}_j))}. \tag{10}$$

Take the negative of the natural log of (10), and set $c_j = -\ln(P(\bar{A}_j)) + \ln(1 - P(\bar{A}_j))$. Then the problem is transformed (ignoring a constant term in the objective) into (11), which is easily recognized as MWFI:

$$\begin{aligned}
\text{minimize} \quad & \sum_{j=1}^{n} c_j \delta(t_j), \\
& x + t \in S, \\
& \delta(t_j) = \begin{cases} 1, & t_j \neq 0, \\ 0, & t_j = 0. \end{cases}
\end{aligned} \tag{11}$$

The advantage of transforming (9) into MWFI is that the latter can be attacked with algorithms based on known mathematical programming methodology (some of which are discussed in the next section). We note that the assumptions required to produce MWFI from such a simple error model were certainly nontrivial, and to our knowledge this exercise has not been attempted for any other error distribution.

4. Minimum weighted fields to impute

4.1. Continuous data and linear constraints

For MWFI with continuous data and linear constraints it is convenient to use the model

$$\min \sum_{j=1}^{n} c_j (\delta(t_j) + \delta(z_j)) \tag{12}$$

subject to

$$\sum_{j=1}^{n} a_{ij}(x_j + t_j - z_j) \leq b_i, \quad i = 1, \ldots, m, \tag{13}$$

$$t_j z_j = 0, \quad j = 1, \ldots, n, \tag{14}$$

$$x_j + t_j - z_j, t_j, z_j \geq 0, \quad j = 1, \ldots, n, \tag{15}$$

where

$$\delta(t_j) = \begin{cases} 1, & \text{if } t_j > 0, \\ 0, & \text{otherwise.} \end{cases} \qquad (16)$$

Thus, t_j and z_j are respectively the positive and negative correction to x_j, and the *complementarity* conditions (14) ensure that at most one correction is used for each x_j. (Note that technically (13) defines *constraints* rather than *edits*. That is, here we deal with the acceptance rather than the rejection region.)

As pointed out by Sande [15] the complementarity conditions (14) are redundant as long as $c_j \geq 0$, which is certainly a reasonable assumption. This follows because any correction $\Delta_j = t_j - z_j$ derived from $t_j > 0$ and $z_j > 0$ yields a contribution of $2c_j$ to the objective. Clearly, a contribution of c_j is realizable by setting t_j or z_j to zero.

By a similar argument, it is also shown that an optimal solution to MWFI can always be found at a vertex of the polyhedron defined by (13) and (15). Sande then solves the problem using a variation of an algorithm due to Chernikova [2] and later modified by Rubin [14]. Chernikova's algorithm is designed to identify all extreme points of the polyhedron defined by a set of linear equations. Fundamentally it works by adding one constraint at a time and by using the extreme points generated at one iteration to identify those for the next. Computational experience with the basic Chernikova algorithm has indicated that storage problems may tend to be burdensome midway through execution (see [10]).

The modifications of the basic algorithm required to enable it to solve MWFI fundamentally consist of elimination of those extreme points which either violate (14) or which are dominated in objective value by a previously generated point. A code has been developed by Statistics Canada, although computational results are not yet available.

Another approach to the solution of the same problem is due to McKeown [11] who solves the problem using a modification of a branch-and-bound code developed for the general fixed charge problem. The code is being applied to survey data of the Bureau of Census involving outputs x_j of n different industries. Constraints are either of the form

$$\sum_{j=1}^{n} x_j = b_i \qquad (17)$$

or of the form

$$x_k - l_i x_j \geq 0, \qquad (18)$$

$$x_k - u_i x_j \leq 0,$$

where the constraints of (17) are "balance tests" and those of (18) are ratio tests. Computational results have not yet been reported.

4.2. Coded data, arithmetic edits

The model (12)–(16) can be applied to this case (see [15]) with an appropriate expansion of the number of variables and constraints. If field j has m possible entries, the binary variables x_j^k, $k = 1,\ldots,m$, are introduced. Then $x_j + t_j - z_j$ in (13) and (15) are replaced with $x_j^k + t_j^k - z_j^k$, where t_j^k and z_j^k are also binary. For a given field j, exactly one of the x_j^k's, say x_j^q, is one. Correspondingly, only z_j^q can be one. This can be represented by:

$$z_j^q = \sum_{k \neq q} y_j^k. \tag{19}$$

In [15] it is shown that the expanded problem can also be handled by Chernikova's algorithm by invoking complementarity conditions.

For problems with many entries per field, it remains to be seen if this approach will introduce too many variables to be practicable.

4.3. Coded data, general edits

Let a given set of edits be $E^* = \{(E^1, E^2,\ldots,E^m\}$. Also, let the set of unilaterally acceptable responses in field j be denoted by $R_j, j = 1,\ldots,n$. Then a record x is acceptable if

$$x_j \in R_j, \quad j = 1,\ldots,n$$

and $\quad\quad\quad\quad\quad\quad\quad\quad\quad\quad\quad\quad\quad\quad\quad\quad\quad\quad\quad\quad\quad\quad$ (20)

$$x \notin E^i, \quad i = 1,\ldots,m.$$

The set of points x satisfying (20) is the *acceptance region* S.

Before discussing solution approaches to MWFI for coded data, we note that it is in the class of NP-complete problems (see, for example, Garey and Johnson [5] for a precise definition of the term). This follows immediately from the fact that the "satisfiability" problem is a special case. The latter is the problem of determining whether there exists a set of truth values which renders each of a set of Boolean expressions true. The equivalence of MWFI follows by considering the complement of each of the Boolean expressions to be an edit, and asking whether a given record can be made acceptable by changing up to n fields.

The classification of MWFI as NP-complete simply means that it is theoretically as difficult as any of a notoriously difficult class of combinatorial problems. This does not prevent us from developing effective techniques for solving the problem. It simply makes it unlikely that we will find a theoretically "good" algorithm (one bounded by a polynomial function of the problem size).

Edits can be written in a number of forms. Perhaps the most natural is

$$E: p_1 \text{ and } p_2, \tag{21}$$

where p_1 and p_2 are propositions concerning the various fields. For instance, if p_1 and p_2 involve the fields "age" and "marital status", respectively, then a valid edit of the form (21) might state

"under 10" and "ever married".

It is easy to see that any general edit like (21) can be written in arithmetic form with the use of x_j^k variables introduced in the previous section. For instance, if the possible entries in the "age" and "marital status" fields are as given in table 1, then (21) would translate into

$$x_1^1 + x_1^2 + x_2^2 + x_2^3 + x_2^4 \geq 2. \tag{22}$$

As an alternative to arithmetization, we describe an approach due to Fellegi and Holt [3] and later modified by Liepins [9] and by Garfinkel [6].

In [3] it is shown that for bounded, discrete spaces any edit E^i can be written in a "normal" form:

$$E^i: F_1^i \times F_2^i \times \ldots \times F_n^i, \tag{23}$$

where $F_j^i \subseteq R_j$. We say that field n "enters" edit E^i if $F_j^i \neq R_j$. Consider a record x which fails the set of edits $E(x)$. One might then suspect that MWFI could be written as the set covering problem

$$\begin{aligned} &\text{minimize} \quad cw, \\ &Aw \geq 1, \\ &w \text{ binary}, \end{aligned} \tag{24}$$

Table 1

Age	Marital status
0–5	Single
6–10	Married
11–15	Divorced
16–20	Widowed
>20	

where $A = (a_{ij})$ and

$$a_{ij} = \begin{cases} 1, & \text{if field } j \text{ enters edit } E^i \in E(x), \\ 0, & \text{otherwise.} \end{cases}$$

In words, the constraints of (24) guarantee that for every failed edit, at least one field that enters is chosen to be changed. It is not hard to see, however, that the formulation (24) is not sufficient to represent MWFI since a solution to (24) may yield a set of fields containing no possible feasible correction.

For example [3], suppose the category "Relationship to head of household" is added to table 1 with possible entries: head, spouse of head, other. Further suppose there are two edits, namely:

E^1: $\{0-5, 6-10\} \times \{\text{married, divorced, widowed}\} \times \{\text{head, spouse, other}\}$;
E^2: $\{0-5, 6-10, 11-15, 16-20, > 20\} \times \{\text{single, divorced, widowed}\} \times \{\text{spouse}\}$.

If the record $x = \{7\} \times \{\text{married}\} \times \{\text{spouse}\}$ is encountered, which fails E^1, the resulting set covering problem is:

$$\min w_1 + w_2 + w_3,$$
$$w_1 + w_2 \geq 1, \tag{25}$$
$$w_1, w_2, w_3 \text{ binary,}$$

assuming $c_1 = c_2 = c_3 = 1$. A solution to (25) is $w_2 = 1$, $w_1 = w_3 = 0$, which indicates that field 2 should be changed. It is easy to see, however, that there is no unilateral change to field 2 which does not fail E^2.

This situation has been addressed by Fellegi–Holt [3], and later by Liepins [8]. The key result is that (24) can be made valid by the inclusion of a subset $\bar{E}$ of the set of "implied" edits, where an edit E is implied by a set of edits T if $E \subset \bigcup_{E^i \in T} E^i$.

Fellegi–Holt have also shown that $\bar{E}$ can be generated by the construction:

$$\hat{F}_j = \bigcap_{i \in Q} F_j^i, \quad j \neq k,$$

$$\hat{F}_k = \bigcup_{i \in Q} F_k^i, \tag{26}$$

as long as $F_j \neq \emptyset$ for any j, where Q is a subset of the edits. Field k is referred to as the "generating field". For instance, in the previous example, using field 2 as the generating field and $Q = \{1, 2\}$ would yield the edit:
E^3: $\{0-5, 6-10\} \times \{\text{single, married, divorced, widowed}\} \times \{\text{spouse}\}$,
which is the only implied edit required to make (24) valid. In order to

guarantee that $\bar{E}$ can be constructed by (26), it is necessary to recursively consider every subset Q of $E^* \cup \bar{E}$, including implied edits.

Liepins [9] has developed a one-pass algorithm for generating a *sufficient set* $\tilde{E} \subset \bar{E}$. The set covering problem (24) can then be solved to find the fields to impute. The actual imputation can be done in a number of ways. One reasonable approach is to search previous records for an acceptable record which agrees with the failing record except in fields to be imputed. The acceptable record can then be used in place of the failed record. This procedure is known as "hot-decking". (Design of efficient search procedures for available acceptable records involve computer science topics which are of interest in their own right.) For many problems the set $\tilde{E}$ may be extremely large causing the edit generation procedure to be lengthy and rendering the set covering problem difficult. Therefore, an alternative "row generation" approach has been developed by Garfinkel [6].

In essence, the algorithm is:

Step 1. Test $x \in S$, if so stop. (Recall that S is the acceptance region.)

Step 2. Solve (24), and attempt to find a feasible imputation over the resulting optimal fields. If successful, stop.

Step 3. Generate an implied edit (by a construction which slightly generalizes (26)). Add the new constraint to (24) and go to step 2.

Note that this algorithm actually supplies an imputation (although it need not be used) as well as solving MWFI. It has the advantage of solving set covering problems for each failed record, at the cost of possibly generating implied edits for each failed record.

It is important to note that the approaches of this section are dependent upon solution techniques for set covering problems. Fortunately, these are among the best-solved class of integer programming problems (see, for example, [7]).

Computational results indicate that the cutting plane approach is the more effective of the two presented in this section. Computer runs are summarized in table 2.

Each row of table 2 denotes a set of runs with 100 random failing records, except for sets 11 and 12 for which only two failing records were run because of time limitations. Computations were performed on the IBM 3031 at the University of Tennessee using the FORTRAN Optimization Compiler. The column headings M, K, and N denote the numbers of edits, data points, and fields, respectively. By the number of data points we mean the total number of possible entries, so that the average number of entries per field is K/N.

The cutting plane algorithm is designated "algorithm 1", and the approach based on generating the sufficient collection is "algorithm 2". Columns *AT1* and *AT2*, *MXT1* and *MXT2* give the average time and maximum time (in seconds) per record for algorithms 1 and 2, respectively (note that the time to

Table 2

No.	M	K	N	AT1	MTX1	AV1	MX1	AT2	MTX2	AVE	MXE	SCT
1	14	50	10	0.09	0.70	0.38	3	0.10	0.23	2.57	9	6.53
2	19	93	21	0.03	0.06	0.01	1	2.90	0.04	1.17	3	294
3	14	53	11	0.06	0.49	0.44	4	0.13	0.21	2.01	9	9.2
4	19	41	13	0.05	0.84	0.16	4	1.02	0.01	2.12	9	98.7
5	10	69	13	0.03	0.13	0.03	2	0.10	0.08	1.53	5	7.38
6	7	69	12	0.03	0.09	0.01	1	0.02	0.05	1.37	4	0.14
7	22	33	14	0.19	1.34	0.76	6	0.36	3.42	7.02	19	21
8	26	38	17	0.04	0.45	0.17	4	3.52	0.15	2.88	13	351
9	22	82	9	1.18	11.73	0.07	1	3.71	0.06	2.57	8	366
10	6	68	13	0.02	0.03	0	0	0.03	0.03	1.32	3	0.5
11	22	500	64	35.80	37.20	0	0	–	–	–	–	*
12	91	740	92	1.89	2.44	0	0	–	–	–	–	*

develop the sufficient collection of edits is included in the calculation of *AT2*). Columns *AV1* and *MX1* give the average and maximum number of implied edits per record generated by algorithm 1, while *AVE* and *MXE* give the average and maximum number of implied edits *failed* per record by algorithm 2. Finally, column *SCT* gives the time required to generate a sufficient collection of implied edits by algorithm 2. For problems 11 and 12 a sufficient collection could not be generated in 30 minutes.

The various problem sets had different "entering field densities" (*EFD*). That is, the average number of fields entering an edit varied from 0.1 to 0.35. In particular problems 3 and 4 had $EFD = 0.35$, problems 5 and 6 had $EFD = 0.25$, problems 9 and 10 had $EFD = 0.2$, while all others had $EFD = 0.1$. The set covering problems were solved by the algorithm given in [1].

5. Conclusions

We have presented a brief survey of the state of the art of error localization. As the reader can readily surmise the field is in its infancy. In particular, very little comprehensive testing of algorithms has been reported, and no systematic analysis of the effect of various error model assumptions on the resulting optimization models has been attempted. In another paper some of these omissions will be addressed.

Acknowledgment

The computer runs were carried out by the third author as part of his dissertation research. The authors benefited from a number of stimulating discussions with Brian Greenberg of the Census Bureau.

References

[1] M. Bellmore and H.D. Ratliff, "Set Covering and Involutory Bases", *Management Science* 18 (1971) 194–206.
[2] N.S. Chernikova, "Algorithm for Finding a General Formula for the Non-Negative Solutions of a System of Linear Equations", *U.S.S.R. Computational Mathematics and Mathematical Physics* 4 (1964) 151–158.
[3] P. Fellegi and D. Holt, "A Systematic Approach to Automatic Edit and Imputation", *JASA* 71 (1976) 17–25.
[4] R.J. Freund and H.O. Hartley, "A Procedure for Automatic Data Editing", *JASA* 62 (1967) 341–352.
[5] M.R. Garey and D.S. Johnson, *Computers and Intractability: A Guide to the Theory of NP-Completeness* (Freeman Press, San Francisco, 1979).

[6] R.S. Garfinkel, "An Algorithm for Optimal Imputation of Erroneous Data", Working Paper 83, College of Business Administration, The University of Tennessee (1979).
[7] R.S. Garfinkel and G.L. Nemhauser, *Integer Programming* (John Wiley & Sons, New York, 1972).
[8] G.E. Liepins, "A Rigorous, Systematic Approach to Automatic Data Editing and its Statistical Basis", ORNL/TM-7126, Oak Ridge National Laboratories (1980).
[9] G.E. Liepins, "Refinement to the Boolean Approach to Automatic Data Editing", Oak Ridge National Laboratories (1979).
[10] J.H. Matheiss and D.S. Rubin, "A Survey and Comparison of Methods for Finding all Vertices of Convex Polyhedral Sets", *Mathematics of Operations Research* 5 (1980) 167–185.
[11] P.G. McKeown, "Solving the Fields to Impute Problems via Mathematical Programming", Working Paper, University of Georgia (1979).
[12] J.I. Naus, T.G. Johnson and R. Montalvo, "A Probabilistic Model for Identifying Errors in Data Editing", *JASA* 67 (1973) 943–950.
[13] L. Pritzker, J. Ogus and M.H. Hansen, "Computer Editing Methods-Some Applications and Results", *Bulletin of the International Statistical Institute* 41 (1965) 442–465.
[14] D.S. Rubin, "Vertex Generation and Cardinality Constrained Linear Programs", *Operations Research* 23 (1975) 555–565.
[15] G. Sande, "An Algorithm for the Fields to Impute Problems of Numerical and Coded Data", Statistics Canada (1978).
[16] M.E. Terry, "The Principles of Statistical Analysis Using Large Electronic Computers", *Bulletin of the International Statistical Institute* 40 (1963) 547–552.

SELECTION OF COST-OPTIMAL FRACTIONAL FACTORIALS, $2^{m-r}3^{n-s}$ SERIES

C.A. MOUNT-CAMPBELL and J.B. NEUHARDT
The Ohio State University

The problem of planning an orthogonal fraction of a factorial experimental design involving a mixture of two- and three-level factors is presented. The cost of experimentation is an explicit consideration in the problem and the objective is to find a balanced fraction that minimizes total cost. A cost model is specified and used to determine which cost structures have uniform costs among the fractions and which do not. The planning process is then illustrated by an example design problem involving computer tests of production planning algorithms. Although the cost model makes it possible manually to determine cost-optimal designs, there is a class of problems that require automatic means for solution. It is suggested that a computer algorithm be fashioned from existing codes. Although the methods used in the paper are related to a specific technique for generating mixed two- and three-level fractions mentioned by Chakravarti, the approach is found to be useful in evaluating alternative costs for those designs cataloged by Connor and Young.

1. Introduction

Experimental design, with its roots in agriculture (e.g. [6,16]) has become an important aid in the controlled laboratory (e.g. [14]), on the industrial production line (e.g. [13]), in computer simulation (e.g. [10], and field studies supporting management decisions related to product selection (e.g. [7]). An important class of designs is the full factorial, allowing qualitative independent variables, and fractional factorials, which have most of the advantages of the full factorial but require fewer total observations.

Many fractional factorial designs require orthogonality, i.e. all estimated effects statistically independent. Recently, an additional constraint was added, that of minimum cost, assuming heterogeneous costs of taking observations. The fractional factorials studied in the minimum cost problem were the two-level [11], and three-level [9]. In this paper a notable extension to the problem is examined, i.e. mixtures of two- and three-level factors (e.g. [5]), and an example is presented involving computer experiments with production scheduling algorithms.

Received September 30, 1980; revised July 31, 1981.

2. Cost model and parameterization

It is assumed that the experiment involves m factors $A_1,\ldots,A_m$, each occuring at three levels, and n factors $B_1,\ldots,B_n$, each occuring at two levels. A particular combination of factors is represented by $(a; b)$, where $a = \{a_i\}$ is an m-tuple with elements 0, 1, 2, and $b = \{b_j\}$ is an n-tuple, with elements 0, 1. Clearly, there are $N = 3^m\, 2^n$ such combinations. Defining μ to be an $N \times 1$ vector of expectations of the dependent variable, or yield, to be measured at all N combinations, the linear model for the complete factorial is assumed:

$$\mu = X\theta, \qquad (1)$$

where $X(N \times M)$, $M \leq N$, consists of orthogonal columns, with one column filled entirely with elements equal to 1, and θ the $M \times 1$ vector of unknown constants.

If a fraction, F, of the factorial is defined (as a subset of the $(a; b)$ consisting of t combinations), the model associated with the resultant μ_F is denoted:

$$\mu_F = X_F \theta,$$

where X_F is the $t \times M$ matrix whose t rows are appropriately chosen from X. An orthogonal fraction, F, is one for which specified columns of X are orthogonal ($M \leq t$ for estimability of all θ).

It is now assumed that with each $(a; b)$ is associated a positive cost $c(a; b)$ and the total cost $TC(F)$, of the fraction F, is given by:

$$TC(F) = \sum_{(a;b) \in F} c(a; b). \qquad (3)$$

We let C be the $N \times 1$ vector of elements $c(a; b)$; then the $P \times 1$ vector, γ, and the $N \times P$ design matrix, Z, are defined to form the cost model:

$$C = Z\gamma \, (P \leq N, \text{rank}(Z) = P). \qquad (4)$$

Thus, Z_F is the $t \times P$ matrix whose row numbers are identical to those of X_F. Now the cost expression may be written:

$$TC(F) = J' Z_F \gamma, \qquad (5)$$

where J' is a row vector with all elements equal to 1 and conformable with Z_F. The total cost is now represented as the inner product of $J'Z_F$, the components of which are column sums of Z_F, and γ, the solution to $C = Z\gamma$.

3. Cartesian product designs

Chakravarti [4] determined the properties of a simple method for determining orthogonal fractions for the mixed two- and three-level series:
(1) determine an orthogonal fraction of 3^m, say 3^{m-r};
(2) determine an orthogonal fraction of the 2^n, say 2^{n-s}; and
(3) The complete $3^{m-r} 2^{n-s}$ fraction is obtained by all $t = 3^{m-r} 2^{n-s}$ combinations of those observations specified in steps (1) and (2).

Let $F1$ represent those factor combinations resulting from step (1). Such an orthogonal fraction is specified by solving r linearly independent equations with coefficient vectors denoted $\boldsymbol{u}_j = (u_{1j}, u_{2j}, \ldots, u_{mj})$ with $u_{ij} = 0$, 1 or 2. Thus,

$$F1 = \left\{ (a;b); \sum_{i=1}^{m} a_i u_{ij} = \phi_j (\mathrm{mod}\, 3), j = 1, 2, \ldots, r \right\}, \tag{6}$$

where the constants ϕ_j are pre-specified (at values 0, 1 or 2). The vectors $\{\boldsymbol{u}_j\}$ are often called the generators of the design, $F1$.

Also, if one lets $F2$ represent the fraction generated in step (2), then it is defined by the solutions to the s linearly independent equations with coefficient vectors $\{\boldsymbol{v}_j = (v_{1j}, v_{2j}, \ldots, v_{mj}); j = 1, 2, \ldots, s\}$ each having component values equal to 0 or 1. Thus,

$$F2 = \left\{ (a;b); \sum_{i=1}^{n} b_i v_{ij} = \delta_j (\mathrm{mod}\, 2), j = 1, 2, \ldots, s \right\}, \tag{7}$$

where δ_j is a pre-specified constant, 0 or 1.
Finally, step (3) forms the complete fraction with all combinations:

$$F = \left\{ (a;b); \sum_{i=i}^{m} a_i u_{ij} = \phi_j, \sum_{i=1}^{n} b_i v_{ik} = \delta_k; j = 1, 2, \ldots, r, k = 1, 2, \ldots, s \right\}. \tag{8}$$

A well-known property of the generators of the designs is that every integer linear combination of the r vectors $\boldsymbol{u}_j, j = 1, \ldots, r$ (reduced modulo 3), forms a finite Abelian group. This group, with the additive identity (an m-tuple with all zero elements) deleted will be designated $G1$. Similarly, every integer linear combination of the s vectors $\boldsymbol{v}_j, j = 1, \ldots, s$ (reduced modulo 2) forms a finite Abelian group. Again, without the identity this group will be designated $G2$. Suppose the cost model is defined so that each parameter in γ is uniquely associated with exactly one of the following:
(1) an element of $G1$;
(2) an element of $G2$;

(3) an element of $G1 \times G2$ (Cartesian product of $G1$, $G2$);
(4) an n-tuple (with elements 0, 1, 2) not in $G1$;
(5) an m-tuple (with elements 0, 1) not in $G2$;
(6) an nm-tuple of all zeros (the concatenated identity); and
(7) an nm-tuple not in $G1 \times G2$ (except the concatenated identity).

If an appropriate coding is used to form the orthogonal columns of Z corresponding to these parameters, then the cost of any Cartesian fraction is computable in a direct and simple manner which facilitates finding the cost-optimal design. Such an appropriate coding scheme for the Z matrix is defined in the appendix. For now, it is sufficient to point out the important property of the Z matrix:

For any Cartesian fraction the matrix Z_F, formed by deleting rows for the observation not in the fraction, has columns which sum to zero (corresponding to estimable parameters), with the remaining columns having constant entries.

The constant columns are those associated with parameters that correspond to elements of $G1$, $G2$, or $G1 \times G2$, and are therefore confounded with the mean vector. The value of the constant in any column depends upon the actual numbers used in the coding of Z.

To discuss the cost model further, the Z matrix is partitioned as follows:

$$Z = \{J, ZA, ZB, ZAB\}, \tag{9}$$

where J is an $N \times 1$ column vector with all components equal to 1, and associated with item (6) above, and the parameter γ_0. ZA is an $N \times (3^m - 1)$ matrix with columns corresponding to the parameters (designated γ_A) that are associated with items (1) and (4) above. ZB is an $N \times (2^n - 1)$ matrix with columns corresponding to the parameters (designated γ_B) that are associated with items (2) and (5) above. ZAB is an $N \times (3^m - 1)(2^n - 1)$ matrix with columns corresponding to the parameters (designated γ_{AB}) that are associated with items (3) and (7) above.

The matrix Z_F of eq. (5) may be similarly partitioned:

$$Z_F = \{J_F, ZA_F, ZB_F, ZAB_F\}. \tag{10}$$

The cost vector may now be represented as:

$$C = J\gamma_0 + (ZA)\gamma_A + (ZB)\gamma_B + (ZAB)\gamma_{AB}, \tag{11}$$

and the total cost of the fraction F may be written:

$$TC(F) = p_1\gamma_0 + p_2 TC(F1) + p_3 TC(F2) + TC(F1, F2), \tag{12}$$

where:

$p_1 = 3^{m-r} 2^{n-s}$,
$p_2 = 2^{n-s}$,
$p_3 = 3^{m-r}$,
$TC(F1) = J'(ZA_F)\gamma_A$,
$TC(F2) = J'(ZB_F)\gamma_B$,
$TC(F1, F2) = J'(ZAB_F)\gamma_{AB}$.

This total cost expression involves the terms $TC(F1)$ and $TC(F2)$ which are the separate costs for both fractions $F1$ and $F2$, respectively. Each can be calculated independently of the other once the cost model (4) has been solved for γ. The term $TC(F1, F2)$ represents the additional cost of joining $F1$ and $F2$ into the fraction, F.

If the parameters are indexed so that γ_{Ai} is associated with the ith element of $G1$, γ_{Bj} is associated with the jth element of $G2$, and γ_{ABk} is associated with the kth element of $G1 \times G2$, then the cost expression (12) may be further simplified by observing that:

$$TC(F1) = 3^{m-r}\left(\sum_{i \in G1} h_{Ai}\gamma_{Ai}\right),$$

$$TC(F2) = 2^{n-s}\left(\sum_{j \in G2} h_{Bj}\gamma_{Bj}\right), \qquad (13)$$

$$TC(F1, F2) = 3^{m-r} 2^{n-s}\left(\sum_{k \in G1 \times G2} h_{ABk}\gamma_{ABk}\right),$$

where

h_{Ai} is $-1, 0, 1,$ or 2,
h_{Bj} is -1 or 1,
h_{ABk} is $-1, 0, 1, 2,$ or -2,

each value depending upon the coding scheme as given in the appendix.

Substituting (13) into (12), the total cost expression reduces to:

$$TC(F) = 3^{m-r} 2^{n-s}\left(\gamma_0 + \sum_{i \in G1} h_{Ai}\gamma_{Ai} + \sum_{j \in G2} h_{Bj}\gamma_{Bj} + \sum_{k \in G1 \times G2} h_{ABk}\gamma_{ABk}\right).$$

(14)

Thus, from (14) one sees that the cost of a fraction is affected by only those cost parameters that would be confounded with the mean in the fractional design. One immediate result of this expression is that if all third-order or higher cost interaction terms are zero, and if one is considering confounding such higher order interaction terms only (e.g. a resolution III design in which the main effects are estimable), then all such fractions have the same cost, namely $3^{m-r} 2^{n-s} \gamma_0$. However, if higher order cost interaction terms are nonzero, then all fractions will not have the same cost and hence the search for the cost-optimal design is warranted. Such a case is illustrated by the example in the next section.

4. An example

A problem recently encountered by Bahl [3] will be modified slightly and used to illustrate the mixed two- and three-level algorithm. Bahl considered a multistage production-inventory system and formulated a unified model for decisions on the master schedule resources requirement planning, component lot sizing, and capacity requirement planning. The model formulation and experimentation with different solution procedures was the main focus of the

Table 1
Factor definitions.

Symbol	Descriptions
A_1	Demand variability; refers to the amount of variance in demand over the planning horizon in the model. It is quantified by the ratio of the standard deviation to the mean and is applied equally to all end items (three levels).
A_2	Number of operations/component; refers to the number of work centers that components must visit with the same number applying to all components (three levels).
A_3	Capacity available; refers to the relationship between capacity and average workload. It is quantified as a percentage of average workload (three levels).
A_4	Problem size; the planning horizon is fixed, but the problem size is changed by varying the number of components and end items (three levels).
B_1	Commonality; refers to the extent to which a component is used in several end items. It is quantified as a ratio of the number of components with duplicate usage to the total number of components (two levels).
B_2	Component inventory carrying cost (two levels).
B_3	End item inventory carrying cost (two levels).
B_4	Overtime cost (two levels).
B_5	Set-up time; expressed as a percent of total time for components (two levels). It is defined in this manner only for experimental purposes. The actual set-up time will depend upon the number of set-ups.
B_6	Lead time offset; refers to the time period between earliest start of production of components and their use in end items (two levels).

research. In particular, Bahl wanted to discover what effects the attributes of the problem had on the success of various heuristic and exact solution procedures.

The factors which Bahl considered are listed in table 1, with the number of levels used in this case to illustrate the design process. The modification for the purpose of an example consisted of increasing the first four factors to the three levels indicated in table 1. Furthermore, subjective estimates of the costs of making n observations were made. These costs were based primarily on envisioned problem complexity and corresponding estimates of computer time. Most costs were increased exponentially with problem size. The objective in the example will be to use the cost model to find the cost optimal Cartesian product fraction that can be used to estimate all main effects (resolution III).

The cost model, described above, was fitted to the estimated costs and resulted in the cost interaction terms (involving three or more factors) given in table 2. Except for interaction terms involving both A and B factors (but less than three of either factor), all other interaction terms involving three or more factors were zero. Since a Cartesian product design with resolution III is desired, none of the interactions involving one or two factors of either type A or B can affect the cost of the fraction. The average cost of all observations was 19.405.

One may begin designing the fraction by first finding the cost-optimal 2^{6-3} fraction. From table 2, $B_2 B_3 B_4 B_5$ has the largest absolute cost, therefore any resolution III design generated from its corresponding six-tuple (0, 1, 1, 1, 1, 0) and a right-hand side value of 1 ($\delta = 1$ in eq. (7)) will be cost optimal and

Table 2
High order cost interaction terms.

Cost interaction terms	Cost interaction effects
$B_3 B_4 B_5$	0.02083
$B_2 B_4 B_5$	0.02083
$B_2 B_3 B_5$	0.10417
$B_2 B_3 B_4$	−0.10417
$B_2 B_3 B_4 B_5$	0.35417
$A_1 A_3 A_4$	−0.66667
$A_1 A_3 A_4^2$	−2.00000
$A_1 A_3^2 A_4$	0.50000
$A_1 A_3^2 A_4^2$	−1.83333
$A_1^2 A_3 A_4$	−0.01852
$A_1^2 A_3 A_4^2$	−1.01850
$A_1^2 A_3^2 A_4$	0.70370
$A_1^2 A_3^2 A_4^2$	−1.29630

result in a cost average of 0.3542 less than average per observation. One such design is based on the following equations:

$$0b_1 + 1b_2 + 1b_3 + 1b_4 + 1b_5 + 0b_6 = 1 (\text{mod} 2),$$

$$0b_1 + 0b_2 + 1b_3 + 0b_4 + 1b_5 + 1b_6 = 0 (\text{mod} 2), \tag{15}$$

$$0b_1 + 0b_2 + 1b_3 + 0b_4 + 1b_5 + 1b_6 = 0 (\text{mod} 2).$$

The next step is to find a cost-optimal 3^{4-2} design, a slightly more complex task, since each interaction term that is nonzero involves both a linear and a quadratic term. Thus, if one were to base a fraction on the term $A_1^2 A_3^2 A_4^2$ with a right-hand side of 2 ($\phi = 2$ in eq. (6)), the average cost would be changed by twice -1.2963, or -2.5926, with, in this case, the linear term $A_1 A_2 A_3$ contributing nothing. This zero contribution is due to the fact that the right-hand side for the linear term is $2*2 \bmod 3 = 1$ ($\phi = 1$ in eq. (6)), which corresponds to code of 0 in the "linear" columns of the X matrix as coded in the appendix. It turns out that the cost-optimal fraction should be based on the linear interaction $A_1 A_3 A_4^2$ with a right-hand side of zero. This will add -2.0 to the average cost, and the corresponding quadratic interaction $A_1^2 A_3^2 A_4$ will therefore also have a right-hand side of 0, causing a subtraction of its cost in the amount of 0.7037. Collectively, this causes a net reduction of 2.7037 in the average cost. Thus, the cost-optimal one-ninth fraction of the three-level factors is based on the following equations:

$$1a_1 + 0a_2 + 1a_3 + 2a_4 = 0 (\text{mod} 3),$$

$$0a_1 + 1a_2 + 1a_3 + 1a_4 = 0 (\text{mod} 3). \tag{16}$$

The complete cost-optimal $3^{4-2} 2^{6-3}$ fraction is the Cartesian product of all solutions to (15) and (16) and the total cost may be computed from the cost model (eq. (14)) as follows:

$$TC(F) = 72(19.405 - 2.0 - 0.7037 - 0.3542) = 1177.$$

The average cost of all fractions of this size is 1,397, so that the cost-optimal fraction represents an average saving of 15.76%. Furthermore, if a fraction is selected without regard for cost, the most expensive fraction (a cost of 1,564) could be inadvertantly selected. This would represent a 32.88% increase over the cost-optimal design.

Table 3 gives the costs for each of the 72 observations in the cost-optimal $3^{4-2} 2^{6-3}$ design, with row and column headings indicating the solution to eqs. (15) and (16).

It should be pointed out that in this case there were relatively few high order

Table 3
The cost-optimal $3^{4-2} 2^{6-3}$ design.

				b_1	0	0	1	1	1	1	0	0
				b_2	1	0	1	0	0	1	0	1
				b_3	0	1	1	0	0	1	1	0
				b_4	0	0	0	0	1	1	1	1
				b_5	0	1	0	1	0	1	0	1
a_1	a_2	a_3	a_4									
2	0	2	1		14	19	13	17	13	17	14	19
1	2	0	1		11	13	11	13	11	14	11	13
0	1	1	1		12	13	12	13	12	13	12	13
0	2	2	2		11	12	11	12	11	12	11	12
2	1	0	2		21	26	20	24	20	24	21	26
1	0	1	2		43	45	43	45	43	46	43	45
1	1	2	0		6	8	6	8	6	9	6	8
0	0	0	0		3	4	3	4	3	4	3	4
2	2	1	0		16	21	15	19	15	19	16	21

cost interaction terms that were nonzero. This made it possible to solve for the cost-optimal fraction by inspection and a minor amount of trial and error. It is anticipated that for most problems involving a large number of factors, and subjective estimates of costs, the need for building a cost structure in a simple and systematic manner will cause this fortuitous situation to occur frequently. However, if large surrogate costs are used on some combinations which, for some reason, are infeasible to observe, then this cost-optimal process can also be used to locate feasible fractions. In this case a large number of high order interaction terms would be nonzero, thus requiring a systematic enumeration strategy such as those proposed in [12] and [9].

5. Other fractional designs

Although the cost expression of (3) is general, advantage was taken of the coding (appendix) of Z, eq. (4), to facilitate calculation of the total cost. Connor and Young [5], motivated by the earlier work of Morrison [8], published a catalog of fractions in the $3^m 2^n$ series [5]. Although some of these designs were not orthogonal, any such design was a collection of Cartesian product designs, and so the methods of cost calculation in this paper become immediately applicable. Whether a simple cost structure could accompany other methods of designing fractions (e.g. [1,2]) remains to be determined.

Appendix

The coding of the matrix ZB is considered first. With each column is associated the symbol $B_1^{\beta_1} B_2^{\beta_2} \ldots B_n^{\beta_n}$, with $\beta_i = 0, 1$ (see, for example, [5]). For example, if only one $\{\beta_i\}$ is nonzero, the column is labelled the "main-effect" of factor $B_i (B_j^0 = 1)$. If exactly two of the $\{\beta_i\}$ are nonzero, the column is labelled a two-factor interaction term. The element in the ZB matrix which is in the column associated with $\beta = (\beta_1, \ldots, \beta_n)$ and row associated with $b = (b_1, \ldots, b_n)$, is coded:

-1, if $\beta'b = 0 \pmod 2$

or

1, if $\beta'b = 1 \pmod 2$.

With each column in ZA is associated the symbol $A_1^{\alpha_1} A_2^{\alpha_2} \ldots A_m^{\alpha_m}$, $\alpha = (\alpha_1, \ldots, \alpha_m)$, not all α_i zero. The coding is then:

$\left.\begin{array}{ll} 1, & \text{if } \alpha'a = 0 \pmod 3 \\ 0, & \text{if } \alpha'a = 1 \pmod 3 \\ -1, & \text{if } \alpha'a = 2 \pmod 3 \end{array}\right\}$, provided $\omega(\alpha) = 1$;

$\left.\begin{array}{ll} -1, & \text{if } \alpha'a = 0 \pmod 3 \\ -1, & \text{if } \alpha'a = 1 \pmod 3 \\ 2, & \text{if } \alpha'a = 2 \pmod 3 \end{array}\right\}$, provided $\omega(\alpha) = 2$;

where $\omega(\alpha)$ = first nonzero element of α, reading left to right. Finally, the ZAB

Table 4
Coding for the Z matrix, 3×2 factorial.

a	b	$(\alpha; \beta)$					
		(0; 0)	(1; 0)	(2; 0)	(0; 1)	(1; 1)	(2; 1)
0	0	1	1	−1	−1	−1	1
1	0	1	0	−1	−1	0	1
2	0	1	−1	2	−1	1	−2
0	1	1	1	−1	1	1	−1
1	1	1	0	−1	1	0	−1
2	1	1	−1	2	1	−1	2

matrix has columns whose components are products, component by component, of every combination of the ZA, ZB matrices, where these columns are now associated with symbols of the form $A_1^{\alpha_1}...A_m^{\alpha_m}B_1^{\beta_1}...B_n^{\beta_n}$ (interaction of A and B factors). Table 4 illustrates a simple coding for a 3×2 factorial Z matrix.

[1] S. Addelman, "Orthogonal Main-effect Plans for Asymmetrical Factorial Experiments", *Technometrics* 4 (1962) 21–46.
[2] S. Addelman and O. Kempthorne, "Orthogonal Main-effect Plans", Aeronautical Research Laboratory Technical Report, 79 (1961).
[3] H. Bahl, "Master Scheduling and Component Lot Sizing Decisions in Resource-Constrained Material Requirements Planning Environments", Ph.D. Thesis, The Ohio State University (1980).
[4] I.M. Chakravarti, "Fractional Replication in Asymmetrical Factorial Designs and Partially Balanced Arrays", *Sankhya* 17, 2 (1956) 143–164.
[5] W.S. Connor and S. Young, "Fractional Designs for Experiments With Factors at Two and Three Levels", *NBS Applied Math Series*, no. 58 (1961).
[6] R.A. Fisher, "The Arrangement of Field Experiments", *Journal Ministry Agriculture* 33 (1926) 503–513.
[7] R. McQuie, "Experimental Design and Simulation in Unloading Ships by Helicopter", *Journal of Operations Research* 5, 17 (1969) 785–799.
[8] M. Morrison, "Fractional Replication for Mixed Series", *Biometrics* 12, 1 (1956) 1–19.
[9] C.A. Mount-Campbell and J.B. Neuhardt, "Selecting Cost-Optimal Main-Effect Fractions of 3^n Factorials", *AIIE Transactions* 12, 1 (1980) 80–86.
[10] Ed. T.H. Naylor, *The Design of Computer Simulation Experiments* (Duke University Press, 1969).
[11] J.B. Neuhardt and C.A. Mount-Campbell, "Selection of Cost-Optimal 2^{n-p} Fractional Factorials", *Communication in Statistics* B7, 4 (1978) 369–383.
[12] J. Pignatiello, "Computational Aspects of Selecting Cost-Optimal Two-Level Fractional Factorial Experimental Designs", Master's Thesis, The Ohio State University (1979).
[13] J.R. Smith and J.M. Beverley, "The Use and Analysis of Staggered Nested Factorial Design", *Journal of Quality Technology* 13, 3 (1981) 166–173.
[14] R.D. Snee, "Developing Blending Models for Gasoline and Other Mixtures", *Technometrics* 23, 2 (1981) 119–130.
[15] S.R. Webb, "Orthogonal Incomplete Factorial Designs and Their Construction Using Linear Programming", Technical Document Report ARL-65-116, Part I, ARL Research Laboratories, Wright-Patterson AFB (1965).
[16] F. Yates, "The Design and Analysis of Factorial Experiments", Rothamsted Experimental Station Technical Communication no. 35 (1937).

PART III

PARAMETER ESTIMATION, RELIABILITY AND QUALITY CONTROL

INTRODUCTION TO CONTRIBUTIONS IN PARAMETER ESTIMATION, RELIABILITY AND QUALITY CONTROL

J.S. RUSTAGI and S.H. ZANAKIS

Statistical inference about parameters of probability models has two basic features: estimation (point and interval) and tests of hypotheses. These procedures are heavily dependent upon the nature of the sample or design of the experiment. An assumption made in statistical inference is that a fixed random sample is available to the decision-maker. There are, however, many situations in practice where such an assumption cannot be made.

In reliability studies it is natural to study a system with several components and wait until a failure occurs. Sometimes early failure data may not be available, or it may be impossible to wait until the whole system fails; rather the experiment is terminated after some fixed maximum amount of time. Such samples are called *censored*. Under various conditions of censoring, the estimation of parameters becomes involved even when one is using classical methods of estimation.

Two of the most common methods of estimation of parameters are maximum likelihood and least squares. The method of maximum likelihood requires maximization of the likelihood function (product of probability density functions evaluated at each sample observation) while the least-squares method requires minimization of the sum of squared differences of the observed frequencies from those expected under the model. Optimization techniques, therefore, naturally enter both estimation processes.

Numerical optimizing techniques play a major role in cases where the solutions of the problems cannot be obtained in closed form. In cases of censored data and those of complicated models, optimization has to be done numerically by means of some iterative procedure. Constrained optimization results in models where parameters must satisfy certain conditions; for example, non-negativity restrictions on the parameters of most distributions, or probabilities summing to one for each row of a Markov transition matrix. Another example here refers to the use of mixtures of distributions as a model for a given phenomenon, in which case the mixture condition is that the probabilities with which individual probability distributions enter into the model add to one. Many of the above optimization problems reduce to those of mathematical programming. A collection of mathematical programming techniques applied to statistical problems can be found in Arthanari and

Dodge [1]. Various numerical methods and algorithms have been provided by Kennedy and Gentle [5] and Bard [2] for statistical problems.

In the paper "Parameter estimation under progressive censoring conditions for a finite mixture of Weibull distributions", Mendelbaum and Harris give algorithms for finding maximum likelihood estimates under different sampling environments. The model chosen is that of a mixture of several Weibull distributions under progressive censoring, most often encountered in reliability applications. Maximum likelihood estimates have desirable asymptotic properties, but such estimators are not available in closed form for several important distributions, including the Weibull. Iterative procedures are therefore needed, which often experience computational difficulties when all three Weibull parameters are unknown (Zanakis, [10,11]). Progressive censoring is concerned with observing objects for an arbitrary interval of time and noting the time of their failure. Therefore two measurements, the arbitrary interval for observation time and time to failure, are obtained for each individual object. The likelihood function in such a case is nonlinear, subject to linear constraints, and thus results in a mathematical programming problem. The applications are made not only to problems in avionic equipment (Mendenhall and Hader [7]) but also to problems in criminal justice program evaluation (Carr-Hill and Carr-Hill [3]). The problems arising from tests of hypotheses of such parameters are still open in the above mixed models.

Data often include observations which are suspicious. The maverick observations may belong naturally to the sample and reflect the distribution as such; or the observations may be genuinely bad ones and should be removed from the data before making any inference. Such observations have been called outliers and there have been several tests in statistics to study outliers, including some developed especially for time-series forecasting and process quality control. Mann, in her paper "Optimal outlier tests for a Weibull model – to identify process changes or to predict failure times", provides a comparison of three different statistics for testing outliers. The properties of the tests are studied and it is shown that one of the statistics proposed gives the most powerful test under certain situations.

Most of the classical tests for outliers have been developed for the normal model. In reliability and some areas of medicine, where Weibull distributions are commonly used, there is a scarcity of such tests. It should be noted that there are distributions which are *outlier-prone*, even though the variance is finite. Several such distributions have been given by Neyman and Scott [8] where they were concerned with studying weather data in cloud-seeding experiments. They have brought out the distinctions between cases where the tendency to suspect and to eliminate outlier observations may be justifiable and those in which it is not. They show that the log normal and gamma families are outlier-prone.

In reliability theory, applications occur frequently where the probability distributions have a decreasing failure rate. Failure (hazard) rate is commonly defined in terms of probability distribution of the time to failure. If $f(x)$ is the probability density function of the time to failure and $F(x)$ its cumulative distribution function, then

$$h(x) = f(x)/[1 - F(x)]$$

defines the failure rate. Then $h(x)\,dx$ is the conditional probability of failure in the interval $(x, x + dx)$ given that the individual has survived till time x. For a Weibull distribution with shape parameter c, the failure rate function $h(x)$ is decreasing rapidly for $c < 1$, constant for $c = 1$, and increasing for $c > 1$, with an increasing ($c > 2$), constant ($c = 2$), or decreasing ($1 < c < 2$) slope. This flexibility of the Weibull model is probably the primary reason for its wide use in practice. For an extensive discussion of reliability theory, the reader is referred to Mann, Schaefer and Singpurwalla [6].

In their paper "Extreme points of the class of discrete decreasing failure rate average life distributions", Langberg, Leon, Lynch and Proschan show that the class of distributions which are discrete with decreasing failure rate average, form a convex set. Such distributions can be represented in terms of the extreme points of the convex set described earlier. Several other aspects of these distributions can also be discussed in terms of the extreme points of the convex set.

Singpurwalla studies the estimation of reliability growth in his paper "A Bayesian scheme for estimating reliability growth under exponential failure times". When the reliability of a system is studied at several stages, the procedure is called *reliability growth*. Using an inverted gamma distribution as the prior distribution in the parameter of the exponential distribution at each stage, the estimate for the mean time to failure is obtained in terms of the posterior distribution. The procedure is especially relevant when the posterior distributions are stochastically ordered, but one situation is also discussed when this is not the case.

Acceptance sampling forms an important area of quality control methodology. In single sampling plans one must determine the sample size and acceptance number (number of defective items in the sample) to achieve a certain level of confidence. In their paper "A bicriterion model for acceptance sampling", Moskowitz, Ravindran, Klein and Eswaran use a criterion with two objectives to solve the above problem; namely to maximize a measure of outgoing lot quality and minimize costs. For a detailed treatment of multiobjective optimization theory and application the reader is referred to Hwang and Masud [4]. See also Starr and Zeleny [9] for a good exposition on multiattribute utility and preference. An implicit enumeration algorithm and an interactive paired-comparisons nonlinear optimization procedure are given by Moskowitz et al. for the sampling problems considered. Simulation experiments are performed to determine the robustness of the sampling plan and the response surfaces of the bicriterion model to various utility functions.

References

[1] T.S. Arthanari and Y. Dodge, *Mathematical Programming in Statistics* (John Wiley and Sons, New York, 1981).
[2] Y. Bard, *Nonlinear Parameter Estimation* (Academic Press, New York, 1974).
[3] G.A. Carr-Hill and R.A. Carr-Hill, "Reconviction as a process", *British Journal of Criminology* 12 (1972) 35–43.
[4] C.L. Hwang and A.S. Masud, *Multiple Objective Decision Making Methods and Applications* (Springer-Verlag, New York, 1979).
[5] W.G. Kennedy, Jr. and J.E. Gentle, *Statistical Computing* (Marcel Dekker, Inc., New York, 1980).
[6] N.R. Mann, R.E. Schafer and N.D. Singpurwalla, *Methods for Statistical Analysis of Reliability and Life Data* (John Wiley and Sons, New York, 1974).
[7] W. Mendenhall and R.J. Hader, "Estimation of parameters of mixed exponentially distributed failure time distributions from censored life test data", *Biometrika* 45 (1958) 504–520.
[8] J. Neyman and E.L. Scott, "Outlier proneness of phenomena and of related distributions", in: J.S. Rustagi, Ed., *Optimizing Methods in Statistics* (Academic Press, New York, 1971) 413–430.
[9] M.K. Starr and M. Zeleny, *Multiple Criteria Decision Making, TIMS Studies in the Management Sciences,* Vol. 6 (North-Holland, Amsterdam, 1977).
[10] S.H. Zanakis, "Computational Experience with Some Nonlinear Optimization Algorithms in Deriving Maximum Likelihood Estimates for the Three-Parameter Weibull Distribution", in M.F. Neuts, ed., *Algorithmic Methods in Probability, TIMS Studies in the Management Sciences*, Vol. 7 (North-Holland Publishing Co., Amsterdam, 1977) 63–77.
[11] S.H. Zanakis, "Extended Pattern Search with Transformations for the Three-Parameter Weibull MLE Problem", *Management Science* 25 (1979) 1149–1161.

PARAMETER ESTIMATION UNDER PROGRESSIVE CENSORING CONDITIONS FOR A FINITE MIXTURE OF WEIBULL DISTRIBUTIONS *

Jay MANDELBAUM
United States Department of Defense

and

Carl M. HARRIS
Center for Management and Policy Research, Inc.

Finite exponential and Weibull mixture models have been found in many fields to represent a wide range of real-world situations. The derivation of maximum-likelihood estimates for the parameters of such models is quite difficult because of the complexity of the likelihood function, particularly when the data arise from a complex sampling regime. This necessitates the development of special likelihood estimation procedures for use on modern computers.

In this work, algorithms have been devised to maximize the likelihood function for random samples from mixed Weibull populations under progressive censoring. This is done for a number of different sampling environments which arose in an extensive number of criminal justice program evaluations. Convergence results have been obtained in each case, and the question of local versus global optimality is explored.

1. Background

In recent years several models featuring mixtures of distributions have been used to describe phenomena in the criminal justice field. Carr-Hill and Carr-Hill [1] introduced a model for recidivism in which they assumed that each released prisoner belongs to either a "quick" or "slow" reconviction group. They defined constant reconviction rates for the two subpopulations and further assumed a random process governing the transfer of members from the "quick" to the "slow" group. These assumptions yield a mixture of two exponential distributions as the probability distribution function of recon-

---

Received September 19, 1980; revised April 2, 1981.

* Prepared under grant no. 79 NI AX 0068 to the Center for Criminal Justice, University of Illinois at Chicago Circle, from the National Institute of Law Enforcement and Criminal Justice, United States Department of Justice.

victions over time. Carr-Hill and Carr-Hill did not, however, directly face the issue of estimation of parameters and mixing proportions. Estimates were found by a bit of guesswork and then partly verified to be reasonable via a chi-square test.

Greenberg [4] proposed a modification of the Carr-Hill–Carr-Hill model to take permanently law-abiding people into consideration. He suggested that the population be viewed as three groups, i.e. strictly law-abiding, potential reconviction, and uncommitted. Once again, however, estimation issues were not resolved.

Another model with two groups in the released population was developed by Maltz and McCleary [9]. Maltz and McCleary treated members of the first group as "successes" and assigned them a zero probability of reconviction. The second group is assumed to fail exponentially with a constant failure rate. Iterative equations to obtain maximum-likelihood estimators for the failure rate and the proportion of the population which failed were derived and illustrated in their work.

The major early efforts on the estimation of parameters for mixture-type models were due to Hasselblad [5] and Mendenhall and Hader [11]. However, as is typical in these kinds of problems, the algorithms developed have very poor convergence properties and are not generally applicable to a wide range of sampling situations which arise in the real world.

To eliminate some of these problems, Kaylan [6] developed and tested a special iterative scheme to calculate maximum-likelihood estimates of mixing proportions and probability-density-function parameters for a finite mixture of exponential or Weibull distributions where all individuals are assumed to fail (complete sampling). Previous work in the exponential area can be found in the aforementioned work of Hasselblad, and Mendenhall and Hader. Kaylan also treated the case of type I censoring, where observation terminates for each item at the same operation time whether or not failure has actually occurred.

There is but brief literature on parameter estimation for Weibull mixtures. The first work published on this was due to Kao [7], as a characterization of the lifetimes of electronic tubes. His was a graphical method for mixtures of two Weibulls and did allow for a location parameter. Much more recently, the paper by Mann [10] in this volume refers to recidivism data of individual drug addicts which appear to be well described by mixtures of two or three well separated IFR Weibulls. Her work identifies three distributions (as a result of patient interviews) resulting from diverse personal motivations driving the drug usage patterns. In addition, the cohort data seem to follow a mixture of three Weibull distributions, at least one of which can be readily identified as exponential. In a sense, then, this becomes a Weibull analog of the three classes offered by Greenberg [4].

2. Introduction

This paper deals with the estimation of mixing proportions and parameters of a finite number of mixed Weibull distributions under conditions of *progressive censoring*. There is considerable literature on the ordinary Weibull under progressive censoring (see for example, Wingo [13] and Cohen [2]). Inclusion of the location parameter makes the optimization more complex, so it is typically estimated by the smallest observation. However, there is literature on the inclusion of the location parameter in the optimization (see Zanakis, [14]). But the problem of a mixture of Weibulls is much more complex, and we stay with the two-parameter Weibull. We choose to omit any consideration of the location parameter because of the added level of difficulty.

By progressive censoring, we mean the following. Observations of objects or individuals may start at an arbitrary time. If there is a failure during the observation period, then the total operating time is recorded and denoted by x_i. However, an individual does not have to fail during the observation period since observations are allowed to be terminated at any point in time. The total observation time for an individual who does not fail is also recorded and denoted by y_j which is the difference between the points in time at which observation was started and ended for that individual.

Our goal is to obtain maximum-likelihood estimators (MLEs) of the parameters and mixing proportions under the assumptions that the data come from a mixed Weibull population. Under fairly general conditions, the MLEs are invariant, consistent, asymptotically unbiased, and best asymptotically normal. They are also functions of sufficient statistics if such exist. The problem here is that it is not possible to obtain an explicit form for the estimator by taking partial derivatives of the likelihood function and setting them equal to zero. In addition, we encounter the constraint that the sum of the mixing proportions must equal unity. The resultant problem can be described as a mathematical program with a nonlinear objective function and linear constraints. Before we attempt to use a mathematical programming algorithm, however, it is advisable to attempt to maximize the log-likelihood function without taking the constraints into account. If the answer is feasible, the problem is solved with much less computational effort. In any event, we utilize iterative numerical procedures to calculate parameter estimates.

Section 3 establishes notation and also present some basic results which will be referenced throughout the paper. Parameter estimates can be made under two different assumptions concerning the failed individuals. After a failure we may assume either: (1) we know the true density in the mixture from which the failure came, or (2) we do not know the true parent. The former case is labelled "post-mortem" and the latter is termed "non-post-mortem". Sections 4 and 5, respectively, present a first- and second-order iterative scheme for parameter estimation under non-post-mortem conditions. Convergence proofs are also given. Section 6 derives an unconstrained method for estimation under special post-mortem conditions. Then, a first-order iterative scheme for the post-mortem case is presented in section 7 along with a convergence proof. Section 8 gives the iterative formulae developed in sections 4 and 5 for the exponential, and then a two-phase second-order method is described in section 9. Finally, sections 10 and 11 offer a detailed computational example and address the issue of global optimality, respectively.

3. Notation

Since we are working with mixtures of K Weibull density functions, the probability density function of the jth Weibull in the mixture will be given as (for $x > 0$; β_j, $\eta_j > 0$):

$$f_j(x; \beta_j, \eta_j) = (\beta_j/\eta_j)(x/\eta_j)^{\beta_j - 1} \exp\left[-(x/\eta_j)^{\beta_j}\right].$$

The mixtures of K Weibulls can thus be expressed as:

$$g(x, \alpha) = \sum_{j=1}^{K} p_j f_j(x) \quad \left(p_j \geq 0, \sum_{j=1}^{K} p_j = 1\right),$$

where α is a vector of the $3K - 1$ unknown parameters.

The cumulative distribution functions for $f_j(x)$ and $g(x, \alpha)$ can also be written respectively as

$$F_j(x) = 1 - \exp\left[-(x/\eta_j)^{\beta_j}\right]$$

and

$$G(x, \alpha) = \sum_{j=1}^{K} p_j F_j(x).$$

It turns out that it is more convenient to work with the complements of the above cumulative distribution functions, namely

$$\bar{F}_j(x) = 1 - F_j(x) = \exp\left[-(x/\eta_j)^{\beta_j}\right]$$

and

$$\bar{G}(x, \alpha) = 1 - G(x, \alpha) = \sum_{j=1}^{K} p_j \bar{F}_j(x).$$

Since we shall constantly be taking partial derivatives of $g(x, \alpha)$ and $\overline{G}(x, \alpha)$ throughout the remainder of the paper, we establish these functions now:

$$\frac{\partial g(x, \alpha)}{\partial \beta_j} = p_j f_j(x) \left[\frac{1}{\beta_j} + \ln \frac{x}{\eta_j} - \left(\ln \frac{x}{\eta_j} \right) \left(\frac{x}{\eta_j} \right)^{\beta_j} \right], \tag{1}$$

$$\frac{\partial \overline{G}(x, \alpha)}{\partial \beta_j} = -p_j \overline{F}_j(x) \left(\ln \frac{x}{\eta_j} \right) \left(\frac{x}{\eta_j} \right)^{\beta_j}, \tag{2}$$

$$\frac{\partial g(x, \alpha)}{\partial \eta_j} = p_j f_j(x) \left[\left(\frac{x}{\eta_j} \right)^{\beta_j} - 1 \right] \frac{\beta_j}{\eta_j}, \tag{3}$$

$$\frac{\partial \overline{G}(x, \alpha)}{\partial \eta_j} = p_j \overline{F}_j(x) \left(\frac{x}{\eta_j} \right)^{\beta_j} \frac{\beta_j}{\eta_j}. \tag{4}$$

Eqs. (1) through (4) hold for $j = 1, 2, \ldots, K$. When we differentiate with respect to p_j, recalling that the $\{p_j\}$ sum to 1, we find that

$$\frac{\partial g(x, \alpha)}{\partial p_j} = f_j(x) - f_K(x) \qquad (j = 1, 2, \ldots, K-1), \tag{5}$$

$$\frac{\partial \overline{G}(x, \alpha)}{\partial p_j} = \overline{F}_j(x) - \overline{F}_K(x) \qquad (j = 1, 2, \ldots, K-1). \tag{6}$$

4. A first-order method – non-post-mortem

Our first method for estimating the mixing proportions and the parameters of the K Weibulls is a first-order iterative method. We assume that there are N observations, R of which are failures during the observation period. We also assume that when there is a failure, we do not know from which of the K Weibull density functions it came. Consistent with the literature, this is called non-post-mortem sampling. Below, we treat a special case of post-mortem analysis with $K = 2$.

The likelihood equation for this problem will be

$$\mathcal{L}(\alpha) = \frac{N!}{(N-R)!} \prod_{i=1}^{R} g(x_i, \alpha) \prod_{l=1}^{N-R} \overline{G}(y_l, \alpha). \tag{7}$$

From this point on we adopt a shorthand notation, by setting $f_j = f_j(x_i)$,

$g = g(x_i, \alpha)$, $\bar{G} = G(y_l, \alpha)$, and $\bar{F}_j = \bar{F}_j(y_l)$. So we have

$$\mathcal{L} = \frac{N!}{(N-R)!} \prod_{i=1}^{R} g \prod_{l=1}^{N-R} \bar{G}.$$

Consistent with standard methods of finding maximum-likelihood estimators, we take logarithms to obtain

$$L = \ln \mathcal{L} = \ln \left[\frac{N!}{(N-R!)} \right] + \sum_{i=1}^{R} \ln g + \sum_{l=1}^{N-R} \ln \bar{G}.$$

We now take partial derivatives with respect to β_j, η_j and p_j and then set them equal to zero. The first term of the log-likelihood is constant and hence does not affect the differentiation. As is shown below, we find that it is not possible to solve explicitly for the parameter; thus, we invoke the following numerical procedure. If there are m equations in m unknowns, we separate each of the m unknowns to the left-hand side of the m equations. Each right-hand side, however, will not be independent of the variable on its corresponding left-hand side. The standard procedure is to solve iteratively for the unknowns with the right-hand sides containing values of the vth iteration and the left-hand sides being the values at the $(v+1)$st iteration. For example, assume there are two parameters a_1 and a_2, such that

$$a_1 = h_1(a_1, a_2) \quad \text{and} \quad a_2 = h_2(a_1, a_2)$$

Then we may solve iteratively for a_1 and a_2. If we use a superscript to index the iterations, then

$$a_1^{v+1} = h_1(a_1^v, a_2^v) \quad \text{and} \quad a_2^{v+1} = h_2(a_1^v, a_2^v).$$

The likelihood analysis thus proceeds as follows:

$$\frac{\partial L}{\partial \beta_j} = \sum_{i=1}^{R} \frac{1}{g} \frac{\partial g}{\partial \beta_j} + \sum_{l=1}^{N-R} \frac{1}{\bar{G}} \frac{\partial \bar{G}}{\partial \beta_j} \quad (j = 1, 2, \ldots, K).$$

From Eqs. (1) and (2) we obtain:

$$\frac{\partial L}{\partial \beta_j} = \sum_{i=1}^{R} \frac{p_j f_j}{g} \left[\frac{1}{\beta_j} + \ln \frac{x_i}{\eta_j} - \left(\ln \frac{x_i}{\eta_j} \right) \left(\frac{x_i}{\eta_j} \right)^{\beta_j} \right]$$

$$- \sum_{l=1}^{N-R} \frac{p_j \bar{F}_j}{\bar{G}} \left(\ln \frac{y_l}{\eta_j} \right) \left(\frac{y_l}{\eta_j} \right)^{\beta_j} \quad (j = 1, 2, \ldots, K). \tag{8}$$

If we now set $\partial L/\partial \beta_j = 0$ for all j, we obtain:

$$\beta_j = \frac{\sum_{i=1}^{R} \left(\frac{f_j}{g}\right)}{\sum_{i=1}^{R} \left(\frac{f_j}{g}\right)\left(\ln \frac{x_i}{\eta_j}\right)\left[\left(\frac{x_i}{\eta_j}\right)^{\beta_j} - 1\right] + \sum_{l=1}^{N-R} \frac{\bar{F}_j}{\bar{G}}\left(\ln \frac{y_l}{\eta_j}\right)\left(\frac{y_l}{\eta_j}\right)^{\beta_j}}. \quad (9)$$

Similarly,

$$\frac{\partial L}{\partial \eta_j} = \sum_{i=1}^{R} \frac{1}{g} \frac{\partial g}{\partial \eta_j} + \sum_{l=1}^{N-R} \frac{1}{\bar{G}} \frac{\partial \bar{G}}{\partial \eta_j} \quad (j = 1, 2, \ldots, K).$$

From Eqs. (3) and (4) we then obtain (for $j = 1, 2, \ldots, K$):

$$\frac{\partial L}{\partial \eta_j} = \sum_{i=1}^{R} \frac{p_j f_j}{g}\left[\left(\frac{x_i}{\eta_j}\right)^{\beta_j} - 1\right]\frac{\beta_j}{\eta_j} + \sum_{l=1}^{N-R} \frac{p_j \bar{F}_j}{\bar{G}}\left(\frac{y_l}{\eta_j}\right)^{\beta_j} \frac{\beta_j}{\eta_j}. \quad (10)$$

When these are set to zero, we find that:

$$\eta_j = \left[\frac{\sum_{i=1}^{R}\left(\frac{f_j}{g}\right) x_i^{\beta_j} + \sum_{l=1}^{N-R}\left(\frac{\bar{F}_j}{\bar{G}}\right) y_l^{\beta_j}}{\sum_{i=1}^{R} \frac{f_j}{g}}\right]^{1/\beta_j}. \quad (11)$$

For the mixing proportions we have:

$$\frac{\partial L}{\partial p_j} = \sum_{i=1}^{R} \frac{1}{g} \frac{\partial g}{\partial p_j} + \sum_{l=1}^{N-R} \frac{1}{\bar{G}} \frac{\partial \bar{G}}{\partial p_j} \quad (j = 1, 2, \ldots, K-1),$$

which becomes, via (4) and (5),

$$\frac{\partial L}{\partial p_j} = \sum_{i=1}^{R} \frac{1}{g}(f_j - f_K) + \sum_{l=1}^{N-R} \frac{1}{\bar{G}}(\bar{F}_j - \bar{F}_K) \quad (j = 1, 2, \ldots, K-1). \quad (12)$$

When we set $\partial L/\partial p_j$ to zero we arrive at

$$\sum_{i=1}^{R} \frac{f_j}{g} + \sum_{l=1}^{N-R} \frac{\bar{F}_j}{\bar{G}} = C \quad (j = 1, 2, \ldots, K), \quad (13)$$

where C is an appropriate constant. If we multiply both sides of (13) by p_j, sum over j, and simplify, then (13) becomes

$$p_j = \frac{p_j}{N}\left[\sum_{i=1}^{R}\frac{f_j}{g} + \sum_{l=1}^{N-R}\frac{\overline{F}_j}{\overline{G}}\right] \quad (j=1,2,\ldots,K). \tag{14}$$

Eqs. (9), (11), and (14) are the basis of the iterative procedure for finding mixing proportions and distribution parameters. The left-hand sides represent their values at the $(v+1)$st iteration. The functions on the right-hand sides contain values at the vth iteration.

The iterative scheme, as it has been developed, can be improved via techniques commonly used in the mathematical programming environment. A mathematical programming algorithm is composed of two main features: the generation of a direction which will lead to improvement in the objective function and the step size or line search problem which indicates how far to move in the prescribed direction. For a nonconcave problem, such a direction is only guaranteed to instantaneously lead to improvement in the objective function. Thus, a step of arbitrary length may or may not be beneficial. Consequently, mathematical programming algorithms commonly generate a step size, s^*, as the solution of

$$\max_s L\left[\alpha^{v+1} + s(\alpha^{v+1} - \alpha^v)\right].$$

As our iterative scheme has been posed, the step size is automatically calculated. We have developed equations which lead to α^{v+1}. In the following section we show that the vector $(\alpha^{v+1} - \alpha^v)$ points in a direction of increasing log-likelihood. We still encounter the possibility that the selection of α^{v+1} does not lead to improvement. In order to ensure such improvement, and at the same time avoid the additional computation required to solve the line search problem, we heuristically bisect the step until an increase is realized. For example, we will first try $\alpha = \alpha^v + (\alpha^{v+1} - \alpha^v)$, then $\alpha = \alpha^v + \frac{1}{2}(\alpha^{v+1} - \alpha^v)$, etc.

Convergence properties

In this section we prove the convergence of the foregoing algorithm. The conditions which constitute global convergence (see Luenberger [8]) form the basis of the proof. We need to show that:

(1) α^v belongs to a compact set;

(2) the algorithm generates a sequence of points such that each new point causes the log-likelihood to increase in value; and

(3) α^v is feasible.

As in Kaylan [6], eqs. (9), (11), and (14) constitute a mapping from α^v to α^{v+1}. Since all of the functions in the equations are continuous, the mapping is closed. Hence, α^v belongs to a compact set.[1]

To show that the algorithm generates a sequence of points so that the log-likelihood increases in value, it is sufficient to show that the following inner product is non-negative:

$$[\alpha^{v+1} - \alpha^v] \cdot \nabla L_\alpha^v \geq 0. \tag{15}$$

We show this in three stages.[2] First,

$$[\beta_j^{v+1} - \beta_j^v] \cdot \nabla L_\beta^v \geq 0 \quad (j = 1, 2, \ldots, K). \tag{16}$$

From (9), after some algebra, we obtain:

$$\beta_j^{v+1} - \beta_j^v = \left[\frac{(\beta_j^{v+1})(\beta_j^v)}{p_j^v \sum_{i=1}^R f_j^v / g^v} \right] \left[\frac{\partial L}{\partial \beta_j} \right]^v \quad (j = 1, 2, \ldots, K).$$

Since the coefficient of $[\partial L / \partial \beta_j]^v$ here is non-negative, it is clear that the condition in (16) is satisfied. The second stage is to show that

$$[\eta_j^{v+1} - \eta_j^v] \cdot \nabla L^v \geq 0 \quad (j = 1, 2, \ldots, K). \tag{17}$$

After some algebra building from (11), we find that

$$\text{sign}(\eta_j^{v+1} - \eta_j^v) = \frac{(\eta_j^v)^{\beta_j^{v+1}}}{p_j^v \beta_j^v \left(\sum_{i=1}^R f_j^v / g^v \right)} \left[\frac{\partial L}{\partial \eta_j} \right]^v \quad (j = 1, 2, \ldots, K).$$

Since the coefficient in the above is non-negative, (17) is satisfied.

The final stage is to show that

$$\sum_{j=1}^K [p_j^{v+1} - p_j^v] \cdot \nabla L_p^v \geq 0. \tag{18}$$

---

[1] The step-size bisection may occasionally violate continuity. To avoid the problem, we imbed a convergent subsequence of points into the principal sequence. This is accomplished through the use of an Armijo step in the gradient direction.

[2] We assume throughout that the log-likelihood is "well-behaved" with respect to the $\{\beta_j\}$ and that good starting points are available.

The proof is patterned after Hasselblad [5]. After substituting for ∇L_p^v from (12), the left-hand side becomes

$$\sum_{j=1}^{K} \left[p_j^{v+1} - p_j^v\right] \cdot \left\{ \sum_{i=1}^{R} \frac{(f_j^v - f_k^v)}{g^v} + \sum_{l=1}^{N-R} \frac{(\bar{F}_j^v - \bar{F}_k^v)}{\bar{G}^v} \right\}.$$

Equation (14) tells us, however, that

$$\frac{Np_j^{v+1}}{p_j^v} = \sum_{i=1}^{R} \frac{f_j^v}{g} + \sum_{l=1}^{N-R} \frac{\bar{F}_j^v}{\bar{G}^v}.$$

Therefore the left-hand side is equal to

$$N \sum_{j=1}^{K} \left[p_j^{v+1} - p_j^v\right] \cdot \left\{ \frac{p_j^{v+1}}{p_j^v} - \frac{p_K^{v+1}}{p_K^v} \right\}$$

$$= N \left[\sum_{j=1}^{K} \frac{(p^{v+1})^2}{p_j^v} - 1 \right]. \tag{19}$$

If we now let

$$p_j^{v+1} = p_j + \delta_j \quad (j=1,2,\ldots,K),$$

where the δ_j sum to zero, then after some algebra (19) becomes

$$N \left[1 + \left(\sum_{j=1}^{K} \frac{\delta_j}{p_j} \right)^2 \right].$$

This is guaranteed to be positive and hence (18) is true. Eqs. (16), (17), and (18) together show that (15) holds and thus that the algorithm does generate a sequence of points such that each new point causes the log-likelihood to increase in value.

The final step in the convergence proof is to show that α^v is feasible. This implies that $p_j \geq 0$, β_j, $\eta_j > 0$ ($j=1,2,\ldots,K$) and that $\Sigma p_j = 1$. If α^0 is such that $p_j \geq 0$, β_j, $\eta_j > 0$ ($j=1,2,\ldots,K$), then this condition will be maintained through all iterations since the right-hand sides of (11) and (14) must be positive.

Under many conditions, the right-hand side of (9) will also be positive, but this is not guaranteed. Thus, we resort to a heuristic method to avoid this difficulty – a bisection of the step size as described in the beginning of this section. We have

$$\alpha^{v+1} = \alpha^v + s(\alpha^{v+1} - \alpha^v).$$

We initially set s to unity. If some β_j is negative, then we will set s to $1/2$, $1/4$, $1/8, \ldots$, until feasibility is achieved. To show that $\Sigma_{j=1}^K p_j = 1$, it is sufficient to show that $\Sigma_{j=1}^{K-1} p_j < 1$. If we take (14) and sum over j, then

$$\sum_{j=1}^{K-1} p_j = \frac{1}{N} \left[\sum_{i=1}^R \frac{\sum_{j=1}^{K-1} p_j f_j}{g} + \sum_{l=1}^{N-R} \frac{\sum_{j=1}^{K-1} p_j \bar{R}_j}{\bar{G}} \right].$$

Since

$$g = \sum_{j=1}^K p_j f_j \quad \text{and} \quad \bar{G} = \sum_{j=1}^K p_j \bar{F}_j,$$

then

$$\sum_{j=1}^{K-1} \frac{p_j f_j}{g} < 1 \quad \text{and} \quad \sum_{j=1}^{K-1} \frac{p_j \bar{F}_j}{g} < 1.$$

Thus,

$$\sum_{j=1}^{K-1} p_j < \frac{1}{N} \left(\sum_{i=1}^R 1 + \sum_{l=1}^{N-R} 1 \right) = 1.$$

5. A second order method – non-post-mortem

Here the log-likelihood equation may be written as:

$$L = \ln \frac{N!}{(N-R)!} + \sum_{i=1}^R \ln \sum_{j=1}^K p_j f_j + \sum_{l=1}^{N-R} \ln \sum_{j=1}^K p_j \bar{F}_j.$$

We want to make use of the fact that a monotone increasing concave function of a concave function is concave. Any function which is linear in the $\{p_j\}$ is concave with respect to the $\{p_j\}$. Also, the logarithm is a monotone increasing concave function. Thus,

$$\ln \sum_{j=1}^{K} p_j f_j \quad \text{and} \quad \ln \sum_{j=1}^{K} p_j \bar{F}_j$$

are both concave functions. Since the sum of concave functions is concave, the log-likelihood function is concave with respect to the $\{p_j\}$.

We next look at the sub-Hessian matrix as:

$$\nabla^2 L_p = \frac{\partial^2 L}{\partial p_{j_1} \partial p_{j_2}} \quad (j_1 = 1, 2, \ldots, K-1; \ j_2 = 1, 2, \ldots, K-1).$$

Eq. (12) defined $\nabla L_p = \partial L / \partial p_j$; therefore

$$\nabla L_p = \sum_{i=1}^{R} \frac{1}{g}(f_{j_1} - f_K) + \sum_{l=1}^{N-R} \frac{1}{\bar{G}}(\bar{F}_{j_1} - \bar{F}_K) \quad (j_1 = 1, 2, \ldots, K-1).$$

Thus (for $j_1, j_2 = 1, 2, \ldots, K-1$),

$$\nabla^2 L = \sum_{i=1}^{R} \frac{(f_{j_1} - f_K)(f_{j_2} - f_K)}{g^2} - \sum_{l=1}^{N-R} \frac{(\bar{F}_{j_1} - \bar{F}_K)(\bar{F}_{j_2} - \bar{F}_K)}{\bar{G}^2}.$$

On the basis of this scheme we may use Newton's method for generating the mixing proportions vector equation:

$$p^{v+1} = p^v - \left(\nabla^2 L_p^v\right)^{-1} \left(\nabla L_p^v\right). \tag{20}$$

Therefore the second-order scheme uses (20) instead of (14). The step-size issue, as discussed within the context of the first-order method, is applicable here as well. In order to guarantee improvement in the log-likelihood at the $(v+1)$st iteration, we will bisect the step size until an increase is realized.

Convergence properties

The properties listed under "convergence properties" in section 4 to show convergence of the first-order method, must also be shown to hold in this second-order method. But we need only examine the properties for the differences between the two algorithms. Clearly, α^v belongs to a compact set since the functions in (20) are continuous. We must next prove that

$$[\alpha^{v+1} - \alpha^v] \cdot \nabla L_\alpha^v \geq 0.$$

Section 4, "convergence properties", showed that the above holds for the β_j and η_j components of α. Thus, we only need to show that:

$$[p^{v+1} - p^v] \cdot \nabla L_p^v \geq 0.$$

From (20) we obtain:

$$(\nabla L_p^v) \cdot [p^{v+1} - p^v] = [\nabla L_p^v][\nabla^2 L_p^v]^{-1}[\nabla L_p^v].$$

Since L is concave with respect to the mixing proportions, $[\nabla^2 L_p]^{-1}$ is negative semidefinite. Therefore the right-hand side of the above is non-negative.

The final step in the convergence proof is to show feasibility. Since the β_j and η_j equations are the same as in the first-order method, we need only show that the $\{p_j\}$ are non-negative and sum to unity. Unfortunately the Newton step does not take the constraints into account. There is no guarantee that the $\{p_j\}$ will sum to unity or remain non-negative. We may however resort to a heuristic method to avoid this problem. If we consider p^v and p^{v+1} to be $(K-1)$-dimensional vectors and let $[p_k = 1 - \Sigma_{j=1}^{K-1} p_j]$, then (20) represents a move in $(K-1)$ space. But (20) is also a step-size equation in which s has been set to one. It can be rewritten as:

$$p^{v+1} = p^v - s(\nabla^2 L_p^v)^{-1}(\nabla L_p)^v.$$

If a value of s implies either

$$\sum_{j=1}^{K-1} p_j > 1 \quad \text{or} \quad p_j < 0, \quad \text{for any } j = 1, 2, \ldots, K-1,$$

then we will bisect s and try again for feasibility. Thus, we will begin with $s = 1$; if the resulting p^{v+1} is infeasible, we will try $s = 1/2$, $s = 1/4$, etc. until feasibility is achieved. Any of these steps is guaranteed to lead to an improvement in the log-likelihood, since the log-likelihood function is concave in the $\{p_j\}$.

6. An unconstrained problem – post-mortem

Up to this point we have been dealing with N observations, R of which fail during the observation period. We also have assumed that the parents of the R failures are unknown. In a "post-mortem" case, we assume instead that the underlying distribution of a failure *is* known. For this problem we need to slightly amend our previous notation. For the R failures, assume that R_j of them were found to belong to the jth parent $f_j(x)$, where $j = 1, 2, \ldots, K$, and $\sum_{j=1}^{K} R_j = R$. Previously, f_j was $f_j(x_i)$ — but now f_j will denote $f_j(x_{ij})$, where x_{ij} is the failure time for the ith object with parent $f_j(x)$.

The likelihood function for the post-mortem problem is thus

$$L = \frac{N!R!}{(N-R)!} \prod_{l=1}^{N-R} \overline{G} \prod_{j=1}^{K} p_j^{R_j} \prod_{j=1}^{K} \left(\frac{f_j^{R_j}}{R_j!} \right),$$

with log-likelihood

$$\ln L = \ln \frac{N!R!}{(N-R)!} + \sum_{l=1}^{N-R} \ln \overline{G} + \sum_{j=1}^{K} R_j \ln p_j + \sum_{j=1}^{K} (R_j \ln f_j - \ln R_j!). \tag{21}$$

We first pose an unconstrained post-mortem problem in which L will be maximized. The optimization is unaffected by the factorial terms so we define

$$L_{\text{mod}} = \sum_{l=1}^{N-R} \ln \overline{G} + \sum_{j=1}^{K} R_j \ln p_j + \sum_{j=1}^{K} \sum_{i=1}^{R_j} \ln f_j.$$

The set of parameter values which maximizes L_{mod} will also maximize L. In the special case where $K = 2$, the following transformations of variables are made:

$$\beta_j = u_j^2, \quad j = 1, 2,$$
$$\eta_j = v_j^2, \quad j = 1, 2.$$

This transformation guarantees that β_j and $\eta_j > 0$. If we also set $p_1 = \sin^2 w$ and $p_2 = \cos^2 w$, then $p_1 + p_2$ must equal unity. When the above is substituted into L_{mod}, an unconstrained maximization problem results since all of the constraints are guaranteed to hold. The solution may be obtained by using any standard unconstrained nonlinear optimization procedure.

7. A first-order method – post-mortem

Let us now draw a parallel to the first-order method of section 4, but for the post-mortem analysis. The restriction ($K = 2$) of the previous section is dropped. We use the procedure of differentiating the log-likelihood function with respect to β_j, η_j and p_j, setting the derivatives to zero and then separating values to form an iterative procedure. The log-likelihood function of (21) is the starting point:

$$\frac{\partial L}{\partial \beta_j} = \sum_{l=1}^{N-R} \frac{1}{\bar{G}} \frac{\partial \bar{G}}{\partial \beta_j} + \sum_{i=1}^{R_j} \frac{1}{f_j} \frac{\partial f_j}{\partial \beta_j} \quad (j=1,2,\ldots,K).$$

After substituting from (1) and (2) this becomes:

$$\frac{\partial L}{\partial \beta_j} = \sum_{l=1}^{N-R} \frac{p_j}{\bar{G}} \left\{ -\bar{F}_j \left(\ln \frac{y_l}{\eta_j} \right) \left(\frac{y_l}{\eta_j} \right)^{\beta_j} \right\}$$

$$+ \sum_{i=1}^{R_j} \left\{ \frac{1}{\beta_j} + \left(\ln \frac{x_{ij}}{\eta_j} \right) \left[1 - \left(\frac{x_{ij}}{\eta_j} \right)^{\beta_j} \right] \right\} \quad (j=1,2,\ldots,K). \tag{22}$$

If $\partial L/\partial \beta_j$ is set to zero, the following holds for all j:

$$\beta_j = \frac{R_j}{\sum_{i=1}^{R_j} \left(\ln \frac{x_{ij}}{\eta_j} \right) \left[\left(\frac{x_{ij}}{\eta_j} \right)^{\beta_j} - 1 \right] + \sum_{l=1}^{N-R} \frac{p_j \bar{F}_j}{\bar{G}} \left(\ln \frac{y_l}{\eta_j} \right) \left(\frac{y_l}{\eta_j} \right)^{\beta_j}}. \tag{23}$$

Similarly,

$$\frac{\partial L}{\partial \eta_j} = \sum_{l=1}^{N-R} \frac{1}{\bar{G}} \frac{\partial \bar{G}}{\partial \eta_j} + \sum_{i=1}^{R_j} \frac{\partial f_j}{\partial \beta_j} \quad (j=1,2,\ldots,K),$$

which via (3) and (4) becomes for all j:

$$\frac{\partial L}{\partial \eta_j} = \sum_{l=1}^{N-R} \frac{p_j \bar{F}_j}{\bar{G}} \left(\frac{y_l}{\eta_j} \right)^{\beta_j} \frac{\beta_j}{\eta_j} + \sum_{i=1}^{R_j} \left[\left(\frac{x_{ij}}{\eta_j} \right)^{\beta_j} - 1 \right] \frac{\beta_j}{\eta_j}. \tag{24}$$

When (24) is set to zero we find that

$$\eta_j = \left[\frac{\sum_{l=1}^{N-R} \frac{p_j \bar{F}_j}{\bar{G}} y_l^{\beta_j} + \sum_{i=1}^{R_j} x_{ij}^{\beta_j}}{R_j} \right]^{1/\beta_j} \quad (j=1,2,\ldots,K). \tag{25}$$

Finally, the mixing proportion equation is:

$$\frac{\partial L}{\partial p_j} = \sum_{l=1}^{N-R} \frac{1}{\bar{G}} \frac{\partial \bar{G}}{\partial p_j} + \frac{R_j}{p_j} - \frac{R_K}{p_K} \quad (j=1,2,\ldots,K-1). \tag{26}$$

Upon substitution of (6) and setting the derivative to zero, we obtain:

$$\sum_{l=1}^{N-R} \frac{\bar{F}_j}{\bar{G}} + \frac{R_j}{P_j} = \text{constant} = C \quad (j=1,2,\ldots,K).$$

If we multiply both sides by p_j and sum, this becomes:

$$\sum_{j=1}^{K} p_j C = C = \sum_{l=1}^{N-R}\sum_{j=1}^{K} \frac{p_j F_j}{\bar{G}} + \sum_{j=1}^{K} R_j = N.$$

Thus, finally,

$$p_j = \frac{R_j}{N} + \frac{1}{N}\sum_{l=1}^{N-R} \frac{p_j \bar{F}_j}{\bar{G}} \quad (j=1,2,\ldots,K). \tag{27}$$

Eqs. (23), (25) and (27) are the basis for finding the distribution parameters and mixing proportions in the post-mortem case. As before, the left-hand sides represent the values at the $(v+1)$st iteration, and the right-hand side functions are evaluated with values at the vth iteration. In addition, we will bisect the step size if no improvement in the log-likelihood function is realized.

Convergence properties

The convergence proof is patterned after section 4 by showing that:
(1) α^v belongs to a compact set;
(2) $[\alpha^{v+1} - \alpha^v] \cdot \nabla L^v \geq 0$; and
(3) α^v is feasible.

Since all functions in (23), (25) and (27) are continuous, the mapping from α^v to α^{v+1} is closed and α^v thus belongs to a compact set.

For the second property we first show

$$\left[\beta_j^{v+1} - \beta_j^v\right] \cdot \nabla L_\beta^v \geq 0 \quad (j=1,2,\ldots,K). \tag{28}$$

Eq. (23) implies for all j that:

$$\frac{1}{\beta_j^{v+1}} - \frac{1}{\beta_j^v} = \sum_{l=1}^{N-R} \frac{p_j^v \bar{F}_j^v}{\bar{G}^v R_j}\left(\ln\frac{y_l}{\eta_j^v}\right)\left(\frac{y_l}{\eta_j^v}\right)^{\beta_j^v} - \frac{R_j}{R_j \beta_j}$$

$$- \sum_{i=1}^{R_j} \frac{\left(\ln\frac{x_{ij}}{\eta_j^v}\right)\left[1 - \left(\frac{x_{ij}}{\eta_j^v}\right)^{\beta_j^v}\right]}{R_j} = -\frac{1}{R_j}\frac{\partial L^v}{\partial \beta_j} \quad \text{(via (20))}.$$

After some algebra, this becomes:

$$\beta_j^{v+1} - \beta_j^v = \frac{\beta_j^v \beta_j^{v+1}}{R_j} \nabla L_\beta^v \quad (j=1,2,\ldots,K).$$

Since the coefficient of ∇L_β^v is positive, (28) holds.

We next need to show that

$$\left[\eta_j^{v+1} - \eta_j^v\right] \cdot \nabla L_\eta^v \geq 0 \quad (j=1,2,\ldots,K). \tag{29}$$

Eq. (25) implies for all j that:

$$\eta_j^{v+1} - \eta_j^v = \left[\frac{\sum_{l=1}^{N-R} \frac{p_j^v \bar{F}_j^v}{\bar{G}^v} y_l^{\beta_j^v} + \sum_{i=1}^{R_j} x_{ij}^{\beta_j^v}}{R_j}\right]^{1/\beta_j^v} - \eta_j^v.$$

Since A-B has the same sign as A^x-B^x, the right-hand side has the same sign as:

$$\frac{\sum_{l=1}^{N-R} \frac{p_j^v \bar{F}_j^v}{\bar{G}^v} y_l^{\beta_j^v} + \sum_{i=1}^{R_j} x_{ij}^{\beta_j^v} - \sum_{i=1}^{R_j} \eta_j^{\beta_j^v}}{R_j}.$$

which, via (24), equals

$$\frac{\eta_j^v}{R_j} \frac{\eta_j^v}{\beta_j^v} \frac{\partial L^v}{\partial \eta_j^v} \quad (j=1,2,\ldots,K).$$

Since the coefficient of $[\partial L/\partial \eta_j]^v$ is positive, (29) holds.

In order to demonstrate the second property, it only remains to show that:

$$\sum_{j=1}^{K} \left[p_j^{v+1} - p_j^v\right] \cdot \nabla L_p^v \geq 0. \tag{30}$$

After substituting for ∇L_p^v from (26), the left-hand side of (30) becomes:

$$\sum_{j=1}^{K} \left[p_j^{v+1} - p_j^v\right] \cdot \left\{\sum_{l=1}^{N-R} \frac{\bar{F}_j^v - \bar{F}_K^v}{\bar{G}^v} + \frac{R_j}{p_j^v} - \frac{R_K}{p_K^v}\right\}.$$

Eq. (27) may be rewritten as:

$$\frac{Np_j^{v+1}}{p_j^v} = \frac{R_j}{p_j^v} + \sum_{l=1}^{N-R} \frac{\bar{F}_j^v}{\bar{G}^v}.$$

Thus, the left-hand side is equal to:

$$N \sum_{j=1}^{K} \left[p_j^{v+1} - p_j^v \right] \cdot \left\{ \frac{p_j^{v+1}}{p_j^v} - \frac{p_K^{v+1}}{p_K^v} \right\}.$$

This expression is identical to (19); thus the arguments of section 4 apply and the expression is positive, completing the proof of (30).

The final step is to show that α^v is feasible. Since the right-hand sides of (23) and (25) are positive and (27) is non-negative, then $\beta_j \eta_j > 0$ and $p_j \geq 0$ for all j. We have only to show that $\sum_{j=1}^{K} p_j = 1$. If both sides of (27) are summed over j, then

$$\sum_{j=1}^{K} p_j^{v+1} = \frac{1}{N} \left[\sum_{j=1}^{K} R_j + \sum_{l=1}^{N-R} \sum_{j=1}^{K} \frac{p_j^v \bar{F}_j^v}{\bar{G}^v} \right] = \frac{1}{N} \left[R + \sum_{l=1}^{N-R} 1 \right] = 1.$$

Thus, convergence is assured.

8. The exponential case

If we deal with a mixture of exponentials rather than a mixture of Weibulls, the results are quite similar. The probability density function and complementary CDF for the exponential are

$$f_j(x) = \frac{1}{\eta_j} \exp(-x/\eta_j) \quad \text{and} \quad \bar{F}_j(x) = \exp(-x/\eta_j).$$

Since these forms are equivalent to the Weibull when β_j is one, the iterative equations for the exponential case are found by setting β_j to unity in each of the various algorithms. The results are as follows.

8.1. First-order method – non-post-mortem

$$\eta_j^{v+1} = \frac{\sum_{i=1}^{R} (f_j^v/g^v) x_i + \sum_{l=1}^{N-R} (\bar{F}_j^v/\bar{G}^v) y_l}{\sum_{i=1}^{R} f_j^v/g^v} \quad (j=1,2,\ldots,K),$$

$$p_j^{v+1} = \frac{p_j^v}{N} \left[\sum_{i=1}^{R} \frac{f_j^v}{g^v} + \sum_{l=1}^{N-R} \frac{\overline{F}_j^v}{\overline{G}^v} \right] \qquad (j=1,2,\ldots,K).$$

8.2. Second-order method – non-post-mortem

$$\eta_j^{v+1} = \frac{\sum_{i=1}^{R} (f_j^v/g^v) x_i + \sum_{l=1}^{N-R} (\overline{F}_j^v/\overline{G}) y_l}{\sum_{i=1}^{R} f_j^v/g^v} \qquad (j=1,2,\ldots,K),$$

$$p_j^{v+1} = p_j^v - (\nabla^2 L_p^v)^{-1} (\nabla L_p^v) \qquad (j=1,2,\ldots,K).$$

First-order method – post-mortem:

$$\eta_j^{v+1} = \frac{\sum_{l=1}^{N-R} \frac{p_j^v \overline{F}_j^v}{\overline{G}^v} y_l + \sum_{i=1}^{R_j} x_{ij}}{R_j} \qquad (j=1,2,\ldots,K),$$

$$p_j^{v+1} = \frac{R_j}{N} + \frac{1}{N} \sum_{l=1}^{N-R} \frac{p_j^v \overline{F}_j^v}{\overline{G}^v} \qquad (j=1,2,\ldots,K).$$

9. A two-phase method

In section 4 we presented a first-order method in the non-post-mortem case. Eqs. (9), (10) and (14) form the basis of this approach. In section 7 analogous equations were developed in circumstances where a post-mortem was performed. We were able to take advantage of the fact that the log-likelihood function is concave with respect to the mixing proportions in the non-post-mortem case in section 5, and consequently were able to use second-order convergence by replacing (14) with (20).

If, in either of the first-order methods, we are in a neighborhood of a local maximum where concavity is guaranteed, then a Newton step can be made in the vector (α) of all parameters as $\alpha^{v+1} = \alpha^v - s(\nabla^2 L_\alpha^v)^{-1} \nabla L_\alpha^v$, where s is the step size, initially set to unity. Convergence will occur since $\nabla^2 L_\alpha^v$ is assumed to be negative semi-definite. If α^{v+1} is not feasible, the step size will be bisected, as was previously the case. Computationally we test the closeness to a solution by the absolute value of the gradient of the log-likelihood being arbitrarily small.

More specifically, when $K=2$, we will define the vector as $\alpha = (\beta_1, \beta_2, \eta_1, \eta_2, p_1)$. Thus, the (i, j) element of $\nabla^2 L_\alpha^v$ will be the second partial of L with respect to α_i and α_j of the vector α. Lengthy formulae can now be derived to enable us to write all terms which may be encountered in a $\nabla^2 L_\alpha$ matrix. The differentiation is straightforward but messy, so the detailed results are not offered here.

10. Question of local vs. global maxima

Clearly, the issue of possible multiple maxima for the likelihood is very important in the case of finite mixtures of distributions. First, it should be noted that there is an inherent symmetry whereby apparently different solutions can be made to be really the same by simply reordering the terms of the mixture. Secondly, it has been our computational experience never to find any distinctly different maxima. Theoretically, it seems possible that multiple solutions may exist if we happen to be on a p boundary. But these are pathological counter-examples.

To expand on the point of symmetry, consider the following observation for the exponential case. If $p_1 = p$ and $p_2 = 1 - p$, $\lambda_1 = \lambda$ and $\lambda_2 = \lambda'$ is a maximizing point, then so is $p_1 = 1 - p$ and $p_2 = p$, $\lambda_1 = \lambda'$ and $\lambda_2 = \lambda$, since it gives the same value to the likelihood function. There must be probabilistic convergence to one or the other of these.

Let us add reference here to some slightly related work by Peters and Walker [12]. They have considered mixtures where the subdensities are completely known so that the MLE optimization is only concerned with the mixing proportions. Such a problem is much better behaved, with a concave log-likelihood function. This leads to a natural convex programming problem and nice, necessary and sufficient conditions for global optimality.

11. Computational example

For purposes of illustration, a data set given by Mendenhall and Hader [11] for some avionic equipment has been further analyzed. Their complete lifetimes were divided into two groups as confirmed and unconfirmed failures. Out of 369 units originally put on test, 107 and 218 of them failed from the first and second subpopulations, respectively. The remaining 44 devices operated for the same amount of time, with observation terminated before failure, without giving any post-mortem information on their subgroup membership.

Mendenhall and Hader assumed their mixture to be exponential and then calculated the MLE (with help form an early desk calculator). We selected their estimates as starting points, namely,

$\beta_1 = \beta_2 = 1;$ $\eta_1 = 0.3718;$ $\eta_2 = 0.5328;$ $p = 0.3098,$

with likelihood value 377.6.

The application of the algorithm of section 6 was run on an IBM 370 machine, and performed quite efficiently. Total CPU time was 29.2 seconds, ultimately giving:

$\beta_1 = 1.1249;$ $\beta_2 = 1.2654;$ $\eta_1 = 0.5606;$ $\eta_2 = 0.3514;$ $p = 0.7028,$

with likelihood 285.2.

To gain additional computational experience, different starting points were used. We kept the shape parameters at 1 and put the value of p at 1 for each of three new trials. For the first example, both scale parameters were started at 1. This took 25.2 seconds and gave:

$\beta_1 = 1.1247;$ $\beta_2 = 1.2657;$ $\eta_1 = 0.5614;$ $\eta_2 = 0.3518;$ $p = 0.7025,$

with likelihood of 285.2 also. We view these answers to differ from the first set only by roundoff error. The second case began from scale parameters equal to 0.75, while the third stepped the scales further down to 0.5. Run times for these were 25.5 and 25.9, respectively, with answers once again essentially the same as the first test.

Related computational experience is documented in Kaylan [6]. That work includes a discussion of the computational aspects of a class of algorithms somewhat like those presented here (though significantly less complex in view of the introduction of censoring).

12. Directions for future research

Several methods for finding the parameters of the mixture model have been presented. The next step is to test these methods under a wide variety of conditions to determine the most appropriate choice corresponding to the particular circumstances. Once the parameters have been estimated, one would want to make statistical statements about them. Thus, another research direction will involve the testing of hypotheses concerning the parameters. Two other issues for consideration are the handling of local solutions and goodness-of-fit testing.

Acknowledgment

The authors wish to express their sincere gratitude to the referees for many helpful suggestions.

References

[1] G.A. Carr-Hill and R.A. Carr-Hill, "Reconviction as a Process", *British Journal of Criminology* 12 (1972) 35–43.
[2] A.C. Cohen, "Multi-Censored Sampling in the 3-Parameter Weibull", *Technometrics* 17 (1975) 347–351.
[3] W. Feller, *Theory of Probability*, Vol. II (John Wiley and Sons, New York, 1966).
[4] David F. Greenberg, "Recidivism as Radioactive Decay", *Journal of Research in Crime and Delinquency* 15 (1978) 124–125.
[5] Victor Hasselblad, "Estimation of Finite Mixtures of Distributions from the Exponential Family", *Journal of the American Statistical Association* 84 (1969) 1459–1471.
[6] A.R. Kaylan, "Statistical Analysis of Finite Mixtures of Exponential and Weibull Distributions", Ph.D. Dissertation, Department of Industrial Engineering and Operations Research, Syracuse University (1978).
[7] J.H.K. Kao, "A Graphical Estimation of Mixed Weibull Parameters in Life Testing of Electron Tubes", *Technometrics* 1 (1959) 389–407.
[8] D.G. Leunberger, *Introduction to Linear and Nonlinear Programming* (Addison-Wesley, Reading, Massachusetts, 1973).
[9] Michael D. Maltz and Richard McClearly, "The Mathematics of Behavioral Change: Recidivism and Construct Validity", *Evaluation Quarterly* 1 (1977) 421–438.
[10] N.R. Mann, "Optimal Outlier Tests for a Weibull Model", *TIMS Studies in the Management Sciences*, this issue.
[11] W. Mendenhall, and R.J. Hader, "Estimation of Parameters for Fixed Exponentially Distributed Failure Time Distributions from Censored Life Test Data", *Biometrika* 45 (1958) 504–520.
[12] B. Charles Peters, Jr. and Homer F. Walker, "The Numerical Evaluation of the Maximum-Likelihood Estimate of a Subset of Mixture Proportions", Report 50, Department of Mathematics, University of Houston, Texas (1976).
[13] D. Wingo, "Solution of 3-Parameter Weibull Equations by Constrained Modified Quasi-linearization (progressively censored samples)", *IEEE Transactions on Reliability* R-22 (1973) 96–102.
[14] Stelios Zanakis, "Computational Experience with Some Nonlinear Optimization Algorithms in Deriving MLE for the 3-Parameter Weibull Distribution", *TIMS Studies in the Management Sciences* 7 (1977) 63–79.

OPTIMAL OUTLIER TESTS FOR A WEIBULL MODEL – TO IDENTIFY PROCESS CHANGES OR TO PREDICT FAILURE TIMES

Nancy R. MANN *

UCLA

In this paper Weibull outlier tests based on three different statistics are investigated with respect to their power optimality under various alternative models. Two of the statistics are new in the context of outlier statistics; and one of these is shown to provide a more powerful test in certain situations than other more classical outlier test statistics. Critical values of the three statistics were computer-generated and are tabulated. The tabulated values allow one to identify "treatment effects" resulting from unsuspected modifications to a process or to predict failure times in a life test. Numerical examples are given.

1. Introduction

The results described in this paper pertain to the detection of Weibull outliers and to the prediction of a future ordered observation in an ongoing life test. The motivation for the research described herein, however, is the need for a method of determining whether or not, in a retrospective study, inordinately long times to failure are statistically significant and thus possible results of "treatment" effects caused by unsuspected modifications to a process.

Detection of outliers (spurious observations) is a problem that has long concerned experimenters and data analysts. An historical survey dealing with outliers was given as early as 1981 by Czuber [5]. A more up-to-date expository review of methods for detection of spurious observations was presented by Grubbs [11]. The latter paper is a modification of one "prepared primarily for the American Society for Testing Materials and represents a rather extensive revision of an earlier Tentative Recommended Practice...".

Grubbs points out that "almost all criteria for outliers are based on an assumed underlying normal (Gaussian) population" and Anscombe [1] in an extensive 1960 survey of the subject of outliers makes an initial assumption of normality for the data. (Discussion of the Anscombe paper and a paper by

---

Received October 25, 1980; revised March 24 and June 2, 1981.

* The research documented herein was supported by the U.S. Office of Naval Research under contract nos. N00014-76-C-0723 and N00014-80-C-0684.

Cuthbert Daniel [6], dealing with outliers in factorial experiments, is given by William Kruskal, Thomas S. Ferguson, John Tukey and E.G. Gumbel [15] and stresses the importance of the outlier problem.)

Most types of *life data* are such that a transformation cannot be made to impose normality on the underlying distribution. Thus, the traditional tests for and methods of treatment of outliers are inappropriate for most data arising from life tests. A statistic for testing for outliers in general location-scale families was recently proposed by Tiku [31] and shown to be more powerful than various other statistics under Tiku's [32, p. 1418] outlier models ("labelled slippage" models of Barnett [2] and Barnett and Lewis [3]), although slightly less powerful under Dixon's [7] contamination models; see Tiku [31,32], Hawkins [14] and Tiku [33, p. 139]. The null distribution of Tiku's statistic is exactly beta for the uniform and exponential populations and approximately beta for the normal population (Tiku [31,32]); the percentage points are not available for any other distribution.

In the study described in the sequel, critical values were generated and have been tabulated for a variation of Tiku's statistic for a type-I extreme-value model (one in which the observations are logarithms of two-parameter Weibull variates). Critical values of two other statistics, shown under certain alternatives to be superior or essentially equivalent in terms of power, are also given. Analysis of optimality of power of tests is given in section 3.3, and numerical examples are provided in section 4.

2. Motivation

Often, during a life test, an experimenter has a need for an upper confidence bound (a prediction interval) for the time of the last (nth) failure in a size-n sample of test items. If the experimenter's data are two-parameter Weibull, table 1 can be used to provide such a prediction interval for sample size $n = 5(1)25$, provided the first $n-1$, $n-2$, or $n-3$ failure times are known. On the basis of the first k failures times, with $n-k = 1, 2, 3$, one can also use table 1 to obtain an upper confidence bound for the time of the $(k+1)$st failure. By use of an approximation described in section 3.2, it is also possible to obtain upper prediction bounds for the $(j+1)$st failure based on the first j failure times, with $n-j = 2, 3, \ldots, n-2$. This approximation can be applied for sample sizes ranging from 3 to as large as required.

Notwithstanding the usefulness of the results herein for obtaining certain prediction intervals, the primary motivation for the research described in the following was precipitated by analysis of data resulting from a large-scale retrospective longitudinal study of times of individuals relapsing to undesirable habitual behavior. Results of Mann and Rothberg [26] and Mann [21,22] appear to indicate that either a two-parameter Weibull model or a mixture of

Table 1
100(gamma) percent points of statistics V, Q, and W for sample size N and $N-K$ outliers.

Stat.	N	$N-K$	Gamma=1-alpha				Stat.	N	$N-K$	Gamma=1-alpha			
			0.80	0.90	0.95	0.99				0.80	0.90	0.95	0.99
V	5	1	1.41	1.78	2.22	3.84	V	8	2	1.58	1.58	1.83	2.56
Q	5	1	1.52	2.37	3.39	7.14	Q	8	2	1.66	2.27	2.92	4.80
W	5	1	1.52	2.37	3.39	7.14	W	8	2	0.73	1.06	1.43	2.27
V	5	2	2.12	3.04	4.40	9.86	V	8	3	1.56	1.92	2.36	3.73
Q	5	2	3.70	5.84	9.04	21.71	Q	8	3	2.50	3.41	4.49	7.96
W	5	2	1.72	2.82	4.28	9.96	W	8	3	0.76	1.14	1.58	2.89
V	5	3	5.77	11.28	21.71	81.80	V	9	1	1.16	1.28	1.40	1.72
Q	5	3	14.04	28.27	55.12	210.75	Q	9	1	0.87	1.23	1.60	2.54
W	5	3	4.49	9.40	18.56	84.65	W	9	1	0.87	1.23	1.60	2.54
V	6	1	1.29	1.53	1.80	2.56	V	9	2	1.27	1.46	1.66	2.22
Q	6	1	1.21	1.80	2.48	4.34	Q	9	2	1.46	1.95	2.48	3.96
W	6	1	1.21	1.80	2.48	4.34	W	9	2	0.64	0.92	1.23	2.06
V	6	2	1.63	2.11	2.71	4.64	V	9	3	1.41	1.69	1.99	2.84
Q	6	2	2.44	3.58	4.99	9.53	Q	9	3	2.09	2.79	3.58	5.69
W	6	2	1.12	1.71	2.43	4.75	W	9	3	0.63	0.92	1.26	2.20
V	6	3	2.38	3.53	5.16	11.20	V	10	1	1.13	1.23	1.34	1.63
Q	6	3	4.84	7.64	11.73	26.80	Q	10	1	0.80	1.12	1.45	1.63
W	6	3	1.57	2.60	4.00	9.55	W	10	1	0.80	1.12	1.45	2.33
V	7	1	1.22	1.40	1.59	2.13	V	10	2	1.22	1.38	1.54	1.94
Q	7	1	1.04	1.52	2.02	3.43	Q	10	2	1.32	1.76	2.20	3.31
W	7	1	1.04	1.52	2.02	3.43	W	10	2	0.57	0.82	1.09	1.80
V	7	2	1.44	1.75	2.12	3.13	V	10	3	1.33	1.54	1.76	2.36
Q	7	2	1.94	2.70	3.61	6.08	Q	10	3	1.85	2.39	2.98	4.56
W	7	2	0.87	1.29	1.77	3.19	W	10	3	0.54	0.79	1.06	1.81
V	7	3	1.77	2.34	2.99	5.35	V	11	1	1.11	1.21	1.31	1.54
Q	7	3	3.11	4.52	6.17	11.86	Q	11	1	0.75	1.05	1.38	2.13
W	7	3	1.00	1.54	2.21	4.40	W	11	1	0.75	1.05	1.38	2.13
V	8	1	1.18	1.33	1.49	1.93	V	11	2	1.20	1.33	1.47	1.81
Q	8	1	0.93	1.35	1.80	3.02	Q	11	2	1.23	1.62	2.01	3.02
W	8	1	0.93	1.35	1.80	3.02	W	11	2	0.53	0.75	0.99	1.61

Table 1 (continued)

Stat.	N	N−K	Gamma=1-alpha				Stat.	N	N−K	Gamma=1-alpha			
			0.80	0.90	0.95	0.99				0.80	0.90	0.95	0.99
V	11	3	1.28	1.47	1.66	2.12	W	18	3	0.32	0.45	0.58	0.92
Q	11	3	1.68	2.16	2.70	3.99	V	19	1	1.06	1.11	1.15	1.26
W	11	3	0.48	0.70	0.94	1.56	Q	19	1	0.57	0.78	0.99	1.45
V	12	1	1.10	1.19	1.27	1.46	W	19	1	0.57	0.78	0.99	1.45
Q	12	1	0.71	0.99	1.27	1.93	V	19	2	1.09	1.15	1.22	1.36
W	12	1	0.71	0.99	1.27	1.93	Q	19	2	0.87	1.12	1.35	1.88
V	12	2	1.17	1.29	1.41	1.70	W	19	2	0.37	0.51	0.66	1.04
Q	12	2	1.14	1.50	1.86	2.75	V	19	3	1.12	1.20	1.27	1.45
W	12	2	0.49	0.70	0.92	1.48	Q	19	3	1.11	1.39	1.66	2.23
V	12	3	1.24	1.40	1.56	1.95	W	19	3	0.30	0.43	0.56	0.89
Q	12	3	1.55	1.99	2.41	3.55	V	20	1	1.05	1.10	1.14	1.24
W	12	3	0.44	0.64	0.84	1.39	Q	20	1	0.56	0.76	0.95	1.40
V	13	1	1.09	1.17	1.25	1.43	W	20	1	0.56	0.76	0.95	1.40
Q	13	1	0.69	0.96	1.23	1.88	V	20	2	1.09	1.15	1.20	1.34
W	13	1	0.69	0.96	1.23	1.88	Q	20	2	0.85	1.08	1.31	1.80
V	13	2	1.15	1.26	1.37	1.62	W	20	2	0.36	0.50	0.64	1.00
Q	13	2	1.09	1.44	1.76	2.55	V	20	3	1.11	1.19	1.26	1.41
W	13	2	0.46	0.66	0.86	1.37	Q	20	3	1.09	1.36	1.60	2.12
V	13	3	1.21	1.35	1.48	1.81	W	20	3	0.29	0.42	0.54	0.85
Q	13	3	1.45	1.85	2.24	3.18	V	21	1	1.05	1.09	1.14	1.23
W	13	3	0.41	0.59	0.77	1.26	Q	21	1	0.54	0.74	0.94	1.38
V	14	1	1.08	1.15	1.22	1.36	W	21	1	0.54	0.74	0.94	1.38
Q	14	1	0.66	0.90	1.15	1.68	V	21	2	1.08	1.14	1.19	1.31
W	14	1	0.66	0.90	1.15	1.68	Q	21	2	0.83	1.06	1.27	1.76
V	14	2	1.14	1.23	1.32	1.53	W	21	2	0.35	0.49	0.63	0.97
Q	14	2	1.04	1.34	1.64	2.28	V	21	3	1.11	1.17	1.24	1.39
W	14	2	0.44	0.62	0.81	1.29	Q	21	3	1.05	1.30	1.54	2.08
V	14	3	1.19	1.30	1.42	1.69	W	21	3	0.29	0.40	0.52	0.82
Q	14	3	1.37	1.71	2.04	2.86	V	22	1	1.05	1.09	1.13	1.22
W	14	3	0.38	0.55	0.72	1.16	Q	22	1	0.54	0.74	0.92	1.38

N.R. Mann, Outlier tests for a Weibull model

Stat	n	k					Stat	n	k				
V	15	1	1.08	1.14	1.20	1.34	W	22	1	0.54	0.74	0.92	1.38
Q	15	1	0.63	0.89	1.11	1.65	V	22	2	1.08	1.13	1.18	1.29
W	15	1	0.63	0.89	1.11	1.65	Q	22	2	0.81	1.03	1.24	1.71
V	15	2	1.12	1.21	1.30	1.48	W	22	2	0.34	0.47	0.61	0.95
Q	15	2	1.00	1.28	1.57	2.21	V	22	3	1.10	1.16	1.22	1.36
W	15	2	0.59	0.59	0.77	1.22	Q	22	3	1.02	1.27	1.50	1.99
V	15	2	1.17	1.28	1.38	1.62	W	22	3	0.28	0.39	0.51	0.80
Q	15	3	1.30	1.63	1.96	2.70	V	23	1	1.05	1.09	1.13	1.21
W	15	3	0.36	0.52	0.68	1.09	Q	23	1	0.53	0.72	0.92	1.33
V	16	1	1.07	1.13	1.18	1.33	W	23	1	0.53	0.72	0.92	1.33
Q	16	1	0.61	0.85	1.07	1.63	V	23	2	1.07	1.13	1.17	1.28
W	16	1	0.61	0.85	1.07	1.63	Q	23	2	0.80	1.03	1.23	1.71
V	16	2	1.11	1.19	1.27	1.44	W	23	2	0.33	0.46	0.60	0.93
Q	16	2	0.96	1.22	1.50	2.10	V	23	3	1.10	1.16	1.22	1.35
W	16	2	0.40	0.57	0.74	1.16	Q	23	3	1.01	1.26	1.48	1.98
V	16	3	1.15	1.25	1.35	1.57	W	23	3	0.27	0.38	0.50	0.78
Q	16	3	1.25	1.56	1.87	2.55	V	24	1	1.04	1.08	1.12	1.29
W	16	3	0.34	0.49	0.64	1.02	Q	24	1	0.51	0.70	0.88	1.26
V	17	1	1.07	1.12	1.18	1.30	W	24	1	0.51	0.70	0.88	1.26
Q	17	1	0.60	0.82	1.05	1.57	V	24	2	1.07	1.12	1.16	1.26
W	17	1	0.60	0.82	1.05	1.57	Q	24	2	0.78	0.99	1.19	1.60
V	17	2	1.11	1.18	1.25	1.42	W	24	2	0.32	0.45	0.59	0.91
Q	17	2	0.93	1.19	1.44	2.03	V	24	3	1.09	1.15	1.20	1.32
W	17	2	0.39	0.55	0.71	1.11	Q	24	3	0.98	1.21	1.42	1.90
V	17	3	1.14	1.23	1.32	1.51	W	24	3	0.26	0.37	0.48	0.75
Q	17	3	1.19	1.49	1.77	2.43	V	25	1	1.04	1.08	1.11	1.19
W	17	3	0.33	0.47	0.61	0.97	Q	25	1	0.51	0.69	0.88	1.27
V	18	1	1.06	1.11	1.16	1.28	W	25	1	0.51	0.69	0.88	1.27
Q	18	1	0.58	0.79	1.00	1.51	V	25	2	1.07	1.11	1.16	1.25
W	18	1	0.58	0.79	1.00	1.51	Q	25	2	0.77	0.97	1.17	1.61
V	18	2	1.10	1.17	1.23	1.38	W	25	2	0.32	0.44	0.57	0.89
Q	18	2	0.89	1.15	1.39	1.94	V	25	3	1.09	1.14	1.19	1.31
W	18	2	0.38	0.53	0.68	1.07	Q	25	3	0.97	1.19	1.40	1.84
V	18	3	1.13	1.22	1.30	1.49	W	25	3	0.26	0.36	0.47	0.74
Q	18	3	1.16	1.45	1.71	2.36							

two-parameter Weibulls is appropriate for "time-to-failure" or return to addictive or other undesirable habitual behavior for longitudinal studies made on individuals. Here, it is convenient to conceptualize independent intentions to abstain from the behavior that wear out or otherwise fail in time. (Time to first failure in a cohort has been studied in the case of prison recidivism by Carr-Hill and Carr-Hill [4], who found that a mixture of exponentials provided a good fit for the data.)

What one is attempting to determine in applying an outlier test to retrospective longitudinal time-to-failure data is whether or not "treatment effects" may have resulted in specified instances. If the Weibull outlier test indicates that a number of seemingly inordinately long times to failure are significantly different from other failure times of an individual, then one can attempt to correlate the instances involving suspected treatment effects with various potential causal factors.

Such an outlier test can also be used to identify treatment effects in hardware on the basis of life-test data. In such situations, identification of an outlier will potentially allow one to discover inadvertent and/or unsuspected modifications that may have been made to a manufacturing process. Note that the immediate goal is not parameter estimation, as in many situations, and also that rather large numbers of outliers are a definite possibility.

3. Determination of appropriate test

3.1. Earlier results

Tiku [31] defined $\hat{\sigma}$ to be the (size-n) maximum-likelihood estimator of the scale parameter of a location-scale-parameter distribution (i.e. a distribution $F_X(x)$ that is of the form $G[(x-\mu)/\sigma]$ for some G). He defined σ_c to be the maximum-likelihood estimator of σ, or an estimator with the asymptotic properties of the maximum likelihood estimator of σ, calculated from all the $k < n$ ordered observations felt not to be outliers (considered together as a censored size-n sample); i.e. σ_c is consistent, asymptotically unbiased, efficient and asymptotically normal for the cases he considered and for the case considered here.

Tiku then proposed

$$T = h(\sigma_c/\hat{\sigma}) \tag{1}$$

(where h is a suitable constant) as a statistic for testing the hypothesis that the sample contains no outliers versus the hypothesis that the suspect observations are all outliers. He demonstrated empirically, for 1, 2 and 4 outliers, $n = 10, 20$ and 40, that the statistic T has higher power than certain other well-known statistics (see Grubbs [11], Tietjen and Moore [37], Shapiro and Wilk [30] and Ferguson [10]) under Tiku's [31,32] labelled slippage models (models A and B of section 3.3). Note that Tiku's statistic is versatile: (i) it can be used to test any specified number of outliers on either side of an ordered sample, and (ii) it can be used to test whether the sample contains outliers, irrspective of how many [32, p. 1420]. A multivariate generalization of Tiku's statistic is also available (Tiku and Singh [35]).

Outliers on the left are not generally of interest in our analyses. They often arise because inspections of hardware or tests for abstinence (such as urinalyses to test for opiates and other drugs) are made at discrete time intervals, perhaps weekly. Thus, small values are relatively more displaced then larger values. Because of the logarithmic transformation, any displacement of small values is magnified as well.

Now, consider a sample with a single large suspected outlier from a one-parameter exponential distribution with parameter σ. Here σ_c and $\hat{\sigma}$ are equal to $S_{n-1}/(n-1)$ and S_n/n, respectively, where

$$S_j = \sum_{i=1}^{j} X_{(i)} + (n-j)X_j,$$

with $X_{(i)}$ the ith exponential order statistic. Thus, for this distribution (in which σ is both a location and scale parameter), the statistic T is proportional to $(n-1)\sigma_c/(n\hat{\sigma}) = S_{n-1}/S_n$, which is equal to $S_{n-1}/[S_{n-1} + (X_{(n)} - X_{(n-1)})]$. If U_k is defined to be $(X_{(n)} - X_{(k)})/S_k$, then $(n-1)\sigma_c/(n\hat{\sigma}) = (1 + U_{n-1})^{-1}$ in this single outlier case.

Lawless [16] proposed the use of U_k for obtaining a prediction interval on $X_{(n)}$, the nth ordered observation, from the first k observations in a life test in which the data are exponential with parameter σ; and he demonstrated that for (one-parameter) exponential data, $(n-1)U_{n-1}$ is distributed as Snedecor's F with 2 and $2n-2$ degrees of freedom.

Monte Carlo results exhibited in table 2 demonstrate similarly that for data from an extreme-value distribution (data that are ordered logarithms ($X_{(1)} < \cdots < X_{(n)}$) of sample observations from a two-parameter Weinbull distribution), the power of a test based on T is equivalent to the power baesd on the ratio of $(X_{(n)} - X_{(n-1)})$ and an estimate equivalent to the maximum likelihood estimate of the extreme-value scale parameter (the Weibull shape parameter) obtained from the first $n-1$ observations.

For more than a single large outlier, the statistic T defined above involves observations that are not available in the prediction interval situation. Hence, for any distribution, using a statistic similar to U_k, i.e. proportional to $Q_{l-k} = (X_{(l)} - X_{(k)})/\sigma_c$, $k < l \leq n$, for testing for $n-k$ outliers would seem to be inefficient for $n > k+1$. It will shown in Section 3.3 that this is not necessarily so.

Table 2
Powers of tests based on V, Q and W for significance level alpha.

Alternative – model A – delta=1

Stat.	N	$N-K$	Alpha 0.20	0.10	0.05	0.01
V	5	1	65	32	17	3
Q	5	1	65	32	17	3
W	5	1	65	32	17	3
V	5	2	41	20	10	3
Q	5	2	40	21	10	3
W	5	2	65	37	17	2
V	5	3	35	16	10	2
Q	5	3	30	18	10	2
W	5	3	42	21	10	1
V	10	1	97	80	55	16
Q	10	1	96	80	54	16
W	10	1	96	80	54	16
V	10	2	87	66	45	14
Q	10	2	81	58	38	11
W	10	2	99	92	76	29
V	10	3	74	53	33	9
Q	10	3	68	47	28	7
W	10	3	99	94	77	31
V	20	1	100	99	91	42
Q	20	1	100	99	91	41
W	20	1	100	99	91	41
V	20	2	99	97	88	45
Q	20	2	98	89	70	30
W	20	2	100	100	99	83
V	20	3	99	94	82	45
Q	20	3	94	79	59	26
W	20	3	100	100	100	89

Alternative – model B – lambda=2.5

Stat.	N	$N-K$	Alpha 0.20	0.10	0.05	0.01
V	10	1	84	69	49	19
Q	10	1	84	69	49	20
W	10	1	84	69	49	20
V	10	2	75	53	35	13
Q	10	2	73	51	34	12
W	10	2	67	48	31	8
V	10	3	60	39	22	8
Q	10	3	61	38	24	9
W	10	3	47	31	19	4

Models A&B – delta=1, lambda=2.5

Stat.	N	$N-K$	Alpha 0.20	0.10	0.05	0.01
V	10	1	100	99	94	59
Q	10	1	100	99	94	59
W	10	1	100	99	94	59
V	10	2	99	92	79	50
Q	10	2	97	88	72	43
W	10	2	100	97	89	47
V	10	3	93	79	61	27
Q	10	3	91	75	57	26
W	10	3	98	93	83	45

3.2. Test statistics for Weibull data

We consider now the variate X, the logarithm of a Weibull variate with

$$F_X(x) = \begin{cases} 1, & 1-\exp[-\exp\{(x-\mu)/\sigma\}], \quad x>0, \\ 0, & \text{otherwise;} \quad \sigma>0. \end{cases} \quad (2)$$

The parameter μ is a location parameter, the mode of the distribution of X (the first asymptotic distribution of the smallest extreme) and is the logarithm of the Weibull scale parameter. The parameter σ, which determines the shape of the Weibull distribution, is a scale parameter of the distribution of X, with $\Pi^2\sigma^2/6$ the variance of X.

Since X has a location-scale parameter distribution, it is to be expected that for the labelled slippage model of Tiku (see section 3.3), an efficient test statistic for testing for large outliers can be provided by $T = h(\sigma_c/\hat{\sigma})$. One might also consider statistics proportional to Q_{l-k}, $k < l \leq n$.

Results of Lawless [16], Thoman, Bain and Antle [36], and Mann and Fertig [23] show that for Weibull data, maximum-likelihood and best linear invariant estimators yield very nearly equal numerical results and their small- and large-sample properties (bias, mean squared error, etc.) are very nearly equivalent. Thus, for testing that the largest $n-k$ of n sample observations are outliers, using T is essentially equivalent to using as a test statistic $\tilde{\sigma}_{k,n}/\tilde{\sigma}_{n,n}$, the ratio of the best linear invariant estimators of σ based on the smallest k and on all n sample observations, respectively. The power is obviously unchanged if one uses $\sigma^*_{k,n}/\sigma^*_{n,n}$, the ratio of the best linear unbiased estimators of σ based on the smallest k and on all n-sample observations, respectively. This is true since best linear invarient and best linear unbiased estimators of σ differ only by a constant factor. See, for example, Mann [19].

In this study we considered specifically:

$$V_{n-k} \equiv hT^{-1} = \tilde{\sigma}_{n,n}/\tilde{\sigma}_{k,n},$$

$$Q_{n-k} = (X_{(n)} - X_{(k)})/\tilde{\sigma}_{k,n}$$

and

$$W_{n-k} = Q_{(k+1)-k} = (X_{(k+1)} - X_{(k)})/\tilde{\sigma}_{k,n}.$$

Note that Q_{n-k} and W_{n-k} yield gap tests somewhat similar to some suggested by Dixon [7]. Critical values of these statistics for testing for large outliers, or predicting later failure times, at 0.20, 0.10, 0.05 and 0.01 significance levels for $n = 5(1)25$, $n-k = 1, 2, 3$, are displayed in table 1 above, and an example of their use is given in section 4.

The values shown for V_{n-k} and Q_{n-k} were generated simultaneously by means of 20,000 Monte Carlo simulations. The exhibited values of W_{n-k} were generated by making use of the fact that, for $k \leq n-2$ (the restriction having been discovered in this research),

$$F_k = \left[(X_{(k+1)} - X_{(k)})/E(X_{(k+1)} - X_{(k)})\right] / \tilde{\sigma}_{k,n} / E(\tilde{\sigma}_{k,n})$$

has approximately a classical F distribution. This is discussed in Mann, Schafer and Singpurwalla [27, pp. 255–256].

In order to generate the tabulated values of W_{n-k}, using the F approximation, it was necessary to use stored values of the expectations of the reduced order statistic $Y_{i,n} = (X_{(i)} - \mu)/\sigma$, $i = k, k+1$, and of $C_{k,n}$, where $\sigma/(1 + C_{k,n})$ is the expectation of $\tilde{\sigma}_{k,n}$ and $C_{k,n}\sigma^2$ is the variance of

$$\sigma^*_{k,n} = (1 + C_{k,n})\tilde{\sigma}_{k,n},$$

the best linear unbiased estimator of σ, based on the smallest k observations of X. Thus,

$$F_k = (X_{(k+1)} - X_{(k)})/[E(Y_{k+1,n}) - E(Y_{k,n})]/[\tilde{\sigma}_{k,n}(1 + C_{k,n})].$$

The degrees of freedom for the approximate variate are based on the result of Patnaik [24], which specifies for ϕ, with $E(\phi) = m$, $\text{var}(\phi) = v$, and m^2 proportional to v, that $2m\phi/v$, is approximately a chi-squared variate with $2m^2/v$ degrees of freedom. Thus, we have for F_k: $\nu_1 = 2 \, \text{var}[(Y_{k+1,n} - Y_{k,n})/E(Y_{k+1,n} - Y_{k,n})] \sim 2$ and $\nu_2 = 2/C_{k,n}$ degrees of freedom. Values of $\text{var}(Y_{k+1,n} - Y_{k,n})$, $n - k = 2, 3$; $n = 5(1)25$, were calculated from stored values, along with the other constants needed for the computations. (See below for the origin of these constants.)

The values obtained from the F approximation were compared with trial simulations having a Monte Carlo sample size of 20,000 to ensure that the tabulated values are sufficiently precise. The agreement increases as significance level α decreases. That is, higher percentile values are more precise. Also, precision increases as sample size n increases and as k decreases. Examples of comparison with Monte Carlo values are shown in table 3.

The method used for obtaining the F values with noninteger degrees of freedom is described in Mann, Schafer and Singpurwalla [27, pp. 172, 173]. This method, along with values of

$$E(Y_{k+1,n} - Y_{k,n}), \quad k = 2, n-1,$$

tabulated in Mann et al. [27, pp. 342–347] for $n = 3(1)16$, and Mann, Scheuer and Fertig [28], for $n = 3(1)25$, and values of $C_{k,n}$, which can be obtained from

Table 3
Comparison of approximate and Monte Carlo generated values of 100(gamma) percent points of W.

Value type	N	$N-K$	Gamma = 1-alpha			
			0.80	0.90	0.95	0.99
Monte Carlo	5	2	1.79	2.88	4.58	9.62
F approx.	5	2	1.72	2.82	4.28	9.96
Monte Carlo	5	3	4.82	10.06	19.36	85.85
F approx.	5	3	4.49	9.40	18.56	84.65
Monte Carlo	10	2	0.60	0.85	1.11	1.81
F approx.	10	2	0.57	0.82	1.09	1.80
Monte Carlo	10	3	0.57	0.81	1.08	1.81
F approx.	10	3	0.54	0.79	1.06	1.81
Monte Carlo	20	2	0.39	0.53	0.65	0.99
F approx.	20	2	0.36	0.50	0.64	1.00
Monte Carlo	20	3	0.32	0.44	0.57	0.86
F approx.	20	3	0.29	0.42	0.54	0.85

values appearing in Mann et al. [27, pp. 194–207], for $n = 2(1)13$ and in Mann [19], for $n = 2(1)25$, can be used to estimate the critical values of W_{n-k} for $n - k > 3$. In these cases one can use $\nu_1 = 2$ along with $\nu_2 = 2/C_{k,n}$ for the degrees of freedom, or can calculate ν_1 more precisely using values of the variances and the covariance of $Y_{k+1,n}$ and $Y_{k,n}$ available in Mann [20].

For samples larger than 25 and $n - k > 1$, one can use the approximation with asymptotic expressions for expectations, variances and covariances of the order statistics available in Mann et al. [27, p. 218], and an asymptotic expression for $C_{k,n}$ available in Harter and Moore [13].

As noted above, maximum-likelihood estimates can be substituted for $\tilde{\sigma}_{k,n}$ and for $\tilde{\sigma}_{n,n}$, and the values in table 1 can be used directly with these estimates without any modification required. One can also $\sigma^*_{k,n}$ and $\sigma^*_{n,n}$, best linear unbiased estimates (see Mann [19]) or simplified linear estimates (see Mann et al. [27, pp. 210–212], Mann and Fertig [24] or Engelhardt and Bain [8,9]), in place of the best linear invariant estimates. In this case, the modified statistics Q_{n-k} and W_{n-k} need to be multiplied by the factor $CQW = (1 + C_{k,n})$ and the modification of V_{n-k} needs to be multiplied by $CV = (1 + C_{k,n})/(1 + C_{n,n})$ before comparison with critical values. In other words, $\sigma^*_{k,n}$ and $\sigma^*_{n,n}$ need to be divided by $(1 + C_{k,n})$ and $(1 + C_{n,n})$, respectively, to convert them to $\tilde{\sigma}_{k,n}$ and $\tilde{\sigma}_{n,n}$. Values of the constants CQW and CV appear in table 4 for $n = 5(1)25$, $n - k = 1, 2, 3$.

Approximations to Q_{n-k} and V_{n-k} can be calculated by using probability plots such as those shown in figs. 1 and 2. Here, the inverse of the slope of the

Table 4
Factors for converting statistics to V, Q or W.

N	N−K	CQW	CV	CQP	CWP	N	N−W	CQW	CV	CQP	CWP
5	1	1.254	1.167	0.731	1.367	15	1	1.052	1.045	0.362	2.760
5	2	1.417	1.167	1.581	0.754	15	2	1.060	1.045	0.619	0.254
5	3	1.895	1.167	3.337	1.223	15	3	1.068	1.045	0.840	0.217
6	1	1.186	1.132	0.620	1.613	16	1	1.048	1.042	0.353	2.836
6	2	1.270	1.132	1.226	0.562	16	2	1.055	1.042	0.600	0.245
6	3	1.432	1.132	2.062	0.679	16	3	1.062	1.042	0.811	0.207
7	1	1.146	1.109	0.551	1.815	17	1	1.045	1.040	0.344	2.908
7	2	1.198	1.109	1.040	0.464	17	2	1.051	1.040	0.583	0.237
7	2	1.280	1.109	1.607	0.496	17	3	1.057	1.040	0.785	0.199
8	1	1.120	1.093	0.503	1.987	18	1	1.042	1.037	0.336	2.975
8	2	1.155	1.093	0.923	0.404	18	2	1.047	1.037	0.568	0.230
8	3	1.205	1.093	1.366	0.403	18	3	1.052	1.035	0.762	0.191
9	1	1.102	1.081	0.468	2.136	19	1	1.039	1.035	0.329	3.038
9	2	1.127	1.081	0.842	0.363	19	2	1.044	1.035	0.554	0.224
9	3	1.161	1.081	1.214	0.346	19	3	1.049	1.035	0.742	0.185
10	1	1.088	1.072	0.441	2.267	20	1	1.037	1.033	0.323	3.098
10	2	1.107	1.072	0.782	0.333	20	2	1.041	1.033	0.542	0.218
10	3	1.132	1.072	1.107	0.308	20	3	1.045	1.033	0.724	0.180
11	1	1.078	1.064	0.419	2.385	21	1	1.035	1.031	0.317	3.154
11	2	1.093	1.064	0.735	0.310	21	2	1.039	1.031	0.531	0.213
11	3	1.112	1.064	1.028	0.280	21	3	1.042	1.031	0.708	0.175
12	1	1.069	1.058	0.401	2.491	22	1	1.033	1.030	0.312	3.208
12	2	1.082	1.058	0.698	0.292	22	2	1.036	1.030	0.521	0.209
12	3	1.096	1.058	0.966	0.258	22	3	1.040	1.030	0.693	0.170
13	1	1.063	1.053	0.386	2.588	23	1	1.032	1.029	0.307	3.259
13	2	1.073	1.053	0.667	0.277	23	2	1.034	1.029	0.512	0.204
13	3	1.085	1.053	0.916	0.242	23	3	1.038	1.029	0.680	0.166
14	1	1.057	1.049	0.374	2.677	24	1	1.030	1.027	0.302	3.308
14	2	1.066	1.049	0.641	0.265	24	2	1.033	1.027	0.504	0.201
14	3	1.076	1.049	0.875	0.228	24	3	1.036	1.026	0.667	0.162
						25	1	1.029	1.026	0.298	3.355
						25	2	1.031	1.026	0.496	0.197
						25	3	1.034	1.026	0.656	0.159

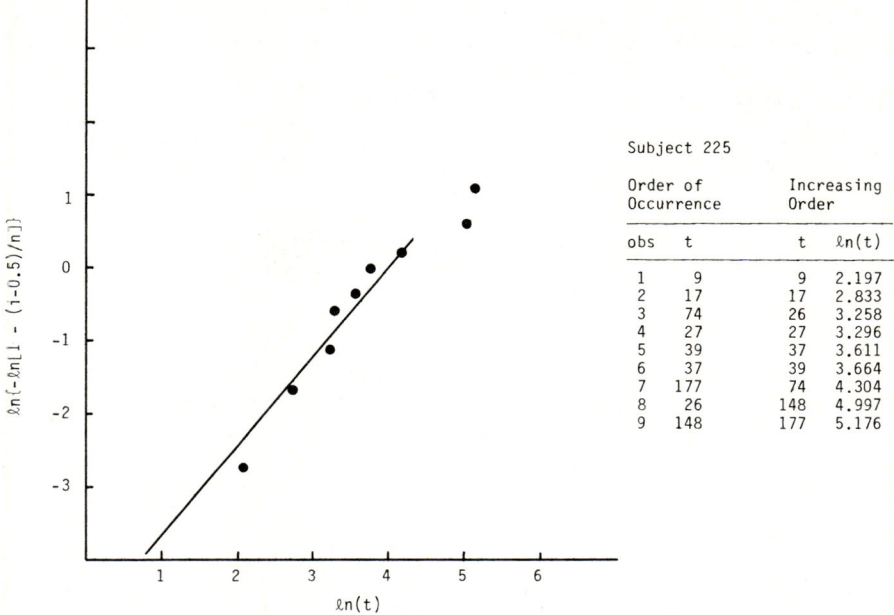

Fig. 1. Weibull probability plot of days until drug relapse.

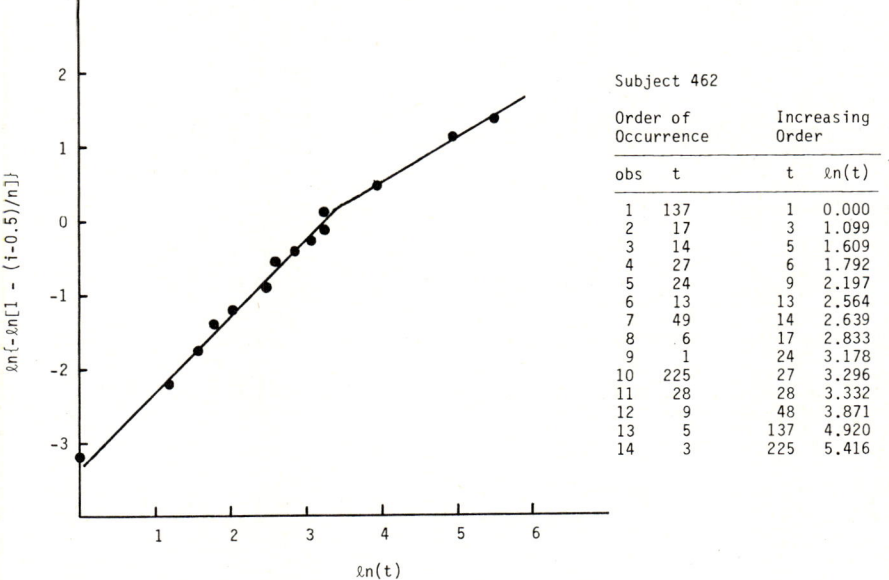

Fig. 2. Weibull probability plot of days until drug relapse.

line plotted on the basis of the smallest k observations gives an approximation to $\sigma^*_{k,n}$; and the inverse of the slopes of the line formed by the nth and kth points and by the line formed by the $(k+1)$st and kth points give approximations to $(X_{(n)} - X_k)/E(Y_{n,n} - Y_{k,n})$, and $(X_{(k+1)} - X_k)/E(Y_{k+1,n} - Y_{k,n})$, respectively.

If the inverse of the slope in the probability plots is used, then the constant factor

$$CQP = E(Y_{n,n} - Y_{k,n})(1 + C_{k,n})$$

or

$$CWP = E(Y_{k+1,n} - Y_{k,n})(1 + C_{k,n})$$

must be used to multiply the value obtained to convert it to one that can be compared with the critical factors for Q_{n-k} or W_{n-k}, respectively. Values of CQP and CWP are given in table 4 for $n = 5(1)25$; $n - k = 1, 2, 3$.

Special probability papers, each one applicable to a specified sample size, have been designed (see [18]), so that individuals without technical training can plot failure times of interest. Without making such plots, one will usually find it very difficult to have much feeling for what might be moderately large values for time-to-failure when the data are Weibull. Using the plots with some minimal instruction, a nontechnical person should be able to determine slopes of lines formed by $x_1, \ldots, x_k$ and by x_k and x_n or x_k and x_{k+1}. This assists a spouse, a "significant other" or a counsellor of a subject engaging in undesirable habitoral behavior to gain insight into what might be, for this subject, motivation for long-term abstinence.

3.3. Optimality of power under the two alternatives

For a Weibull model, the hypothesis H_0 to be tested is:

$X_{(1)}, \ldots, X_{(n)}$ are order statistics from

$$f_X(x) = (1/\sigma)g[(x - \mu)/\sigma], \tag{3}$$

where $f_X(x)$ is the density function corresponding to the distribution function (2). Model A and model B are given, respectively, by:

A: $X_{(1)}, \ldots, X_{(k)}$ are the smallest k-order statistics from (3) and $X_{(k+1)}, \ldots, X_{(n)}$ are the largest $n - k$ order statistics from $f_X(x) = (1/\sigma)f\{[x - (\mu + \delta\sigma)]/\sigma\}$, and

B: $X_{(1)},\ldots,X_{(k)}$ are the smallest k-order statistics from (3) and $X_{(k+1)},\ldots,X_{(n)}$ are the largest $n-k$ order statistics from $f_X(x)=(1/\lambda\sigma)g[(x-\mu)/\lambda\sigma[$.

These models may not correspond to the manner in which data are generated for the situation described. Nonetheless, a mixture of any two specified Weibull distributions can be represented by a mixture of models A and B if the "outliers" are larger than other values and the number of outliers is only one or two. Models A and B can be combined also to approximate very well almost any model that is a mixture of a Weibull sample of small values and a Weibull sample of larger values (the "outliers").

Examples of model A and model B are shown as probability plots (on Weibull probability paper) in figs. 1 and 2, respectively. It was the object of the research described in this paper to determine test statistics that are optimal, in terms of power considerations, for testing for outliers, in general, and for testing against model A or model B, or a mixture of these, in particular. To this end, the power of the various test statistics under consideration was calculated by 2,000 Monte Carlo simulations (in addition to the 20,000 used to generate critical values for the test statistics). These power calculations were made for each critical value generated for V_{n-k}, Q_{n-k} and for selected sample sizes for W_{n-k} for model A: $\delta=0.5, 1$; model B: $\lambda=2, 5$; and mixed models: $\delta=1$; $\lambda=2, 5$. Illustrative examples are exhibited in table 2. Note that only for the test statistic W_{n-k} (under model A with $n\geqslant 10$) does the power increase as $n-k$ increases. This is probably due to the fact that observations near to μ are closer together than observations near the tail. Hence, displacement of 1σ is less critical near the tail.

On the basis of the many similations that were made, it has been well established that when one is testing H_0 versus a single outlier, a test based on $Q_{n-k}\equiv Q_{n-(n-1)}\equiv W_{(k+1)-k}\equiv W_{n-(n-1)}$ has power essentially identical to that of one based on $V_{n-(n-1)}\equiv hT^{-1}$. This was pointed out in section 3.1.

Tiku [27] has demonstrated for Gaussian families that a test based on the statistic T has higher power in more general situations (more than a single outlier) than other classical outlier tests under his labelled slippage model. As the number of outliers, $n-k$, increases, however, the ratio of the power of W_{n-k} relative to the power of V_{n-k} increases under model A (shift in location). That is to say, under model A a test based on a measure of the gap $(X_{(k+1)}-X_{(k)})$ between the smallest suspected outlier and the largest observation thought not to be an outlier, relative to a measure of the dispersion ($\tilde{\sigma}_{k,n}$) of the observations thought not to be outliers is more powerful than one based on T (see table 2). It is clear from fig. 1 that for model A it is essentially this quantity, i.e. the size of the gap relative to the dispersion of the smaller observations, that is the critical factor in establishing the suspicion of outliers.

Thus, it is not unlikely that a test based on statistic, such as W_{n-k} involving $X_{(k+1)} - X_{(k)}$, is optimal for alternative models resembling model A.

If it were established that model A was precisely the alternative (which it usually will not be), then using in the denominator of W_{n-k} an estimator of σ that involves all differences of successive order statistics except $X_{k+1} - X_k$ would be more powerful than W_{n-k} as it is defined. Such a test would be equivalent, in terms of power, to one having this statistic in the denominator and $\tilde{\sigma}_{n,n}$ in the numerator and should be optimal for the labelled slippage model with model A as the alternative. Note that Mann and Fertig [25] demonstrate that for a goodness-of-fit test, involving gaps (which all estimates of σ in location-scale families involve), the important consideration in determining optimality is *which* gaps are involved in the test and in *what position*, rather than how the gaps are combined. That is to say, an optimal estimator of σ based on the first $k-1$ gaps performs no better than one which is the sum of each of the $k-1$ gaps divided by its (the gap's) expectation.

In this context it is noted that the statistic $V^* = \tilde{\sigma}_{k+1,n}/\tilde{\sigma}_{k,n}$ for $k = n-2$, $n-3,\ldots$, has the same functional relationship with W_{n-k} that $V_{n-(n-1)}$ has with $Q_{n-(n-1)}$. Therefore, the statistic W_{n-k} also has essentially the same power as V^*. The inverse of V^* is a special case of Z, a statistic proposed by Tiku [34, eq. 1.4] for testing goodness of fit when H_0 is exponentiality. The statistic Z is equivalent to V^* when the exponential censored sample of size n consists only of the smallest $k+1$ observations. Thus, Z stresses the difference of the two largest observed order statistics.

It should also be pointed out (see [24]) that μ^* and $\tilde{\mu}$ based on $X_{(1)},\ldots,X_{(k)}$, $k < n$, from an extreme-value distribution are of the approximate form, $X_{(k)} + c\tilde{\sigma}_{k,n}$, where c is an appropriate constant. Thus, a test of form $(X_{(k+1)} - \tilde{\mu}_{k,n})/\tilde{\sigma}_{k,n}$ is essentially the test W_{n-k}.

For model B, the critical factor is the ratio of the slopes of the plots of the smallest k and the largest $n-k+1$ observations. For this model, W_{n-k} performs poorly relative to V_{n-k}, as one might suspect, but Q_{n-k}, which is proportional to the ratio of estimates of these two slopes, approximates V_{n-k} very well, i.e. powers of V_{n-k} and Q_{n-k} are very nearly equivalent (see table 2). The statistic Q_{n-k} is shown (in table 2) to perform very poorly, in terms of power, under model A, however.

For a mixture of the models, results shown in table 2 indicate that a test based on W_{n-k} tends to be most powerful, with Q_{n-k} performing most poorly. Again, as with model A, the gap $X_{(k+1)} - X_{(k)}$ relative to $\tilde{\sigma}_{k,n}$ appears to be the most critical factor.

It seems clear from this study that in considering whether or not to test for outliers, one should, if possible, plot the data on probability paper. Plotting is useful in providing perspective even though there is a single suspected outlier. For more than a single outlier, plotting is essential if one is to know whether to use W_{n-k} (for model A or mixed models) or either V_{n-k} or Q_{n-k} (for model B) or V_{n-k} (for a more general alternative model). In this way no one can insure using a test with what appears to be optimal power.

Clearly, the power of the outlier test is affected by the a priori analysis, as is always the case to some extent in looking at the data before performing an outlier test. However, in this context it is important to identify *large* outliers in order to determine if treatment effects (extending life or, for human subjects, extending periods of abstinence) have resulted and what might have caused such effects. The goal is not primarily one of estimation of parameters, but rather of exploration. This point is discussed by Barnett and Lewis [3, pp. 5–6].

Finally, it is to be noted that the results obtained here are likely to extend to other location-scale families. Thus, an analog of W_{n-k} involving the gap $X_{(k+1)} - X_{(k)}$ will possibly tend to be more powerful for any location-scale family (including Gaussian distributions) for testing H_0 under model A than is the statistic T.

4. Numerical examples

The data in the probability plots (figs. 1 and 2) are used here to provide examples of the use of the various test statistics.

First, we consider fig. 1, which exhibits two possible outliers from a mixed model with $\lambda < 1$. Here n is equal to 9, so that the tables in [22] can be used to obtain the weights to calculate $\tilde{\sigma}_{7,9} = 0.709$ and $\tilde{\sigma}_{9,9} = 0.884$. Also, $x_{(9)} - x_{(7)} = 0.872$ and $x_{(8)} - x_{(7)} = 0.693$. Thus, $v_{9-7} = 1.245$, $q_{9-7} = 1.228$ and $w_{9-7} = 0.976$. Comparing these values with the tabulated critical values, one finds that if the specified significance level is 0.10, then only the test statistic w_{9-7}, involving $x_{(8)} - x_{(7)}$, rejects the hypothesis of no outliers.

The plotted line drawn (by hand) in fig. 1 gives highest weight to the kth or, in this case, the seventh value, as do the weights for optimal linear estimates of σ, such as $\tilde{\sigma}$ and σ^*. Also note that horizontal rather than vertical distances from points should be minimized. The slope of the line is about 1.20 so that an approximation to $\sigma_{7,9}^*$ is about 0.833. This gives 0.717 as an approximation to $\tilde{\sigma}_{7,9}$ with the use of $CQW = 1.161$ (found in table 4) as a divisor.

The plot in fig 2 suggests three large outliers of the general type specified by model B. Thus, using tabulated values in [14], one finds $\tilde{\sigma}_{11,14} = 0.84226$, $\tilde{\sigma}_{14,14} = 1.353$, $x_{(14)} - x_{(11)} = 2.084$ and $x_{(12)} - x_{(11)} = 0.560$, so that $v_{14-11} = 1.595$, $q_{14-11} = 2.457$ and $w_{14-11} = 0.660$. In comparing these values with the critical values of table 1, one finds that if the specified significance level is 0.10, all three test statistics reject a "no outliers" hypothesis. The statistics v_{14-11} and q_{14-11} reject also at the 0.05 significance level, while w_{14-11} does not. This is to be expected since the probability plot demonstrates that the appropriate test statistic is v_{14-11} or q_{14-11}.

The slope of the line plotted in fig. 2 is about 1.1, giving an approximation of about $0.91/CQW = 0.91/1.076 = 0.845$ for $\tilde{\sigma}_{11,14} = 0.842$. Note that again the kth value has been weighted most heavily.

References

[1] F.J. Anscombe, "Rejection of Outliers", *Technometrics* 2 (1960) 123–147.
[2] V. Barnett, "The Study of Outliers: Purpose and Model", *Appl. Statist.* (1978) 242–250.
[3] V. Barnett and T. Lewis, *Outliers in Statistical Data* (John Wiley, New York, 1978).
[4] G.A. Carr-Hill and R.A. Carr-Hill, "Reconviction as a Process", *British J. of Criminology* 12 (1972) 35–43.
[5] E. Czuber, *Theorie der Beobachtungsfehler* (Leipzig, 1981).
[6] C. Daniel, "Locating Outliers in Factorial Experiments", *Technometrics* 2 (1960) 149–156.
[7] W.J. Dixon, "Analysis of Extreme Values", *Ann. Math. Statist.* 21 (1950) 488–506.
[8] M. Engelhardt and L.J. Bain, "Some Complete and Censored Sampling Results for the Weibull or Extreme Value Distribution", *Technometrics* 15 (1973) 541–549.
[9] M. Engelhardt and L.J. Bain, "Simplified Statistical Procedures for the Weibull or Extreme-Value Distribution", *Technometrics* 19 (1977) 323–331.
[10] T.S. Ferguson, "Rules for Rejection of Outliers", *Rev. Int. Statist. Inst.* 3 (1961) 29–43.
[11] F.E. Grubbs, "Sample Criteria for Testing Outlying Observations", *Ann. Math. Statist.* 21 (1950) 27–58.
[12] F.E. Grubbs, "Procedures for Detecting Outlying Observations", *Technometrics* 11 (1969) 1–21.
[13] H.L. Harter and A.H. Moore, "Maximum Likelihood Estimation, from Doubly Censored Samples, of the Parameters of the First Asymptotic Distribution of Extreme Values", *J. Amer. Statist. Assoc.* 63 (1968) 889–901.
[14] D.M. Hawkins, "Comment on 'A New Statistic for Testing Suspected Outliers'", *Commun. in Statist.* A6 (1977) 435–438.
[15] W.H. Kruskal, T.S. Ferguson, J.W. Tukey and E.G. Gumbel, "Discussion of the Papers of Messrs. Anscombe and Daniel", *Technometrics* 2 (1960) 157–166.
[16] J.F. Lawless, "A Prediction Problem Concerning Samples from the Exponential Distribution, with Application in Life Testing", *Technometrics* 13 (1971) 725–730.
[17] J.F. Lawless, "Construction of Tolerance Bounds for the Extreme-Value and Weibull Distributions", *Technometrics* 17 (1975) 255–261.
[18] N.R. Mann, "Results on Location and Scale Parameter Estimation with Application to the Extreme-Value Distribution", Aerospace Research Laboratories Report ARL 67-0023, Office of Aerospace Research, U.S. Air Force, Wright-Patterson Air Force Base, Ohio (1967).
[19] N.R. Mann, "Tables for Obtaining the Best Linear Invariant Estimates of Parameters of the Weibull Distribution", *Technometrics* 9 (1967) 629–645.
[20] N.R. Mann, "Results on Statistical Estimation and Hypothesis Testing with Application to the Weibull and Extreme-Value Distributions", Aerospace Research Laboratories Report ARL 68-0068, Office of Aerospace Research, U.S. Air Force, Wright-Patterson Air Force Base, Ohio (1968).
[21] N.R. Mann, "Use of Life-Test Data Analysis Methodology for Analyzing Undesirable Habitual Behavior", Serial T-406, The George Washington University Institute for Management Science and Engineering, Washington, D.C. (1979). Also in *Proceedings of the 14th Annual Conference on Statistics, Computer Science and Operations Research*, Cairo University, Cairo, Egypt (March 1979).
[22] N.R. Mann, "Insights to Addictive Behavior Based on the Application of Life-Test Data-Analysis Methodology to Longitudinal Drug-Use Recidivism Data", *International Journal of Addiction*, to appear, 1983.

[23] N.R. Mann and K.W. Fertig, "Tables for Obtaining Confidence Bounds and Tolerance Bounds Based on Best Linear Inviarant Estimates of the Extreme-Value Distribution", *Technometrics* 15 (1973) 86–100.
[24] N.R. Mann and K.W. Fertig, "Simplified Efficient Point and Interval Estimators for Weibull Parameters", *Technometrics* 17 (1975) 361–368.
[25] N.R. Mann and K.W. Fertig, "A Goodness-of-Fit for the Two-Parameter Weibull Distribution Against Three-Parameter Weibull Alternatives", *Technometrics* 17 (1975) 237–245.
[26] N.R. Mann and J. Rothberg, "Using Reliability Methodology to Think About Human Behavior", presented at the Joint National TIMS/ORSA Meeting, Washington, D.C., 1980.
[27] N.R. Mann, R.E. Schafer and N.D. Singpurwalla, *Methods for Statistical Analysis of Reliability and Life Data* (John Wiley, New York, 1974).
[28] N.R. Mann, E.M. Scheuer and K.W. Fertig, "A New Goodness-of-Fit Test for Two-Parameter Weibull or Extreme-Value Distribution with Unknown Parameters", *Commun. in Statist.* 2 (1973) 383–400.
[29] P.B. Patnaik, "The Non-Central χ^2 and F Distributions and their Applications", *Biometrika* 36 (1949) 202–232.
[30] S.S. Shapiro and M.B. Wilk, "An Analysis of Variance Test for Normality (complete samples)", *Biometrika* 52 (1965) 591–611.
[31] M.L. Tiku, "A New Statistic for Testing for Suspected Outliers", *Commun. in Statist.* 4 (1975) 737–752.
[32] M.L. Tiku, "Rejoinder: Comment on a New Statistic for Testing Suspected Outliers", *Commun. in Statist.* A6 (14) (1977) 1417–1422.
[33] M.L. Tiku, "Robustness of MML Estimators Based on Censored Samples and Robust Test Statistics", *J. Statistical Planning and Inference* 4 (1980) 123–143.
[34] M.L. Tiku, "Goodness-of-Fit Statistics Based on the Spacings of Complete or Censored Samples", *Austral. J. Statist.* 22 (1980) 260–275.
[35] M.L. Tiku and M. Singh, "Testing Outliers in Multivariate Data", in: G.P. Patil, C. Taillie and B. Baldessari, eds., *Multivariate distributions in Scientific Work*, (North-Holland, Amsterdam, 1980).
[36] D.R. Thoman, L.J. Bain and C.E. Antle, "Reliability and Tolerance Limits in the Weibull Distribution", *Technometrics* 12 (1970) 363–371.
[37] G.L. Tietjen and R.H. Moore, "Some Grubbs-type Statistics for the Detection of Outliers", *Technometrics* 14 (1972) 583–597.

A BAYESIAN SCHEME FOR ESTIMATING RELIABILITY GROWTH UNDER EXPONENTIAL FAILURE TIMES

Nozer D. SINGPURWALLA
The George Washington University

In this expository paper we consider an adaptive approach for estimating reliability growth, based on prior information which is motivated from practical considerations. We discuss two situations: in the first, both the prior distribution and the posterior distributions of the mean time to failure of an exponential distribution are stochastically ordered; in the second situation, the prior distribution is stochastically ordered with respect to the last posterior distribution. The former situation leads us to procedure which is not fully Bayesian, and is therefore termed "pseudo-Bayesian". Since we do not know the properties of this pseudo-Bayesian approach, we can best describe our work here as being a "pseudo-Bayesian scheme". The second situation leads us to an approach which is fully Bayesian under certain assumptions. Our work along the lines indicated in this paper is still in progress, and we invite the attention of other researchers to some of the problems we have posed, and the questions we have raised.

1. Introduction

A complex, newly designed system generally undergoes several stages of testing before it is put into operation. After each stage of testing changes are made to the design (or the operating conditions are respecified) with the hope that the new design would lead to a longer period of performance. This procedure is referred to as *reliability growth*; this is because a longer period of performance implies an improvement in reliability. Often, in practice a change in the system design may result in a deterioration of the system performance, so that the term reliability growth may not be an appropriate description of what is actually happening to the system. However, since the intent of a design change is to improve the reliability, we shall continue to use the term reliability growth throughout this paper.

Suppose that the system has been tested at stages 0, 1, 2,.... At the end of each stage stage an evaluation of the reliability of the system is made using any of the conventional procedures which are used to measure reliability. At each stage of testing, we may test either a single copy of the system, or several copies of the system, depending on the particular situation being considered.

Received June 6, 1980; revised July 29, 1981.

Having tested the system over, say, i, $i = 0, 1, 2,\ldots$, stages, we would like to obtain the best (in a sense to be made precise) estimate of the reliability of the system at the ith stage; in doing so we would like to use all the testing information that we have acquired over the various stages.

Information of the type indicated above will enable a decision-maker to determine if the design changes do indeed result in an improvement in reliability, and perhaps determine the rate (with respect to the stages) at which there is a growth in reliability. It will also enable a decision-maker to decide if the system is ready to be put into operation, and to arrive at a suitable cost–warranty agreement with the user.

Since 1966 there have been several papers written under the heading of reliability growth, each emphasizing a different point of view. For a recent survey of these papers we refer the reader to Balaban [1]. The approach that we shall take in this paper is suggested by Bayesian considerations, with the analysis being undertaken in a manner which will be described in section 2. Others who have adopted a Bayesian point of view, but with a direction different from ours, are Smith [5], Weinrich and Gross [6], and Jewell [3].

As stated in the abstract, our paper is expository in nature; we do not claim to offer a definitive solution to the problem of estimating reliability growth. What we give here is an outline of a few potential approaches and discuss their ramifications, both from a philosophical and a practical point of view. We raise several statistical issues pertinent to each approach, and invite the attention of other researchers to the questions that we have raised. Toward the end of the paper we also describe some additional practical problems pertinent to reliability growth which are of interest, especially to management scientists and operations research analysts, since they involve cost considerations and optimization. The contribution of this paper, then, is the introduction of a comprehensive and unified Bayesian view of the problem of reliability growth, and an outline of some possible approaches that we can take with their ramifications. Our hope is that the text of this paper will stimulate more thought along the lines we described.

2. A Bayesian scheme for estimating reliability growth

We start by considering the following model.

Suppose that the failure distribution of the system after the ith design change, $i = 0, 1,\ldots$, is exponential with mean θ_i. That is, $f(t;\ \theta_i) = \theta_i^{-1} \exp(-t/\theta_i)$, $\theta_i > 0$, $t > 0$. In particular, at stage 0, when the system is newly built, its time to failure has an exponential distribution with mean θ_0. Based upon our knowledge of the performance of similarly designed systems, we assign a prior distribution to θ_0, say $G(\theta_0;\ \cdot)$. Without any loss in generality, we let $G(\theta_0;\ \cdot)$ be an *inverted gamma* distribution with parameters α_0 and

β_0; this is the natural conjugate prior distribution for θ_0. Thus, we have $dG(\theta_0; \cdot) = g(\theta_0, \alpha_0, \beta_0)$ as

$$\left(\alpha_0^{\beta_0}/\Gamma(\beta_0)\right)\theta_0^{-(\beta_0+1)} \exp(-\alpha_0/\theta_0), \quad \text{for } \alpha_0, \beta_0, \theta_0 > 0. \tag{1}$$

It is easy to verify (Mann, Schafer and Singpurwalla [4, p. 112]) that the mean and the variance of θ_0 with respect to the above prior distribution are $E(\theta_0) = \alpha_0/(\beta_0 - 1)$ and $\text{var}(\theta_0) = \alpha_0^2/(\beta_0 - 1)(\beta_0 - 2)^2$, respectively. Thus, to be assured that both the prior mean and variance of θ_0 are finite, we must have $\beta_0 > 2$.

One way of choosing α_0 and β_0, the prior parameters, is to set $E(\theta_0)$ equal to our best guess of the mean time to failure θ_0, and to set $\text{var}(\theta_0)$ equal to a number which reflects the quality of our guess about θ_0.

Having assigned our prior distribution, $G(\theta_0; \cdot)$, we test n_0 (≥ 1) copies of the system, and observe their times to failure, $t_{0,i}$, $i = 1, \ldots, n_0$. Let $T_0 = \Sigma_1^{n_0} t_{0,i}$ be the *total time on test*. Then conditional on T_0, the posterior distribution of θ_0 is also an inverted gamma distribution; that is, $g(\theta_0 | T_0; \alpha_0, \beta_0, n_0)$ is

$$\left((\alpha_0 + T_0)^{\beta_0 + n_0}/\Gamma(\beta_0 + n_0)\right)\theta_0^{-(\beta_0 + n_0 + 1)} \exp(-(\alpha_0 + T_0)/\theta_0). \tag{2}$$

It is well known that under the assumption of a squared error loss function, the Bayesian estimate of the mean time to failure at stage 0 is the mean of the posterior distribution of θ_0. Thus,

$$E(\theta_0 | T_0) = (\alpha_0 + T_0)/(\beta_0 + n_0 - 1) \tag{3}$$

is our estimate of the mean time to failure at stage 0, given the failure data T_0. To express our uncertainty about this estimate, we compute the *posterior* $(1 - \gamma)$ *credibility interval* for θ_0. This interval is provided by specifying two numbers $(\theta_0 | T_0)_L$ and $(\theta_0 | T_0)_U$, $(\theta_0 | T_0)_L < (\theta_0 | T_0)_U$, such that

$$I[(\theta_0 | T_0)_L, (\theta_0 | T_0)_U; g(\theta_0 | T_0; \alpha_0, \beta_0, n_0)] = 1 - \gamma, \tag{4}$$

where

$$I[a, b; g(s)] = \int_a^b g(s) \, ds.$$

The parameter θ_0, with its posterior distribution, is a measure of the reliability of the system at stage 0. Having observed $E(\theta_0 | T_0)$, we may want to increase it either by making design changes or by respecifying the operating conditions of the system. Suppose that we do decide to make these changes; let θ_1 denote the mean time to failure after the changes have been made. The system is now at stage 1, and we are ready to test the system and verify if the changes have resulted in an improvement of the system.

Since we have adopted a Bayesian point of view, our first task is to assign a prior distribution for θ_1, say $G(\theta_1; \cdot)$. The novelty of our approach pertains to the manner in which we go about choosing $G(\theta_1; \cdot)$, and this is motivated by the following consideration.

Even though the changes that we have instituted have been undertaken with a view towards increasing the reliability of the system, there is a possibility that the changes could be deleterious to the system. In order to account for this possibility we shall choose $G(\theta_1; \cdot)$ in such a manner that θ_1 is *stochastically larger* than $(\theta_0|T_0)$; this is written as

$$\theta_1 \overset{st}{\geq} (\theta_0|T_0).$$

Thus, we must have:

$$P(\theta_1 \geq x) \geq P((\theta_0|T_0) \geq x), \quad \text{for all } x \geq 0.$$

The usual approach for situations involving reliability growth of the type described above is to take $\theta_1 \geq (\theta_0|T_0)$ with probability 1, as has been done by Barlow, Bartholomew, Bremner and Brunk [2] and by Smith [5]. However, a prior chosen to satisfy the above condition does not place any mass in the region $\theta_1 < (\theta_0|T_0)$, and therefore would exclude from our analysis the possibility of an adverse effect of the changes. Hence, the requirement that

$$\theta_1 \overset{st}{\geq} (\theta_0|T_0)$$

appears to be more realistic.

Following our previous discussion, suppose that the prior distribution $G(\theta_1; \cdot)$ is also an inverted gamma with parameters α_1 and β_1. That is, $g(\theta_1; \alpha_1, \beta_1)$ is

$$(\alpha_1^{\beta_1}/\Gamma(\beta_1))\theta_1^{-(\beta_1+1)} \exp(-\alpha_1/\theta_1), \quad \text{for } \alpha_1, \beta_1, \theta_1, > 0. \tag{5}$$

A sufficient condition for ensuring that

$$\theta_1 \overset{st}{\geq} (\theta_0|T_0)$$

is to have $\alpha_1 \geq \alpha_0 + T_0$, and $\beta_1 = \beta_0 + n_0$. We can verify that under these conditions $E(\theta_1) \geq E(\theta_0|T_0)$, an inequality we would hope to achieve under the hypothesis that a change is beneficial to the system. One possibility is to let $\alpha_1 = \alpha_0 + T_0 + a_1$, where a_1 is a measure of our prior belief about the magnitude of the improvement in the mean as a result of the changes. A possible strategy for choosing a_1 is to set $E(\theta_1) = (\alpha_0 + T_0 + a_1)/(\beta_1 - 1)$ equal to our best guess of the mean time to failure θ_1. For reasons to be given in section 2.1, we would also want $a_1 \geq \alpha_0$. With α_1 thus chosen, it is important that at stage 1 we treat T_0 as being a *constant*. If this is not done, then the posterior distribution of θ_1 will have to be obtained by treating α_1 as a hyperparameter with a prior distribution which is related to the unconditional distribution of T_0. Under these circumstances the posterior distribution of θ_1 will not be an inverted gamma, and our procedure will become computationally quite involved. Furthermore, we shall also assume that $g(\theta_1; \alpha_1, \beta_1)$ is *independent* of $g(\theta_0|T_0; \cdot)$.

Having chosen $g(\theta_1; \alpha_1, \beta_1)$, we test $n_1 \geq 1$ copies of the system (under stage 1) and observe T_1, the corresponding total time on test. Since we now treat T_0 as a constant, the posterior distribution of θ_1 conditioned on T_1, $g(\theta_1|T_1; \cdot)$, is:

$$((\alpha_0 + T_0 + a_1 + T_1)^{\beta_0 + n_0 + n_1}/\Gamma(\beta_0 + n_0 + n_1))(1/\theta_1)^{\beta_0 + n_0 + n_1 - 1}$$

$$\exp(-(\alpha_0 + T_0 + a_1 + T_1)/\theta_0). \tag{6}$$

Analogous to (3), the Bayes estimator of θ_1 is:

$$E(\theta_1|T_1) = (\alpha_0 + T_0 + a_1 + T_1)/(\beta_0 + n_0 + n_1 - 1). \tag{7}$$

Under our postulate of reliability growth, we would also want to have

$$(\theta_1|T_1) \stackrel{st}{\geq} (\theta_0|T_0);$$

a *necessary* (but not sufficient) condition for the above inequality is that:

$$(\alpha_0 + T_0 + a_1 + T_1)/(\beta_0 + n_0 + n_1 - 1) \geq (\alpha_0 + T_0)/(\beta_0 + n_0 - 1), \tag{8}$$

which reduces to the requirement that:

$$(a_1 + T_1)/n_1 \geq (\alpha_0 + T_0)/(\beta_0 + n_0 - 1). \tag{9}$$

If (9) is *not satisfied*, we proceed to section 2.1 wherein we indicate a pooling procedure which gives us a desired inequality. If (9) is satisfied, then $E(\theta_1|T_1)$, the Bayes estimator of θ_1, with a $(1-\gamma)$ posterior credibility interval given by

$$I[(\theta_1|T_1)_L, (\theta_1|T_1)_U; g(\theta_1|T_1; \cdot)] = 1 - \gamma,$$

represents our evaluation of the reliability of the system at stages 0 and 1.

If the value of $E(\theta_1|T_1)$ and the credibility interval $[(\theta_1|T_1)_L, (\theta_1|T_1)_U]$ meet our desired reliability goals, then we choose not to make any further improvements to the system, and stop the testing procedure. If not, we proceed to stage 2 by making appropriate modifications to the system.

If we need to proceed to stage 2, then the prior distribution of θ_2 must be such that

$$\theta_2 \overset{st}{\geq} (\theta_1|T_1). \tag{10}$$

One way of achieving the requirement specified by (10) is to choose α_2 and β_2, the parameters of the inverted gamma prior distribution of θ_2, in such a manner that $\alpha_2 = \alpha_0 + T_0 + a_1 + T_1 + a_2$ and $\beta_2 = \beta_0 + n_1 + n_2$. Once again, a_2 reflects our belief about the magnitude of the improvement in the mean as a consequence of the changes at stage 2, and now *both* T_0 and T_1 are treated as constants. Furthermore, for reasons given in section 2.2, we shall want to have $a_2 \geq \alpha_0 + T_0 + a_1$. We shall continue with our discussion of this procedure in section 2.2

2.1. Procedures when the posterior distributions at stages 0 and 1 are not stochastically ordered

If (9) is not satisfied, i.e. if

$$(a_1 + T_1)/n_1 < (\alpha_0 + T_0)/(\beta_0 + n_0 - 1), \tag{11}$$

and if we have no prior reason to believe that the reliability growth postulate is *not* true, then we consider that the randomness of T_0 and T_1 are causes for (11), and propose the following two strategies as a means of overcoming this.

Strategy 1. We essentially ignore (11), and *for the time being* do not concern ourselves with the fact that $E(\theta_1|T_1) < E(\theta_0|T_0)$. Thus, we proceed to stage 2 by choosing the prior distribution of θ_2 in such a manner that (10) is satisfied. After completing the testing over all the stages, we will perform an *isotonic regression* of the posterior means $E(\theta_i|T_i)$, $i = 0, 1, \ldots$. This will represent our final evaluation of the reliability of the system based upon tests performed at the various stages of testing. More about this strategy will be said in section 3.

Strategy 2. By ignoring (11) and directly proceeding to stage 2 we will, through the prior distribution of θ_2, allow the effects of (11) to perpetuate over the succeeding stages. We can avoid this by pooling (the violators) T_0 and T_1 and n_0 and n_1. That is, we replace
 (a) both T_0 and T_1 by T_{01}, where $T_{01} = (T_0 + T_1)/2$, and
 (b) both n_0 and n_1 by n_{01}, where $n_{01} = (n_0 + n_1)/2$.
When we pool as indicated, we must test to see if

$$(a_1 + T_{01})/n_{01} \geq (\alpha_0 + T_{01})/(\beta_0 + n_{01} - 1), \tag{12}$$

a condition analogous to (9) except that the n_i and the T_i, $i = 0, 1$, have been replaced by their pooled values. Since $(\beta_0 - 1) > 0$, a sufficient condition for (12) is that $a_1 \geq \alpha_0$.

To summarize, we must choose $a_1 \geq \alpha_0$, and when (9) is violated, we shall pool as indicated above, and be assured that:

$$(\alpha_0 + 2T_{01} + a_1)/(\beta_0 + 2n_{01} - 1) \geq (\alpha_0 + T_{01})/(\beta_0 + n_{01} - 1), \tag{13}$$

a condition motivated by (8).

Note that since (13) can also be written as

$$(\alpha_0 + T_0 + T_1 + a_1)/(\beta_0 + n_0 + n_1 - 1) \geq (2\alpha_0 + T_0 + T_1)/(2\beta_0 + n_0 + n_1 - 2),$$

and since the left-hand side of the above equation is identical to the left-hand side of (8), the effect of pooling is to lower the magnitude of the right-hand side of (8), the posterior mean of θ_0 given T_0. Thus, the effect of pooling is to lower the posterior mean at the previous stage in the light of information obtained at the current stage, the previous stage, and the prior conviction that reliability growth has indeed taken place.

From (3) replacing T_0 by T_{01}, and n_0 and n_{01}, we see that our *revised* Bayes estimator of θ_0 is:

$$E(\theta_0 | T_0, T_1) = (2\alpha_0 + T_0 + T_1)/(2\beta_0 + n_0 + n_1 - 2), \tag{14}$$

and the *revised* $(1 - \gamma)$ posterior credibility intervals for θ_0 are given by $(\theta_0 | T_0, T_1)_L$ an $(\theta_0 | T_0, T_1)_U$, where

$$I[(\theta_0 | T_0, T_1)_L, (\theta_0 | T_0, T_1)_U; g(\theta_0 | T_0, T_1; \cdot)] = 1 - \gamma,$$

with $g(\theta_0 | T_0, T_1; \cdot)$ being an inverted gamma distribution with scale parameter $(2\alpha_0 + T_0 + T_1)/2$ and shape parameter $(2\beta_0 + n_0 + n_1)/2$. To obtain a revised estimator of θ_1, the one which incorporates the effect of pooling, we replace, in (7), T_0, T_1, n_0, and n_1 by their pooled values. We can now immediately verify that the Bayes estimator of θ_1 remains unchanges and is therefore given by (7). Consequently, the credibility interval for θ_1 is also unchanged. We can now either stop at this point or proceed to stage 2, our choice being determined by the criteria described prior to and following (10). If we choose to stop, then our assessment of the reliability of the system at stages 0 and 1 is given by (14) and (7), respectively.

We remark that instead of pooling the T_i's and the n_i's, we could have pooled the quantities T_i/n_i, $i = 0, 1$. The reason we choose the former is that the resulting expressions are much simpler to work with, and are also easier to interpret.

2.2. Procedures for analysis at stage 2

In the previous section we saw that whether (9) is satisfied or not (i.e. irrespective of pooling), the prior distribution of θ_2 could be an inverted gamma with parameters α_2 and β_2, where $\alpha_2 = \alpha_0 + T_0 + a_1 + T_1 + a_2$ and $\beta_2 = \beta_0 + n_0 + n_1$.

Having chosen $g(\theta_2; \alpha_2, \beta_2)$, we test $n_2 \geq 1$ copies of the system (under stage 2) and observe T_2, the corresponding total time on test. If $g(\theta_2; \alpha_2, \beta_2)$ is assumed to be independent of $g(\theta_1 | T_1; \cdot)$, then the posterior distribution of θ_2 given T_2 is also an inverted gamma with parameters $(\alpha_0 + T_0 + a_1 + T_1 + a_2 + T_2)$ and $(\beta_0 + n_0 + n_1 + n_2)$.

Here again, in order to be assured that

$$(\theta_2 | T_2) \stackrel{st}{\geq} (\theta_1 | T_1),$$

we need, as a necessary condition,

$$(a_2 + T_2)/n_2 \geq (\alpha_0 + T_0 + a_1 + T_1)/(\beta_0 + n_0 + n_1 - 1). \tag{15}$$

If the above condition is satisfied, then we can either stop and (because of the independence assumption) use $(\alpha_0 + T_0 + a_1 + T_1 + a_2 + T_2)/(\beta_0 + n_0 + n_1 + n_2 - 1)$ as our Bayes estimator of θ_2, with a $100(1-\gamma)$ credibility interval for θ_2 given by the two numbers $(\theta_2 | T_2)_L$ and $(\theta_2 | T_2)_U$, where

$$I[(\theta_2|T_2)_L, (\theta_2|T_2)_U; g(\theta_2|T_2; \cdot)] = 1 - \gamma,$$

or we can proceed to stage 3 by making the appropriate modifications to the system. If we proceed to stage 3, we repeat our cycle so that the prior distribution of θ_3 satisfies the condition

$$\theta_3 \stackrel{st}{\geq} (\theta_2 | T_2).$$

If condition (15) is *not satisfied*, then we can, following strategy 1 of section 2.1, either ignore the inequality (15) and proceed directly to stage 3, or follow strategy 2 and pool T_1 and T_2 to obtain $T_{12} = (T_1 + T_2)/2$ and $n_{12} = (n_1 + n_2)/2$. If we choose to pool then, analogous to (12), we must have

$$(a_2 + T_{12})/n_{12} \geq (\alpha_0 + T_0 + a_1 + T_{12})/(\beta_0 + n_0 + n_{12} - 1). \tag{16}$$

A *sufficient* condition for the above is that $a_2 \geq \alpha_0 + T_0 + a_1$. Thus, at this stage, the prior parameter a_2 has to be related to T_0 the total time on test at stage 0.

As we emphasized in section 2.1, the effect of pooling is a lowering of the posterior mean at the previous stage. In the present case, we have changed the posterior mean at stage 1 from $(\alpha_0 + T_0 + a_1 + T_1)/(\beta_0 + n_0 + n_1 - 1)$ (eq. (7)) to

$$E(\theta_1|T_0, T_1, T_2) = \frac{\alpha_0 + T_0 + a_1 + T_{12}}{\beta_0 + n_0 + n_{12} - 1} = \frac{2(\alpha_0 + T_0 + a_1) + T_1 + T_2}{2(\beta_0 + n_0) + n_1 + n_2 - 2}. \quad (17)$$

Furthermore, our credibility intervals for θ_1 will now be given by the equation:

$$I[(\theta_1|T_0, T_1, T_2)_L, (\theta_1|T_0, T_1, T_2)_U; g(\theta_1|T_0, T_1, T_2; \cdot)] = 1 - \gamma,$$

where $g(\theta_1|T_0, T_1, T_2)$ is an inverted gamma distribution with a scale parameter $(2(\alpha_0 + T_0 + a_1) + T_1 + T_2)/2$ and a shape parameter $(2(\beta_0 + n_0) + n_1 + n_2 - 2)/2$. Of course, our Bayes estimator of θ_2 remains unchanged as $(\alpha_0 + T_0 + a_1 + T_1 + a_2 + T_2)/(\beta_0 + n_0 + n_1 + n_2 - 1)$.

Since we have revised our estimator of θ_1 from that given by (7) to that given by (17), we will have to see if

$$(\theta_1|T_0, T_1, T_2) \overset{st}{\geqslant} (\theta_0|T_0), \quad (18)$$

if we *did not* have to pool at stage 1, or

$$(\theta_1|T_0, T_1, T_2) \overset{st}{\geqslant} (\theta_0|T_0, T_1), \quad (19)$$

if we had to pool at stage 1. A necessary condition for (18) is that

$$(\alpha_0 + T_0 + a_1 + T_{12})/(\beta_0 + n_0 + n_{12} - 1) \geqslant (\alpha_0 + T_0)/(\beta_0 + n_0 - 1), \quad (20)$$

which reduces to the requirement that

$$(a_1 + T_{12})/n_{12} \geqslant (\alpha_0 + T_0)/(\beta_0 + n_0 - 1). \quad (21)$$

If (21) is not violated, we proceed to stage 3. If (21) is violated, then we shall pool T_0 and T_{12} and n_0 and n_{12} to form

$$T_{0,12} = (T_0 + T_{12})/2 \quad \text{and} \quad n_{0,12} = (n_0 + n_{12})/2,$$

and replace the appropriate quantities in (20) by their pooled values. Having done this, we shall need to have

$$\frac{\alpha_0 + T_{0,12} + a_1 + T_{0,12}}{\beta_0 + n_{0,12} + n_{0,12} - 1} \geqslant \frac{\alpha_0 + T_{0,12}}{\beta_0 + n_{0,12} - 1} \quad (22)$$

which, because of $a_1 \geq \alpha_0$, is always true.

Since $T_{0,12} = (1/2)(T_0 + (1/2)(T_1 + T_2))$ and $n_{0,12} = (1/2)(n_0 + (1/2)(n_1 + n_2))$, condition (22) reduces to

$$\frac{\alpha_0 + T_0 + a_1 + T_{12}}{\beta_0 + n_0 + n_{12} - 1} \geq \frac{4\alpha_0 + 2T_0 + T_1 + T_2}{4\beta_0 + 2n_0 + n_1 + n_2 - 4}.$$

Thus, if we did not have to pool at stage 1, and if condition (21) is violated, our Bayes estimator of θ_0 conditioned on T_0, T_1, and T_2 is

$$E(\theta_0 | T_0, T_1, T_2) = (4\alpha_0 + 2T_0 + T_1 + T_2)/(4\beta_0 + 2n_0 + n_1 + n_2 - 4). \tag{23}$$

The credibility intervals for θ_0 are now given by an inverted gamma distribution with scale parameter $(4\alpha_0 + 2T_0 + T_1 + T_2)/4$ and shape parameter $(4\beta_0 + 2n_0 + n_1 + n_2)/4$. We can now either stop or proceed to stage 3.

Reverting to (19), we note that a necessary condition for satisfying this equation is that

$$(\alpha_0 + T_0 + a_1 + T_{12})/(\beta_0 + n_0 + n_{12} - 1)$$
$$\geq (2\alpha_0 + T_0 + T_1)/(2\beta_0 + n_0 + n_1 - 1). \tag{24}$$

If condition (24) is satisfied, we proceed to stage 3. Note that for $n_1 \geq n_2$ condition (24) reduces to

$$(a_1 + T_{12})/(\beta_0 + n_0 + n_{12} - 1) \geq (\alpha_0 + T_1)/(2\beta_0 + n_0 + n_1 - 1); \tag{25}$$

note that when $n_1 \geq n_2$, $2\beta_0 + n_0 + n_1 - 1 \geq \beta_0 + n_0 + n_{12} - 1$. Clearly, condition (25) is satisfied whenever $T_{12} \geq T_1$, i.e. when $T_2 > T_1$.

Thus, in view of the above arguments, whenever condition (24) is violated, i.e. whenever $n_1 < n_2$ or $T_2 < T_1$, or both, we shall replace T_1 and n_1 by their pooled values T_{12} and n_{12}. Thus, after pooling, (24) becomes

$$(\alpha_0 + T_0 + a_1 + T_{12})/(\beta_0 + n_0 + n_{12} - 1)$$
$$\geq (4\alpha_0 + 2T_0 + T_1 + T_2)/(4\beta_0 + 2n_0 + n_1 + n_2 - 2)$$

which, because $a_1 \geq \alpha_0$, is always true.

To summarize, if we had to pool at stage 1, and if condition (24) is violated, our Bayes estimator of θ_0 conditioned on T_0, T_1, an T_2 is

$$E(\theta_0 | T_0, T_1, T_2) = (4\alpha_0 + 2T_0 + T_1 + T_2)/(4\beta_0 + 2n_0 + n_1 + n_2 - 2).$$

Note that its estimator is identical to the one given by eq. (23), which was based on the fact that there was no pooling at stage 1.

However, in the present case the credibility intervals for θ_0 are given by an inverted gamma distribution with scale paramter $(4\alpha_0 + 2T_0 + T_1 + T_2)/2$ and shape parameter $(4\beta_0 + 2n_0 + n_1 + n_2)/2$. We can now either stop or proceed to stage 3.

We contrast these to the credibility intervals for θ_0 given after (23); these pertained to the case of no pooling at stage 1. We note that pooling at both stages 1 and 0 has a tendency to make the credibility intervals wider than those obtained when there is pooling at stage 0 only. *Thus, based on the above analysis, we claim that excessive pooling results in wider credibility intervals.*

Our analysis of the failure data at the succeeding stages follows along the lines mentioned above. Our evaluation of the reliability of the system at stage i, $i = 1, 2, \ldots$, is given by the appropriately conditioned Bayes estimator of θ_i and its associated credibility interval.

2.3. Some remarks on the pooling procedure

It is fairly clear that condition (7) is likely to be violated whenever (T_1/n_1) is not much larger than (T_0/n_0). Note that (T_i/n_i), $i = 0, 1$, is the (classical) maximum likelihood estimator of θ_i, $i = 0, 1$. Thus, (7) will be violated if the improvement in reliability in going from stage 0 to stage 1 is not significantly large. Hence, pooling will be necessary whenever the effect of the design changes is not substantial (or if the design changes have produced a significant deterioration).

The pooling procedure advocated here is one among several others that can be used. For example, we could have pooled the estimated mean times to failure, (T_0/n_0) and (T_1/n_1), or we could have just pooled the observed total time on test, T_0 and T_1. However we pool, the important question is whether pooling the data is a legitimate Bayesian procedure.

An orthodox Bayesian might argue that by pooling we have violated the "likelihood principle" of statistical inference. He will object on the grounds that our decision rule is not based on the information provided to us by the true posterior distribution, but is instead based on a posterior distribution which is modified to suit our hypothesis. He would recommend that instead of pooling, it would be better to choose $a_1 \gg \alpha_0$, so that condition (7) will always be satisfied, or to choose a joint prior distribution on θ_0 and θ_1 in such a manner that there is no probability mass in the region $\theta_1 < \theta_0$ (as was done by Barlow et al. [2]).

Our response to the above arguments is that not allowing any prior or posterior probability in the region $\theta_1 < \theta_0$ is too strong, and is perhaps an unreasonable requirement, and that pooling is necessitated by the randomness of the data. Thus, whenever the posterior distributions violate our requirement, i.e.,

$$(\theta_1|T_1) \stackrel{st}{\geq} (\theta_0|T_0),$$

it is preferable to pool the variables rather than to change the prior parameters in order to make $a_1 \gg \alpha_0$. As a compromise, we may want to delete the requirement that

$$\theta_1 \stackrel{st}{\geq} \theta_0$$

with respect to the posterior distributions, and just work with the requirement that

$$(\theta_1) \stackrel{st}{\geq} (\theta_0|T_0)$$

(see section 4).

Another comment about our procedure pertains to our rationale for requiring that subsequent to pooling, conditions of the type given by (13) be satisfied. Note that (12) is analogous to the necessary condition (8). However, the terms which comprise condition (13) are not the means of the true posterior distribution after pooling. For instance, after we replace T_0 by T_{01}, and n_0 by n_{01}, the mean of the posterior distribution of θ_0 conditioned on T_{01} is *not* $(\alpha_0 + T_{01})/(\beta_0 + n_{01} - 1)$, as is implied by the right-hand side of (13). The actual mean of the true posterior distribution is quite complicated, and in view of the resulting computational difficulties we choose $(\alpha_0 + T_{01})/(\beta_0 + n_{01} - 1)$ as being *analogous* to the mean of the posterior distribution of θ_0 given T_{01}. We approximate the mean of the posterior distribution of θ_1 given T_{01} in a similar manner, and thus write condition (13). Since the above approximations are motivated by the arguments which lead us to pool, we feel that they are inherently satisfactory.

We close this section by stating that in the light of the above discussions, our approach should be called a "pseudo-Bayesian approach".

3. An isotonic regression of the raw posterior means

Our strategy 1 of section 2.1 specifies that the inequality (11) and similar inequalities be ignored whenever the posterior means do not have the correct order. As a result, we will have at the end of testing over, say $(\tau + 1)$ stages, the $(\tau + 1)$ posterior means

$$E(\theta_0|T_0), E(\theta_1|T_0, T_1), \ldots, E(\theta_\tau|T_0, T_1, \ldots, T_\tau),$$

where

$$E(\theta_0|T_0) = (\alpha_0 + T_0)/(\beta_0 + n_0 - 1)$$

and

$$E(\theta_i|T_0, T_1, \ldots, T_i) = \frac{\alpha_0 + T_0 + \sum_{j=1}^{i}(a_j + T_j)}{\beta_0 + n_0 + \sum_{j=0}^{i} n_j - 1}, \quad i = 1, 2, \ldots, \tau.$$

Under our postulate of reliability growth, we would need to have (as a necessary condition)

$$E(\theta_i|T_0, \ldots, T_i) \leq E(\theta_{i+1}|T_0, \ldots, T_{i+1}), \quad i = 0, 1, \ldots, \tau - 1. \tag{26}$$

If condition (26) is satisfied, then our evaluation of the reliability growth curve is given by these posterior means, and our Bayes estimator of the reliability at stage τ is simply $E(\theta_\tau|T_0, T_1, \ldots, T_\tau)$. Note that because of the adaptive nature of our scheme, $E(\theta_\tau|T_0, \ldots, T_\tau)$ is based on the failure data over all the previous and present stages of testing, and our prior knowledge about the magnitude of the improvement over each stage.

If condition (26) is violated by any one or more of the indices i, $i = 0, 1, \ldots, \tau$, then we shall, following Barlow et al. [2], pool the adjacent violators to obtain the isotonic regression of $E(\theta_i|T_0, \ldots, T_i)$, $i = 0, \ldots, \tau$, say $E^*(\theta_i|T_0, \ldots, T_i)$. We shall use the $E^*(\theta_i|T_0, \ldots, T_i)$, $i = 0, \ldots, \tau$, as our evaluation of the reliability growth, and $E^*(\theta_\tau|T_0, \ldots, T_\tau)$ as our estimate of the reliablity at stage τ. Note that like $E(\theta_\tau|T_0, \ldots, T_\tau)$, $E^*(\theta_\tau|T_0, \ldots, T_\tau)$ is based on the failure data over all the previous stages, our prior knowledge about the magnitude of the improvements at each stage, and the postulate of reliability growth.

The remarks of section 2.3 are also appropriate for the isotonic regression estimators $E^*(\theta_i|T_0, \ldots, T_i)$, since

(a) by performing an isotonic regression of the true posterior means we have violated the likelihood principle; and

(b) the estimators $E^*(\theta_i|T_0, \ldots, T_i)$ are not the true posterior means of the θ_i, $i = 0, \ldots, \tau$, and thus are not fully Bayesian.

4. Estimation when the ordering is with respect to the priors only

In section 2 we considered the case when the mean lifetimes were stochastically ordered with respect to both the prior and the posterior distributions. In this section we delete the requirement that the means be ordered with respect to the posterior distribution. When this is done we will not have to pool the violators, nor will we have to perform an isotonic regression of the posterior means should we choose not to pool.

We start by choosing a prior distribution of θ_0, $g(\theta_0; \alpha_0, \beta_0)$, as given by (1). The posterior distribution of θ_0 conditioned on T_0, $g(\theta_0|T_0; \alpha_0, \beta_0 n_0)$, is given by (2). Assuming a squared error loss, the Bayes estimator of θ_0 is $E(\theta_0|T_0)$, and this is given by (3); the credibility intervals for θ_0 are given by (4).

We now choose a prior distribution of θ_1, $g(\theta_1; \alpha_1, \beta_1)$, in such a manner that

$$\theta_1 \overset{st}{\geqslant} (\theta_0|T_0).$$

Following the discussion of section 2, we choose $\alpha_1 = \alpha_0 + T_0 + a_1$ and $\beta_1 = \beta_0 + n_0$, where a_1 has the same interpretation as in section 2. If at stage 1 we look upon T_0 as a constant, and assume that $g(\theta_1; \alpha_1, \beta_1)$ is independent of $g(\theta_0|T_0; \alpha_0, \beta_0), n_0)$, then under the assumption of a squared error loss function the Bayes estimator of θ_1, conditioned on T_1, is $E(\theta_1|T_1)$, given by (7). The posterior distribution of θ_1 given T_1, $g(\theta_1|T_1; \cdot)$, is given by (6), and the credibility intervals for θ_1 follow in the usual manner. Note that the above statements are only true if T_0 is viewed as a constant at stage 1, and the prior distribution at stage 1 is assumed to be independent of the posterior distribution at stage 0.

Once we obtain $E(\theta_1|T_1)$ we do not care to compare it with $E(\theta_0|T_0)$, since we have not imposed any requirements on our parameters with respect to posterior distributions.

We now proceed to stage 2 by choosing our prior distribution of θ_2 in such a manner that

$$\theta_2 \overset{st}{\geqslant} (\theta_1|T_1).$$

Following our discussion in section 2, we take $g(\theta_2; \alpha_2, \beta_2)$ as our prior distribution of θ_2 with $\alpha_2 = \alpha_0 + T_0 + a_1 + T_1 + a_2$ and $\beta_2 = \beta_0 + n_0 + n_1$; here again we treat T_0 and T_1 as constants. As before, if $g(\theta_2; \alpha_2, \beta_2)$ is taken to be independent of $g(\theta_1|T_1; \cdot)$, then the mean of the posterior distribution of θ_2 conditioned on T_2 is our Bayes estimator of θ_2. We continue in this manner going from one stage to the next, obtaining at each stage the Bayes estimator of θ_i, $i = 3, \ldots, \tau$.

5. Summary and conclusions

In this paper we have considered an adaptive approach for estimating reliability growth based on prior information.

In section 2 we imposed a strong requirement on our approach by requiring that the mean times to failure at the various stages be stochastically ordered with respect to *both* the prior and the posterior distributions. The latter requirement can be satisfied if we pool the violators; however, pooling results in a violation of the likelihood principle and creates other computational difficulties. Even though the computational difficulties can be avoided by using some approximations (see section 2.3), the pooling makes our procedure not fully Bayesian. Thus, what we present in section 2 can best be described as a *pseudo-Bayesian scheme* for estimating reliability growth. A formal investigation of the properties of our scheme, despite the fact that it is not fully Bayesian, is a matter which needs further attention. Our scheme, however, does produce results which are reasonable and intuitively satisfying.

Our review of the literature in Bayesian statistics indicates that there is no discussion or *even mention* of the problem of estimating parameters which are stochastically ordered. As mentioned before, our strategy of pooling the violators to obtain the stochastic order may be unacceptable to a Bayesian. We therefore hope that this paper can stimulate some basic research into this general problem area.

In view of the difficulties mentioned above, in section 4 we weaken the specifications on our approach by deleting the requirement that the parameters by stochastically ordered with respect to the posterior distributions. This simplification obviates the need for pooling the violators, and thus would make our procedure fully Bayesian and therefore optimal in the usual sense of minimizing the square error loss function. However, the adaptive nature of our problem imposes certain computational difficulties. We circumvent these by treating the observed statistic T_i as constant at stage $(i+1)$, $i = 0, 1, 2, \ldots, \tau - 1$, and by assuming the prior distribution at stage j to be independent of the posterior distribution at stage $(j-1)$, $j = 1, 2, \ldots, \tau$. Then, within the context of the above assumptions, the procedure of section 4 is fully Bayesian.

Future work

There are several other aspects of the reliability growth problem that we plan to address in subsequent work. These are as follows.

(i) An evaluation of the gain in information obtained by considering an adaptive scheme wherein previous data obtained at stages $0, 1, \ldots, i-1$, is used in the estimation at stage i, versus a nonadaptive scheme wherein only the data at stage i is used. It is conceivable that an adaptive procedure will be advantageous whenever the improvement in reliability from one stage to another is small, whereas if there is a drastic change in reliability at a particular stage, then the data from the previous stages will tend to diminish its true effect.

(ii) A cost–benefit analysis of the reliability growth procedure. That is, we would like to evaluate the trade-off between the costs incurred in improving the reliability at stage i versus the actual improvement in reliability at stage $(i+1)$, say $(\theta_{i+1} - \theta_i) = \Delta\theta_i$, $i = 0, 1, \ldots, \tau$. It is conceivable that the $\Delta\theta_i$ will be decreasing in i (there is only so much that one can do to improve a system), whereas the C_i will be either constant or increasing in i. What we need is a stopping rule which tells us when to stop performing the improvements on the system and put the system into operation, based on the costs C_i and our best estimate of $\Delta\theta_i$.

Acknowledgements

The author gratefully acknowledges several helpful conversations with Professor Gerald J. Lieberman of Stanford University and Professor William Jewell of the University of California, Berkeley. This work was performed while the author was a Visiting Professor of Statistics at Stanford University during the academic year 1978–79, and a visitor at the University of California at Berkeley during Summer 1978. It has been partially supported by the U.S. Army Research Office under Grant DAAG 29-77-G-0031 with Stanford University, and by the Air Force Office of Scientific Research under Grant AFOSR-77-3179 with the University of California at Berkeley.

References

[1] H. Balaban, "Reliability Growth Models", *Environmental Sci.* (1978) 11–18.
[2] R.E. Barlow, D.J. Bartholomew, J.M. Bremner and H.D. Brunk, *Statistical Inference Under Order Restrictions* (Wiley, New York, 1972).
[3] W.S. Jewell, "Stochastically Ordered Parameters in Bayesian Prediction", Technical Report ORC 79-12, Operations Research Center, Univ. of California, Berkeley, California (1979).
[4] N.R. Mann, R.E. Schafer and N.D. Singpurwalla, *Methods for Statistical Analysis of Reliability and Life Data* (Wiley, New York, 1974).
[5] A.F.M. Smith, "A Bayesian Note on Reliability Growth During a Development Testing Program", *IEEE Trans. Reliability* R-26 (1977) 346–347.
[6] M.C. Weinrich and A. Gross, "The Barlow–Scheuer Reliability Growth Model from a Bayesian Point of View", *Technometric* 20 (1978) 249–254.

EXTREME POINTS OF THE CLASS OF DISCRETE DECREASING FAILURE RATE AVERAGE LIFE DISTRIBUTIONS *

Naftali A. LANGBERG **, Ramón V. LEÓN †, James LYNCH ‡
and Frank PROSCHAN §

The Florida State University

We show that the class of discrete decreasing failure rate average (discrete DFRA) life distributions is a convex set. We then obtain the extreme points of this class. Finally, we show how to represent any discrete DFRA life distribution as a mixture of these extreme points.

1. Introduction and summary

Distributions with decreasing failure rate (DFR) occur frequently in reliability theory and application. See, for example, Barlow and Proschan [1] and Proschan [5]. Less common is the use of distributions with decreasing failure rate average (DFRA). However, it is easy to formulate models and find real-life examples of DFRA distributions. We list a few, since the literature seems devoid of such cases.

(1) The life length of a retail outfit merchandizing seasonal goods may be DFRA. The failure rate tends to decrease with increased experience, growing capital, public recognition of the product or company name, and other factors monotonic with age. However, the seasonal factor prevents the failure rate from being monotonically decreasing, so that only the *average* failure rate is decreasing.

(2) A device operates 16 hours a day, say. On the ith day the failure rate during operation is λ_i and is 0 during the 8 hours of "rest". Suppose $\lambda_1 > \lambda_2 > \lambda_3 > \ldots$, but, in addition, are such that the cumulative failure rate

---

Received May 12, 1980; revised March 25, 1981.
  * Research supported by the U.S. Army Research Office, Durham, under grant no. DAAGG-24-79-C-0158.
  ** At the Haifa University, Israel.
  † Now at Rutgers University.
  ‡ At Pennsylvania State University, University Park.
  § Research supported by the Air Force Office of Scientific Research, AFSC, USAF, under grant AFOSR-78-3678.

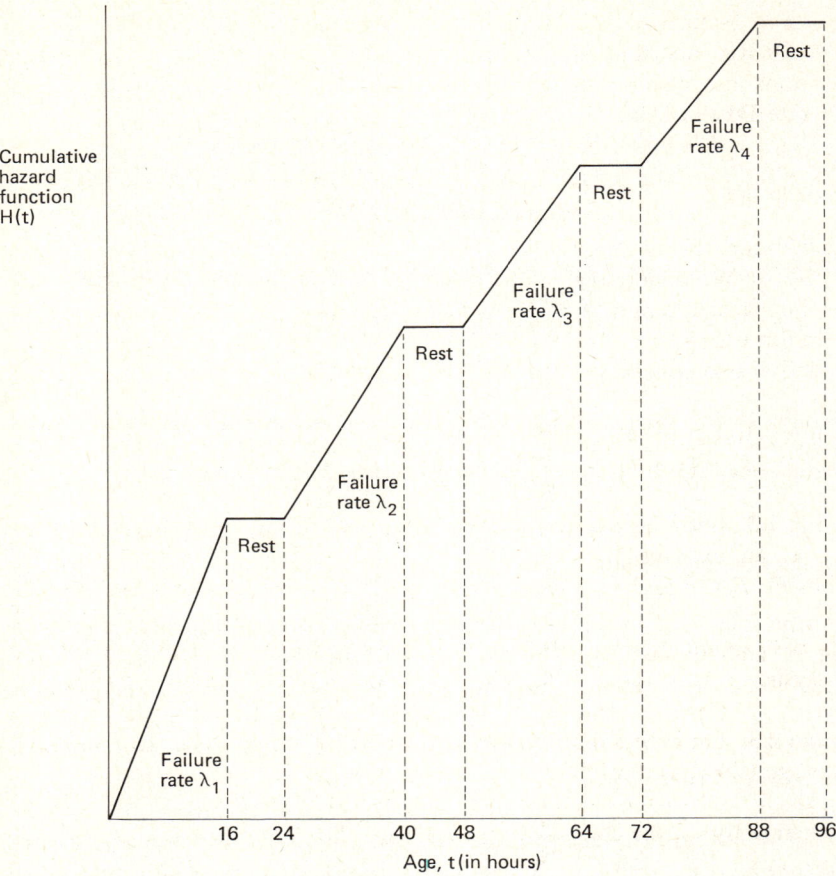

Fig. 1. Cumulative hazard function of a DFRA distribution.

$H(t)$ over time (including rest periods) satisfies: $(1/t)H(t)$ is decreasing. The graph in fig. 1 displays the situation.

Note that $(1/t)H(t)$ is decreasing, but $H(t)$ is not concave. Thus, the underlying distribution is DFRA but not DFR.

Other practical examples may be listed in the continuous case. In a similar fashion, DFRA life distributions occur in the discrete case, in which age is measured by the number of cycles that have occurred since the unit was initially put into operation.

In this paper we obtain the extreme points of the convex class of discrete DFRA distributions. In addition, we show constructively how to represent any discrete DFRA distribution as a convex combination of extreme points.

As is well known in optimization theory, from a knowledge of the extreme points it may be possible to obtain maxima or minima of certain functionals of discrete DFRA distributions. Bounds and inequalities for discrete DFRA distributions may also be derivable from a knowledge of the extreme points of the discrete DFRA class.

2. Preliminaries

A distribution F is a *discrete life distribution* if its support is contained in the set $\{0, 1, \ldots\}$. We denote the corresponding survival function $1 - F$ by $\bar{F}$. We define the hazard function $H(F, x)$ of a discrete life distribution as $-\ln \bar{F}(k-1)$ for $x \in [k, k+1]$ and $k = 0, 1, \ldots$.

We define two concepts used in the sequel.

Definition 2.1. Let $\mathcal{G}$ be a class of distribution functions. Then $\mathcal{G}$ is a *convex class* if $F \equiv \theta F_1 + (1 - \theta) F_2 \in \mathcal{G}$ whenever $F_1, F_2 \in \mathcal{G}$ and $\theta \in [0, 1]$.

Definition 2.2. Let $\mathcal{G}$ be a convex class of distribution functions, and let $F \in \mathcal{G}$. Then F is an *extreme point* of $\mathcal{G}$ if there are no two distinct distribution functions $F_1, F_2 \in \mathcal{G}$ and a real number $\theta \in (0, 1)$ such that $F \equiv \theta F_1 + (1 - \theta) F_2$.

Next we present the class of discrete life distributions that is the subject of our analysis.

Definition 2.3. Let F be a discrete life distribution. Then F is *decreasing failure rate average* (DFRA) if $x^{-1} H(F, x)$ is nonincreasing for $x \in (0, \infty)$.

We denote by $\mathcal{G}_D$ the class of discrete DFRA life distributions. Throughout we define $(\bar{F}(-1))^{0^{-1}} \equiv 0$.

3. The extreme points of the discrete DFRA class

In this section we identify the extreme points of the class of discrete DFRA life distributions. The function $g(\theta, x, y) = -\ln\{\theta e^{-x} + (1 - \theta) e^{-y}\}$, $\theta \in [0, 1]$, $x, y \in [0, \infty)$, plays a key role in our analysis. First, we prove three properties of $g(\theta, x, y)$ used in the sequel

Lemma 3.1. Let $\theta \in (0, 1)$. Then (i) $g(\theta, x, y)$ is strictly increasing in x and y; (ii) for $\alpha \in [0, 1]$, and $x, y \in [0, \infty)$, $g(\theta, \alpha x, \alpha y) \geq \alpha g(\theta, x, y)$; and (iii) for $\alpha \in (0, 1)$ and $x, y \in [0, \infty)$, $g(\theta, \alpha x, \alpha y) = \alpha g(\theta, x, y)$ iff $x = y$.

Proof. (i) Follows in a straightforward way. (ii) and (iii). Define the random variable Z as follows:

$$Z = \begin{cases} e^{-x}, & \text{with probability } \theta, \\ e^{-y}, & \text{with probability } 1 - \theta. \end{cases}$$

Then $g(\theta, x, y) = -\ln E[Z]$, and $g(\theta, \alpha x, \alpha y) = -\ln E[Z^\alpha]$. Consequently, (ii) and (iii) follow by the Liapounov inequality (Chung [2, p. 47]). □

Next we show that the class $\mathcal{G}_D$ is convex.

Lemma 3.2. *The class of discrete DFRA life distributions is convex.*

Proof. Let $F \equiv \theta F_1 + (1 - \theta) F_2$, where $F_1, F_2 \in \mathcal{G}_D$ and $\theta \in (0, 1)$. We show that $F \in \mathcal{G}_D$.

Let $k \in \{1, 2, \ldots\}$. Then $H(F, k) = g\{\theta, H(F_1, k), H(F_2, k)\}$. By lemma 3.1(i):

$$H(F, k) \geq g\{\theta, k(k+1)^{-1} H(F_1, k+1), k(k+1)^{-1} H(F_2, k+1)\}.$$

By lemma 3.1(ii):

$$g\{\theta, k(k+1)^{-1} H(F_1, k+1), k(k+1)^{-1} H(F_2, k+2)\}$$

$$\geq k(k+1)^{-1} g\{\theta, H(F_1, k+1), H(F_2, k+1)\} = k(k+1)^{-1} H(F, k+1).$$

Consequently the desired result follows. □

Define $\Delta(F, 0) = -\infty$, an $\Delta(F, k) = (k+1)^{-1} H(F, k+1) - k^{-1} H(F, k)$, $k = 1, 2, \ldots$. To accomplish the objective of this section we need the following lemma.

Lemma 3.3. *Let $F \equiv \theta F_1 + (1 - \theta) F_2$, where $F_1, F_2 \in \mathcal{G}_D$ and $\theta \in (0, 1)$. Assume $\Delta(F, k) = 0$ for some positive integer k. Then $H(F_1, k) = H(F_2, k) = H(F, k)$, and $H(F_1, k+1) = H(F_2, k+1) = H(F, k+1)$.*

Proof. First, by lemma 3.1(i) and (ii):

$$k^{-1} H(F, k) = k^{-1} g\{\theta, H(F_1, k), H(F_2, k)\}$$

$$\geq k^{-1} g\{\theta, k(k+1)^{-1} H(F_1, k+1), k(k+1)^{-1} H(F_2, k+1)\}$$

$$\geq (k+1)^{-1} g\{\theta, H(F_1, k+1), H(F_2, k+1)\}$$

$$= (k+1)^{-1} H(F, k+1).$$

Since $\Delta(F, k) = 0$, the extreme values in the preceding chain of inequalities are equal. Thus, by lemma 3.1(i), $\Delta(F_j, k) = 0, j = 1, 2$, and by lemma 3.1(iii), $H(F_1, k+1) = H(F_2, k+1)$. Consequently, the desired results follow. □

To describe the extreme points of $\mathcal{G}_D$ we need the following definition and notation.

Definition 3.4. Let F be a discrete life distribution and $k_1 < k_2$ be two integers in the support of F. Then k_1 and k_2 are *successive support points* if no integer in the interval (k_1, k_2) belongs to the support of F.

Let $\mathcal{G}_{D,e} = \{F: F \in \mathcal{G}_D$, and for every two successive support points of F, k_1, $k_2, \Delta(F, k_1) = 0$ or $\Delta(F, k_2) = 0\}$.

We are ready now to identify the extreme points of $\mathcal{G}_D$.

Theorem 3.5. $\mathcal{G}_{D,e}$ is the class of all extreme points in the class of discrete DFRA life distributions.

Proof. First, we show that all the life distributions in $\mathcal{G}_{D,e}$ are extreme points. Let $F \equiv \theta F + (1 - \theta)F_2$, where $F_1, F_2 \in \mathcal{G}_D$, $F \in \mathcal{G}_{D,e}$, and $\theta \in (0, 1)$. We show that $F \equiv F_2$.

Let $d = \sup\{q: H(F_1, j) = H(F_2, j), j = 0, \ldots, q\}$. To prove that $F_1 \equiv F_2$ it suffices to show that $d = \infty$. Assume $d < \infty$. Then there is a positive integer k such that $d < k$, and d, k are two successive support points of F. By the definition of d and lemma 3.3, $\Delta(F, d) < 0$. Thus, $\Delta(F, k) = 0$. Consequently, by lemma 3.3, $d \geq k + 1 > d$, a contradiction. Hence, $d = \infty$ and F is an extreme point in the class $\mathcal{G}_D$.

To complete the proof of the theorem, we show that a discrete DFRA life distribution that does not belong to $\mathcal{G}_{D,e}$ is not an extreme point in $\mathcal{G}_D$. To show that a discrete DFRA life distribution is not an extreme point, it suffices to prove that the life distribution can be written as a proper convex combination of two distinct discrete DFRA life distributions. Let $F \in \mathcal{G}_D$ such that $F \notin \mathcal{G}_{D,e}$. We show that F is not an extreme point.

Let k_1 and k_2 be two successive support points of F, such that $k_2 > k_1 \geq 0$, and $\Delta(F, k_j) < 0, j = 1, 2$. Let $\bar{F}_1(k) = \bar{F}_2(k) = \bar{F}(k)$ for $k \notin (k_1 - 1, k_2 - 1]$, $\bar{F}_1(k) = \max\{[\bar{F}(k_1 - 1)]^{1+k_1^{-1}}, \bar{F}(k_2)\}$, $\bar{F}_2(k) = \min\{[\bar{F}(k_2)]^{k_2/(k_2+1)}, \bar{F}(k_1 - 1)\}$ for $k \in (k_1 - 1, k_2 - 1]$ and let $\theta = [\bar{F}(k_1) - \bar{F}_1(k_1)][\bar{F}_2(k_1) - \bar{F}_1(k_1)]^{-1}$. For the sake of clarity we recall that $[\bar{F}(-1)]^{0^{-1}}$ is defined as zero. Finally, observe that F_1 and F_2 are two distinct DFRA life distributions, that $\theta \in (0, 1)$, and that $F \equiv \theta F_1 + (1 - \theta)F_2$. Consequently, F is not an extreme point in $\mathcal{G}_D$. □

4. Representation of a discrete DFRA life distribution as a mixture of extreme points

In this section we prove that every discrete DFRA life distribution can be presented as a mixture of the extreme points of the discrete DFRA class. More explicitly, for any discrete DFRA life distribution F, we first construct a probability space $(\Omega_F, \mathcal{B}_F, P_F)$. We then define for each $\omega \in \Omega_F$ a discrete life distribution $G_F(\cdot, \omega)$ that belongs to the class of the discrete DFRA extreme points. Finally, we prove that:

$$F(k) = \int_{\Omega_F} G(k, \omega) \, dP_F(\omega), \quad k = 0, 1, \ldots . \tag{1}$$

Let $F \in \mathcal{G}_{D,e}$. Define $\Omega_F = \{1\}$, $\mathcal{B}_F = \{\phi, \{1\}\}$, $P_F\{1\} = 1$, and $G(\cdot, \{1\}) = F(\cdot)$. Then clearly

$$F(k) = \int_{\Omega_F} G(k, \omega) \, dP_F(\omega), \quad k = 0, 1, \ldots .$$

Thus, to prove statement (1) it suffices to consider discrete DFRA life distributions that are not extreme points.

Let F belong to $\mathcal{G}_D$ but not to $\mathcal{G}_{D,e}$. Next, we construct the probability space $(\Omega_F, \mathcal{B}_F, P_F)$, define the extreme discrete DFRA life distributions $G(\cdot, \omega)$, $\omega \in \Omega_F$, and prove statement (1). Let m be the number (possibly infinite) of all pairs of non-negative integers (k_q, k_{q+1}), $1 \leq q < 2m+1$, such that:

$$k_q, \; 1 \leq q < 2m+1 \text{ is a strictly increasing sequence,} \tag{2}$$

$$\Delta(F, k_q) < 0, \quad 1 \leq q < 2m+1, \tag{3}$$

and

$$K_{2q-1}, k_2 \text{ are successive support points of } F \text{ for } 1 \leq q < m+1. \tag{4}$$

Furthermore, let $\mathcal{I}_{2q} = \{k: k \in (k_{2q-1} - 1, k_{2q} - 1]\}$, $1 \leq q < m+1$:

$$\bar{F}_1(k) = \begin{cases} \max\left\{\left[\bar{F}(k_{2q-1} - 1)\right]^{1+k_{2q}^{-1}}, \bar{F}(k_{2q})\right\}, & k \in \mathcal{I}_2, \; 1 \leq q < m+1, \\ \bar{F}(k), & k \notin \mathcal{I}_{2q}, \; 1 \leq q < m+1, \end{cases} \tag{5}$$

$$\bar{F}_2(k) = \begin{cases} \min\{[\bar{F}(k_{2q})]^{k_2/(k_{2q}+1)}, \bar{F}(k_{2q-1}-1)\}, & k \in \mathcal{I}_{2q}, 1 \leq q < m+1, \\ \bar{F}(k), & k \notin \mathcal{I}_{2q}, 1 \leq q < m+1, \end{cases}$$

and (6)

$$\theta_{2q} = [\bar{F}(k_{2q-1}) - \bar{F}_1(k_{2q-1})][\bar{F}_2(k_{2q-1})]^{-1}, \quad 1 \leq q < m+1. \tag{7}$$

For the sake of clarity, note that $F_1, F_2 \in \mathcal{G}_{D,e}$, that $\theta_{2q} \in (0, 1)$, $1 \leq q < m+1$, and that for $k \in \mathcal{I}_{2q}$, $1 \leq q < m+1$:

$$\bar{F}(k) = \theta_{2q}\bar{F}_2(k) + (1 - \theta_{2q})\bar{F}_1(k). \tag{8}$$

We now construct the probability space $(\Omega_F, \mathcal{B}_F, P_F)$. Let

$$\Omega_F = \{\omega: \omega = (\omega_{2q}, 1 \leq q < m+1), \omega_{2q} \in \{0, 1\}, 1 \leq q < m+1\},$$

$\mathcal{B}_F =$ the set of all subsets of Ω_F,

and let

$$P_F\{\omega: \omega_{2q} = \delta_q, q = 1, \ldots, k\} = \prod_{q=1}^{k} \theta_{2q}^{\delta_q}(1 - \theta_{2q})^{1-\delta_q},$$

where $1 \leq k < m+1$, and $\delta_1, \ldots, \delta_k \in \{0, 1\}$. For the case $m = \infty$, we extend P_F to all subsets of Ω_F by the caratheodory extension theorem (Halmos [3, p. 54]).

Next, we define for each $\omega \in \Omega_F$, a discrete DFRA life distribution that belongs to $\mathcal{G}_{D,e}$:

$$\bar{G}(k, \omega) = \begin{cases} \bar{F}(k), & k \notin \mathcal{I}_{2q}, & 1 \leq q < m+1, \\ \bar{F}_2(k), & k \in \mathcal{I}_{2q}, \omega_{2q} = 1, & 1 \leq q < m+1, \\ \bar{F}_1(k), & k \in \mathcal{I}_{2q}, \omega_{2q} = 0, & 1 \leq q < m+1. \end{cases} \tag{9}$$

Note that for each $\omega \in \Omega_F$, $\bar{G}(\cdot, \omega) \in \mathcal{G}_{D,e}$.

Finally, we prove statement (1).

Theorem 4.1. Let F be a discrete DFRA life distribution that does not belong to $\mathcal{G}_{D,e}$. Then for $k = 0, 1, \ldots$,

$$F(k) = \int_{\Omega_F} G(k, \omega) \, dP_F(\omega).$$

Proof. Let $k \in \{0, 1, \dots\}$, and let I denote the indicator function. Then:

$$\int_{\Omega_F} G(k, \omega) \, dP_F(\omega) = \bar{F}(k) I\left(k \notin \bigcup_{q=1}^{m} \mathcal{I}_q\right)$$

$$+ \sum_{q=1}^{m} I(k \in \mathcal{I}_{2q}) \{\theta_{2q} \bar{F}_2(k) + (1 - \theta_{2q}) \bar{F}_1(k)\}.$$

Consequently, the desired result follows by (8). □

Finally we note that, as is frequently the case, the representation is not unique.

References

[1] R.E. Barlow and F. Proschan, *Statistical Theory of Reliability and Life Testing: Probability Models* (Holt, Rinehart, and Winston, New York, 1975).
[2] K.L. Chung, *A Course in Probability Theory* (Academic Press, New York, 1974).
[3] P.R. Halmos, Measure Theory (D. Van Nostrand Co., Inc., New York, 1965).
[4] N.A. Langberg, R.V. León, J. Lynch and F. Proschan, "Extreme Points of the Class of Discrete Decreasing Failure Rate Life Distributions", *Mathematics of Operations Research* 5 (1980) 35–42.
[5] F. Proschan, "Theoretical Explanation of Observed Decreasing Failure Rate", *Technometrics* 5 (1963) 375–383.

A BICRITERION MODEL FOR ACCEPTANCE SAMPLING

Herbert MOSKOWITZ *, A. RAVINDRAN [†], Gary KLEIN* and
P.K. ESWARAN [†]

Purdue University

A bicriterion utility model for acceptance sampling in quality control is developed, which incorporates a decision-maker's preferences in selecting an optimal plan. Such a model allows explicit consideration of conflicting goals that are currently not considered in acceptance sampling schemes. Two optimization procedures for solving the bicriterion model are developed and illustrated. The first requires the construction of a bicriterion utility function which is optimized by implicit enumeration. The second procedure employs an interactive method which circumvents the problem of constructing an explicit bicriterion utility function. The interactive method employs paired comparisons, poses less cognitive burden and converges more rapidly to an optimal solution than other interactive approaches.

Simulation experiments were run to investigate the robustness of the optimal sampling plan and response surface to the following factors: (1) the convexity/concavity and functional forms of the single criterion utility functions; (2) the scaling constants of the bicriterion utility function associated with each criterion; and (3) the functional form of the bicriterion utility function. The optimal sampling plan was sensitive only to the scaling constants and, to a lesser degree, to the convexity of the single criterion utility functions. Local maxima problems were minimal, indicating a reasonably well-behaved response surface. Existing nonlinear programming codes as well as interactive methods may thus be applied appropriately to determine an optimal sampling plan.

1. Introduction

An interesting and challenging area of research in acceptance sampling concerns the development of efficient computational schemes for determining optimal acceptance sampling plans. This focus becomes increasingly significant as the models developed become more complex and realistic. A great deal of recent attention has been aimed at cost-based models [1,17,18], which incorporate such additional factors as inspection error [4,13], multiple attributes (e.g. multiple type defects) [3,9,16], and multiple criteria [11]. Another additional factor is a decision-maker's (DM's) preferences and risk attitude.

In this paper we consider multiple criteria and a DM's preferences and risk attitude in focusing on development and optimization of a bicriterion model

---

Received October 29, 1980; revised March 17, 1981.

* Krannert Graduate School of Management.
[†] School of Industrial Engineering.

for determining an optimal sample size n^* and acceptance number c^* for single sampling acceptance plans. The model incorporates the two commonly employed criteria of average lot inspection cost ($ALIC$) and average outgoing quality (AOQ), as well as a DM's preferences for these criteria, into a bicriterion utility function. The model departs significantly from previous approaches used in determining optimal sampling plans in two respects: (1) it includes risk preference, i.e. a utility function is explicitly incorporated into a choice of an acceptance plan; and (2) acceptance sampling is treated as a bicriterion problem. Risk preference and multiple objectives are important considerations in decision-making in general, and should be embedded in acceptance sampling models in quality control. Moreover, the conscious consideration of these factors has the advantage of involving the decision maker in model formulation and, hence, the acceptability and implementation of its results.

The use of a bicriterion utility model, however, poses some potential measurement and optimization difficulties which must be addressed, in order to determine the model's implementability. One obvious difficulty concerns the nature of the response surface, i.e. whether a global optimum plan can be found. A more practical difficulty concerns the choice of the multiobjective optimization scheme. At least two general approaches can be taken in obtaining (n^*, c^*). One is to measure a DM's bicriterion utility function a priori, and then optimize the utility function on n and c using standard nonlinear programming (NLP) codes or implicit enumeration methods. A second approach is to employ interactive methods, whereby a DM, in a symbiotic relationship with the computer, progressively articulates his preferences by making tradeoffs, to select an optimal acceptance plan. Such a procedure circumvents the difficult and onerous task of explicitly measuring a DM's multicriteria utility function, thereby avoiding the difficulties and limitations inherent in the construction of a real valued utility function.

To reconcile and resolve the potential measurement and optimization problems, the following issues were addressed.

(1) How robust are the sampling plans generated by the bicriterion model with respect to the nature and type of single criterion and bicriterion utility functions employed?

(2) How well-behaved is the response surface generated by the bicriterion model associated with various types of single criterion and bicriterion utility functions that could be employed? That is, is the bicriterion utility function unimodal or do there exist local optima, which may limit the use of standard nonlinear optimization methods for such functions?

(3) How feasible and effective are the interactive methods (vis-à-vis the more conventional approach of optimizing the decision variables on an expected utility function), which do not require the prior determination of a DM's utility function, in solving the bicriterion optimization problem?

In responding to these issues, we shall proceed as follows. In the next section we present a bicriterion single acceptance sampling model and discuss various common types of bicriterion and single-criterion utility functions that were employed in evaluating the model. In sections 3 and 4 we develop and apply both implicit enumeration and interactive optimization algorithms to the bicriterion model to demonstrate the distinctions in the two approaches. Then in section 5 we discuss a series of simulation experiments used to examine the robustness of the choice of sampling plan and the nature of response surface of the bicriterion model to the utility functions employed. Finally, in section 6 we present our conclusions and outline several areas of future research.

2. Bicriterion model

Two desirable properties of any acceptance sampling scheme are low cost and high outgoing quality. A suitable measure of each provides a reasonable criterion for selecting a sampling plan, and have been used accordingly. AOQ provides a common measure of outgoing lot quality. Implicit in the traditional AOQ measure are costs due to sending defective items to the customer or on to further internal processing. $ALIC$ includes the cost of inspecting the entire sample as well as incorporating the expected cost of inspecting the entire lot upon rejection. The use of AOQ and $ALIC$ as joint criteria thus simultaneously incorporates the desirable characteristics of each criterion in selecting a sampling plan when integrated into a bicriterion model. Assuming identical lots and no inspection error, each of the above criteria is defined as follows [5,10]:

$$AOQ(n,c) = (N-n)P_A P_D / N \tag{1}$$

and

$$ALIC(n,c) = nC_I + (N-n)C_I(1-P_A), \tag{2}$$

where c is the acceptance number, n is the sample size, P_D is the true lot fraction defective, N is the lot size, P_A is the probability of acceptance (a function of n and c as given below), and C_I is the unit cost associated with inspection. $C_I = C_S + P_D C_R$, where C_S is the sampling cost and C_R the replacement cost per defective unit. Assuming random sampling without replacement, P_A can be computed using the hypergeometric probability function:

$$P_A = \sum_{x=0}^{c} h(x;n,a,b) = \sum_{x=0}^{c} \frac{\binom{a}{x}\binom{b}{n-x}}{\binom{a+b}{n}}, \tag{3}$$

where a and b are the number of defective and nondefective items in the lot, respectively, such that $a + b = N$ and $a/(a+b) = P_D$.

We choose (n^*, c^*) by maximizing

$$U\{AOQ(n,c); ALIC(n,c)\} = f\{u_1(AOQ(n,c)); u_2(ALIC(n,c))\}, \quad (4)$$

where U denotes the bicriterion utility function, and u_1 and u_2 are the individual utility functions for each criterion. The various utility functions considered in our model, and examined in our simulation experiments, are discussed below.

2.1. Bicriterion utility functions

The two most common types of bicriterion utility functions employed which are applicable to the bicriterion model are the multilinear and additive forms [7,8]:

(1) *multilinear*

$$U(x_1, x_2) = k_1 u_1(x_1) + k_2 u_2(x_2) + kk_1 k_2 u_1(x_1) u_2(x_2); \quad (5)$$

(2) *additive* (a special case of the multilinear form)

$$U(x_1, x_2) = k_1 u_1(x_1) + k_2 u_2(x_2), \quad (6)$$

where k, k_1 and k_2 are scaling constants, and the $u_i(x_i)$ are individual utility functions for each criterion. The above bicriterion utility functions can be constructed by having a DM assess two single criterion utility functions and two scaling constants. The multilinear and additive forms require that the two criteria be *mutual utility independent* (MUI). Moreover, the additive form also implies that the criteria are *additive independent* (AI). Procedures for testing for MUI and AI, measuring single criterion utility functions, and for determining the scaling constants are described in Keeney and Raiffa [8,Ch. 5].

2.2. Single criterion utility functions

The single criterion utility functions examined are shown in table 1. These functions provide a broad spectrum of shapes, and range from highly risk averse (concave functions) to highly risk seeking (convex functions). All are monotonic in the scaled range from zero to one. The "mild" exponential utility function was included as a contrast to the extreme parameters of the "strong" exponential. $ALIC$ was divided by NC_1 to normalize its scale to $(0, 1)$, while AOQ fell naturally between these values.

Table 1
Single-criterion utility functions investigated.

Type	Functional form [a]
Quadratic	
Concave	$1 - AOQ^2$
Convex	$1 - 2 \cdot AOQ + AOQ^2$
"Strong" exponential	
Concave	$1 - e^{-5(1-AOQ)}$
Convex	$e^{-5(AOQ)}$
"Mild" exponential	
Concave	$1 - (e^{2 \cdot AOQ} - 1)/(e^2 - 1)$
Convex	$1 - (e^{-2 \cdot AOQ} - 1)/(e^{-2} - 1)$
Logarithmic	
Concave	$2 - 2/(\log(10 - 9 \cdot AOQ) + 1)$
Convex	$\log(10 - 9 \cdot AOQ)$
Linear	$1 - AOQ$

[a] Single-criterion functions for $ALIC$ are identical (e.g. for the linear form $U(ALIC) = 1 - ALIC$).

3. Optimization via prior preference articulation: Implicit enumeration

Since the nature of the objective function of the bicriterion model can assume a variety of forms, the decision variables are integer-valued, and constraints such as non-negativity and $c \leq n$ are inherent to the model, gradient techniques are inappropriate. Moreover, complete enumeration is often inefficient. A general implicit enumeration algorithm was thus developed which could evaluate all c for a given n, over all $n \leq N$.

NP_D is the number of defectives in the lot. Hence for any sample size n,

$$c_{\max} = b = \min(n, NP_D), \quad \text{for } n \leq N. \tag{7}$$

When $n > N - NP_D$, then there would be defectives in the sample of at least $n - (N - NP_D)$, so

$$c_{\min} = a = \max[0, n - N(1 - P_D)], \quad \text{for } n \leq N. \tag{8}$$

Hence, for a given n, there are limits, a and b, within which we search for optimal c. Another simple bound, based on the following theorem, is employed to further reduce the number of c values that need be evaluated for a given n.

Theorem 1.
    Let u_1, u_2 and U be functions with the following three conditions:
    (1) $u_1(n, c) \geq 0$ increases as c increases, keeping n fixed;

(2) $u_2(n,c) \geq 0$ decreases as c increases, keeping n fixed; and
(3) $U(u_1, u_2) \geq 0$ increases as u_1 or u_2 or both u_1 and u_2 increase.

Let $(\bar{n}, \bar{c})$ be a specific (n, c) pair, and $(\bar{a}, \bar{b})$ be upper and lower bounds on c for $n = \bar{n}$ so that $\bar{a} \leq \bar{c} \leq \bar{b}$. Finally, let $U_0 = U[u_1(n_0, c_0), u_2(n_0, c_0)]$, where U_0 is the current estimate of the maximum value of total utility U.

(a) If $U[u_1(\bar{n}, \bar{c}), u_2(\bar{n}, \bar{a})] \leq U_0$, then $U_0 \geq U[u_1(\bar{n}, c), u_2(\bar{n}, c)]$ for $\bar{c} \leq c \leq \bar{b}$.
(b) If $U[u_1(\bar{n}, \bar{b}), u_2(\bar{n}, \bar{c})] \leq U_0$, then $U_0 \geq U(u_1(\bar{n}, c), u_2(\bar{n}, c)]$ for $\bar{a} \leq c \leq \bar{c}$.

Proof
(a) Monotonic properties indicate that: $u_1(\bar{n}, \bar{c}) \geq u_1(\bar{n}, c)$ for $\bar{a} \leq c \leq \bar{c}$ and $u_2(\bar{n}, \bar{a}) \geq u_2(\bar{n}, c)$ for $\bar{a} \leq c \leq \bar{b}$.
Hence, $U_0 \geq U[u_1(\bar{n}, \bar{c}), u_2(\bar{n}, \bar{a})] \geq U[u_1(\bar{n}, c), u_2(\bar{n}, c)]$ for $\bar{a} \leq c \leq \bar{c}$.

Proof
(b) Monotonic properties indicate that:
$u_1(\bar{n}, \bar{b}) \geq u_1(\bar{n}, c)$ for $\bar{a} \leq c \leq \bar{b}$ and $u_2(\bar{n}, \bar{c}) \geq u_2(\bar{n}, c)$ for $\bar{c} \leq c \leq \bar{b}$.
Hence, $U_0 \geq U[u_1(\bar{n}, \bar{b}), u_2(\bar{n}, \bar{c})] \geq U[u_1(\bar{n}, c), u_2(\bar{n}, c)]$ for $\bar{c} \leq c \leq \bar{b}$.

Consider the properties of the bicriterion model for acceptance sampling. From (3), as c increases keeping n fixed, P_A increases. Hence, AOQ also increases as c increases, while $ALIC$ decreases as c increases. In addition, the utilities for $ALIC$ and AOQ (given by $u_1(ALIC)$ and $u_2(AOQ)$) increase as the criteria levels decrease. Restricting utility to positive values therefore gives $u_1(ALIC) = u_1(n, c)$ and $u_2(AOQ) = u_2(n, c)$ as conditions (1) and (2) of theorem 1. The total utility $U(u_1, u_2)$ will increase if either or both of the single criterion utilities increase when using the multilinear and additive forms of eqs. (5) and (6). Total utility, therefore, satisfies condition (3) of theorem 1. The listed properties enable the use of theorem 1 to update upper and lower bounds in an appropriate implicit enumeration algorithm.

At a certain stage of the enumeration let the values of (n, c) be $(\bar{n}, \bar{c})$. Let $(\bar{a}, \bar{b})$ be the lower and upper bounds on c for $n = \bar{n}$. Finally, let (n_0, c_0) be the current values which maximize total utility U and $U_0 = U[u_1, u_2]$ at (n_0, c_0).

Using part (a) of theorem 1, if $U[u_1(\bar{n}, \bar{c}), u_2(\bar{n}, \bar{a})] \leq U_0$, then the lower bound $\bar{a}$ may be updated to $\bar{c}$. That is, if $ALIC$ at the enumeration's current sample size (n) and acceptance level (c) and AOQ at the lower bound on c for the current n have less total utility than the best utility yet found, no greater total utility will be found between the current c and current lower bound on c. Therefore, to avoid checking nonoptimal points, the lower bound on the acceptance level for the current sample size should be raised.

Part (b) of theorem 1 indicates how to update the upper bound on the acceptance level. If $U[u_1(\bar{n}, \bar{b}), u_2(\bar{n}, \bar{c})] \leq U_0$, then the upper bound $\bar{b}$ may be lowered to $\bar{c}$. That is, if $ALIC$ at the enumeration's current sample size and upper bound on the acceptance level, and AOQ at the current values have less total utility than the best utility yet found, then the upper bound on c should be lowered to avoid enumerating nonoptimal points, which reduces processing time.

3.1. General steps of the implicit enumeration algorithm

> *Step 1.* Select an $n = \bar{n}$. Determine $(\bar{a}, \bar{b})$ for $\bar{n}$ using relations (7) and (8). Start with $\bar{c} = \bar{a}$.
> *Step 2.* Choose a new $\bar{c} =$ old $\bar{c} + 1$. If $\bar{c} > \bar{b}$ terminate for $n = \bar{n}$.
> *Step 3.* Compute the MAU function value $\bar{U}$ at $(\bar{n}, \bar{c})$.
> *Step 4.* If $\bar{U} > U_0$ update U_0 by $\bar{U}$, and (n_0, c_0) by $(\bar{n}, \bar{c})$.
> *Step 5.* Update bounds $(\bar{a}, \bar{b})$, if possible using theorem 1. Return to step 2.

Example. Assume that $P_D = 0.10$, $C_I = \$1$, and that a DM's MAU function is the following additive form:

$$U\{AOQ(n,c); ALIC(n,c)\} = 0.1\left(1 - \left(\frac{ALIC}{1200}\right)^2\right)$$

$$+ 0.9(1 - 2 \cdot AOQ + AOQ^2), \tag{9}$$

where $k_1 = 0.1$, $k_2 = 0.9$, $u(ALIC(n,c)) = (1 - (\frac{ALIC}{1200})^2)$, and $u(AOQ(n,c)) = (1 - 2 \cdot AOQ + AOQ^2)$. Note from the scaling constants that the criterion of AOQ is deemed to be nine times as "salient" as $ALIC$; also, $ALIC$ and AOQ are respectively represented by a concave (risk averse) and convex (risk prone) quadratic utility function. Assume the implicit enumeration is at $\bar{n} = 115$, then the step-by-step algorithm would proceed as follows.

> *Step 1.* Select $n = \bar{n} = 115$. Eq. (7) gives a starting upper bound $\bar{b} = \min(\bar{n}, NP_D) = \min(115, 120) = 115$. Eq. (8) gives a starting lower bound $\bar{a} = \max[0, \bar{n} - N(1 - P_D)] = \max[0, -1065] = 0$.
> *Step 2.* Assume the previous $\bar{c}$ was 6 with no update to $(\bar{a}, \bar{b})$, then choose a new $\bar{c} =$ old $\bar{c} + 1 = 7$. Since $7 \leq \bar{b}$, we continue.
> *Step 3.* The bicriterion utility value at $(\bar{n}, \bar{c}) = (115, 7)$ is computed using (1), (2) and (9). The utility is found to be 0.90103.
> *Step 4.* The current maximum utility in this example would have been found at $(n_0, c_0) = (110, 7)$ with $U_0 = 0.90110$. No update of U_0 is indicated.
> *Step 5.* In order to update the bound $\bar{a}$, we must compute the MAU value with AOQ at $(\bar{n}, \bar{a})$ and $ALIC$ at $(\bar{n}, \bar{c})$. This gives us a MAU of 0.91563. Since this is greater than $U_0 = 0.90110$, no update of $\bar{a}$ may be made. To test for the update of $\bar{b}$, the MAU value with AOQ at $(\bar{n}, \bar{c})$ and $ALIC$ at $(\bar{n}, \bar{b})$ is computed to be 0.98447. Using theorem 1, we see that $\bar{b}$ cannot be updated. At this point the algorithm specifies a return to step 2. When this problem is

solved completely, the implicit enumeration scheme yields $U^* = 0.9011$ for $(n^*, c^*) = (110, 7)$; $AOQ = 0.01$ and $ALIC$ (scaled) $= 0.891$ which is equivalent to \$1069.20 ($= 0.891 \times 1200$). CPU time on a CDC 6500 computer was 15.38 seconds, which was 16.6% less than the time required for complete enumeration. Over many problems the algorithm on the average reduced CPU time about 40% with a minimum and maximum reduction of 8 and 87%, respectively, relative to complete enumeration.

4. Optimization via an interactive approach: Paired comparison method

Interactive approaches in the more general context of multicriteria problems have been addressed, for example, by Geoffrion, Dyer and Feinberg [6] and Zionts [20]. However, such methods can pose a considerable cognitive burden on a DM [19], converge to the optimal solution rather slowly, generally apply to continuous variables, and only treat linear and sometimes strictly concave MAU functions. In this paper, we describe an interactive procedure, called the *paired comparison method* (PCM), which uses only "soft" interaction, in that it poses less cognitive burden on a DM, rapidly converges to an optimal solution, applies to discrete (as well as continuous) variables, and accommodates nonlinear constraints and objective functions.

Consider the following bicriterion mathematical programming (BCMP) problem:

$$\text{vector max}[f_1(x), f_2(x)] = \max U[f_1, f_2] \text{ subject to } x \in S,$$

where the feasible region S is a compact convex set; the individual criterion functions f_1 and f_2 are concave and the implicit MAU function U is increasing and strictly quasi-concave. The assumptions on the criteria and feasible region are made only to guarantee the attainment of global maxima during single criterion optimization. The quasi-concavity assumption of U ensures that local optima are also global optima. In addition, it is sufficient to restrict ourselves to *effecient* solutions defined below as candidates for the optima:

Definition. A solution $x° \in S$ is said to be *efficient* if $f_i(x) > f_i(x°)$ for some $x \in S \Rightarrow f_k(x°) < f_k(x)$ for at least one other index k.

We shall now give the theory for generating the efficient solutions and then a method of "efficiently" searching among the efficient solutions for the "best compromise" solution of a DM. Define the payoff set $Y = \{y | f(x) = y$ for $x \in S\}$. Consider the following single objective mathematical programs:

P1: $\max f_1(x)$ subject to $x \in S$.

P2: max $f_2(x)$ subject to $x \in S$.

Let the optimal values of P1 and P2 be v^* and w^*, respectively. Now consider the following parametric mathematical programs:

P_v: max $f_2(x)$ subject to $x \in S$, $f_1(x) \geq v$.

Q_w: max $f_1(x)$ subject to $x \in S$, $f_2(x) \geq w$.

Let x_2^* solve Q_w for $w = w^*$. Then, $f_1(x_2^*) = v_l$ is the minimum achievable value for f_1 without sacrificing any achievement on f_2, while v^* is the maximum value of f_1 at the expense of f_2. Hence, the range of achievable values for f_1, denoted by $\bar{v}$, is given by $[v_l, v^*]$ and the best compromise value for f_1 lies in this range.

The following results will enable us to generate efficient solutions to BCMP with a specified level of achievement for f_1 (the proofs are available in Sadagopan and Ravindran [15]).

(1) $\bar{x} \in S$ is efficient iff $\bar{x}$ solves $P_{\bar{v}}$, where $\bar{v} \in [v_l, v^*]$.

(2) In the optimal solution to the program $P_{\bar{v}}$, where $\bar{v} \in [v_l, v^*]$, the constraint $f_1(x) \geq \bar{v}$ will be a binding constraint, i.e. if $\bar{x}$ solves $P_{\bar{v}}$ then $f_1(\bar{x}) = \bar{v}$.

Thus, the entire set of efficient solutions can be generated by parametrically solving $P_{\bar{v}}$ and the generated solutions will have specific levels of attainment of f_1, namely $\bar{v}$. In addition, using the quasi-concave assumption of MAU function $U(f_1, f_2)$, we can devise an efficient parameterization.

Consider the problem of maximizing U over the payoff set Y, keeping f_1 fixed at v:

P3: max $U(v, f_2)$ subject to $(v, f_2) \in Y$.

Note that problem P3 and P_v are equivalent since U is increasing in f_2, and their optimal solutions will be the same. Let $g(v) =$ maximum of $U(v, f_2)$. Then, it can be proved that $g(v)$ is strongly quasi-concave, i.e. unimodal in v.

Note that the BCMP problem has now been reduced to determining the maximum of $g(v)$, where v belongs to the interval $[v_l, v^*]$. However, $g(v)$ is not known explicitly since U is not known. But, using a search technique which requires only *functional comparison* and not function values, we can still solve the BCMP problem using the following region elimination concept.

For a unimodal function $g(v)$ defined over a finite interval (v_l, v^*), let v_A and v_B be two points in the interval such that $v_A < v_B$. Then, $g(v_A) < g(v_B)$ implies that the maximum of $g(v)$ will not lie in the interval (v_l, v_A). On the other hand, $g(v_A) > g(v_B)$ implies that the maximum will not lie in the interval (v_B, v^*)

4.1. General steps of the paired comparison method

Step 1. Solve P1: max $f_1(x)$ subject to $x \in S$. Set max $f_1(x) = v^*$.

Step 2. (a) Solve P2: max $f_2(x)$ subject to $x \in S$. Set max $f_2(x) = w^*$. (b) Solve max $f_1(x)$ subject to $x \in S$, and $f_2(x) \geq w^*$. Set the maximum of $f_1(x) = v_l$. Now the optimal value of f_1 to the BCMP lies between v_l and v^*.

Step 3. Choose two values of f_1, v_A and v_B such that $v_l < v_A < v_B < v^*$. Here we could use the Fibonacci Method or the Golden Section Search to choose the two points effeciently.

Step 4. Solve the problem P_v: Maximize $f_2(x)$ subject to $x \in S$ and $f_1(x) \geq v$ for the two different values of v, $v = v_A$ and $v = v_B$; let $g(v) = \max f_2(x)$ for P_v. Let $Y^{(1)} = [v_A, g(v_A)]$, $Y^{(2)} = [v_B, g(v_B)]$. Given $Y^{(1)}$ and $Y^{(2)}$, the DM is asked to specify whether $Y^{(1)} \succ$ (preferred to) $Y^{(2)}$, or $Y^{(2)} \succ Y^{(1)}$, or indifferent. Note that the DM's preference comparison translated in terms of preference function U reduces to function comparison of $g(v)$. Using the unimodal

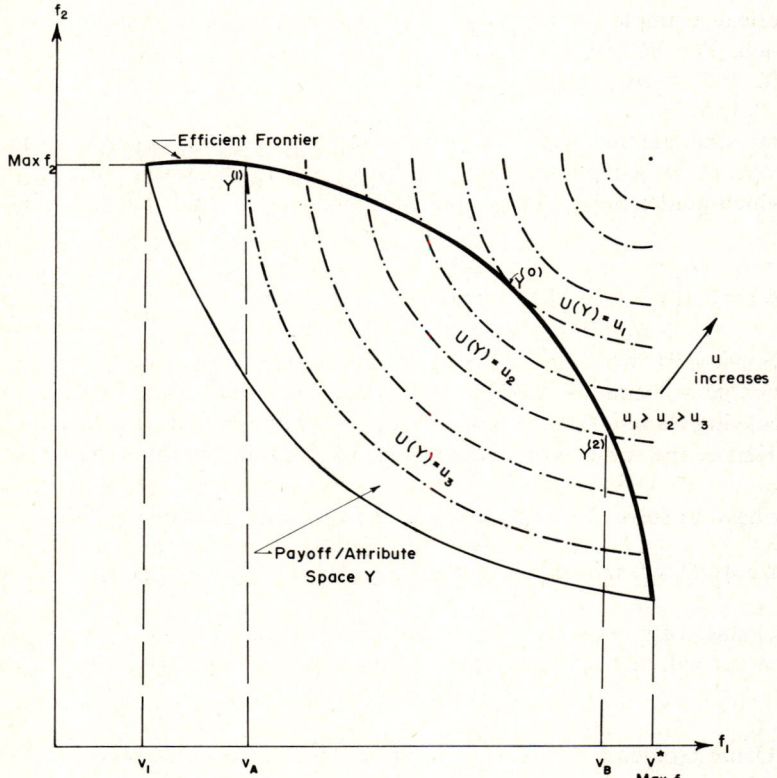

Fig. 1. Paired comparison method.

property of $g(v)$ and the DM's responses, a portion of the efficient set can be eliminated and we progressively converge to the best compromise solution.

Fig. 1 illustrates the paired comparison method. Here, $Y^{(2)} \succ Y^{(1)}$ since $U(Y^{(2)}) > U(Y^{(1)})$. Hence, the maximum of U cannot lie in the interval $[v_l, v_A]$. Eliminating the region (v_l, v_A), the interval of uncertainty where the optimum lies reduces to (v_A, v^*). If Golden Section Search is used to generate two new points between v_A and v^*, one of the new points will turn out to be v_B. The algorithm is continued until the interval of uncertainty becomes very small. With Golden Section Search, we will be able to bracket the best compromise solution Y^0 to less than 10% of the original interval on v with just five paired comparisons from the DM; in ten paired comparisons the optimal solution will be within 1% of the original interval of uncertainty $[v_l, v^*]$. Thus, for a specified level of uncertainty we have a finite, rapidly convergent procedure using only soft interaction, namely paired comparison of two efficient solutions.

Example. To demonstrate the paired comparison method, we employ the same numerical example used previously to illustrate the implicit enumeration approach. The BCMP formulation is a vector minimum problem: minimize $ALIC = f_1 = nC_1 + (N - n)C_1(1 - P_A)$; minimize $AOQ = f_2 = [(N - n)P_A P_D]/N$.

In order to simulate the interaction process and provide a direct comparison of the approaches, we will assume that the DM's implicit bicriterion preference function (which guides him in the paired comparisons) is again that given by (9):

$$\max U(f_1, f_2) = 0.1\left[1 - (f_1/1200)^2\right] + 0.9(1 - 2f_2 + f_2^2). \tag{10}$$

Step 1. Solving P1, which corresponds to minimizing f_1 ignoring f_2, we get $n^* = 1$, $c^* = 1$, $f_1^* = 1$ and $f_2 = 0.0999$.

Step 2. Solving P2 ignoring $ALIC$, we get $n^* = 1200$, $c^* = 1080$, $f_2^* = 0$ and $f_1 = 1200$. Hence, the lower bound v_l on f_1 is $1 and the upper bound v^* is $1200.

We now have to solve the single objective optimization problem:

P_v: Minimize $AOQ = f_2$ subject to $x \in S$ and $f_1 \leq v$,

for various values of v between 1 and 1200. The implicit enumeration scheme discussed earlier will be used to solve the single-objective problems.

Interaction 1.

Step 3. Using Golden Section ratios 0.618 and 0.382, we compute $v_A = 459$ and $v_B = 742$.

Table 2
Iterations of BCMP using PCM

Iteration	v_l	v^*	v_A	v_B	f_{2A} [a]	f_{2B}	$U(f_A)$	$U(f_B)$	Remaining interval (%)
1	1	1200	459	742	0.064	0.039	0.8753	0.8937	61.8
2	459	1200	742	917	0.039	0.024	0.8937	0.8996	38.2
3	742	1200	917	1025	0.024	0.015	0.8996	0.9005	23.6
4	917	1200	1025	1092	0.015	0.009	0.9005	0.9013	14.6
5	1025	1200	1092	1133	0.009	0.006	0.9013	0.9003	9.0
6	1025	1133	1066	1092	0.011	0.009	0.9015	0.9013	5.5
7	1025	1092	1051	1066	0.013	0.011	0.9008	0.9015	3.4
8	1051	1092	1066	1076	0.011	0.011	0.9015	0.9003	2.1
9	1051	1076	1061	1066	0.012	0.011	0.9010	0.9015	1.3
10	1061	1076	1066	1070	0.011	0.011	0.9015	0.9015	0.33

[a] f_{2A} corresponds to minimum AOQ when $ALIC$ is set to v_A.

Step 4. Solve P_v for $v = v_A$ and $v = v_B$. The optimal solution $v = v_A$ has objective values $f_A = (f_{1A}, f_{2A}) = (459, 0.064)$; similarly, the solution for $v = v_B$ gives $f_B = (f_{1B}, f_{2B}) = (742, 0.0390)$.

The solutions f_A and f_B are presented to the DM and a paired comparison is made between (459, 0.064) and (742, 0.039). Assuming that the DM is guided by the implicit preference function in (10), $U[f_B] = 0.8937$ is greater than $U[f_A] = 0.875$. Hence, the DM would prefer f_B over f_A. Consequently, the interval (1,459) can be eliminated and the lower bound v_l is updated to 459.

Assume that the algorithm will be terminated when $(v^* - v_l) < 1\%$ of the original interval. For brevity of presentation, the results of further iterations are summarized in table 2. The optimal integer sampling plan for both v_A and v_B is given by $n^* = 110$, $c^* = 7$. This plan has an $ALIC = \$1065$ and $AOQ = 0.011$, which agrees with the solution obtained by the implicit enumeration approach using (10) as an explicit utility function.

5. Simulation experiments

The purpose of the simulation experiments was to conduct sensitivity analysis on the bicriterion and individual criterion utility functions given by (5) and (6) and table 1 to ascertain their effects on the choice of the sampling plan and the response surface.

5.1. Design

The dependent variables were: (1) the frequency of changes to the optimal sampling plan (n^*, c^*) based on maximizing $(ALIC(n, c); AOQ(n, c))$ over the feasible region, or change in the $(ALIC, AOQ)$ vector, and (2) the frequency of existence of local maxima in the feasible solution region.

The independent variables in the simulation experiment were: (1) the scaling constants: (2) the convexity, concavity, and linearity of the single criterion utility functions; (3) the logarithmic, quadratic, and exponential forms for the single criterion utility function; and (4) the multilinear and additive forms of the bicriterion utility function. True fraction defectives were randomly varied over the realistic range from 0.05 to 0.30 [12,13]. The scaling constants ("importance weights") employed varied from equal weights to the nearly extreme case where almost all weight was assigned to a given criterion (e.g. $k_1 = 0.1, k_2 = 0.9$). C_I was equal to \$1, N was equal to 1200, and n was varied in increments of 10.

5.2. Effect on optimal sampling plan (n^*, c^*), ALIC, and AOQ

A MANOVA was performed using a $5 \times 4 \times 3 \times 3$ factorial design. The vector of criteria (ALIC, AOQ) was treated as the dependent variable set. The bicriterion utility models considered did not affect ALIC and AOQ significantly and thus were not considered in the MANOVA to reduce problem size. The MANOVA yielded the following significant main and interaction effects based on the Hotelling T^2 statistic and Wilk's Lambda (λ) criterion (table 3). The optimal vector (ALIC*, AOQ*), and hence (n^*, c^*), were sensitive to the magnitudes of scaling constants and convexity/concavity of the single-criterion utility functions in the bicriterion case. The only significant interaction effect was between scaling constants and convexity/concavity of U(AOQ).

For the quadratic, logarithmic, linear, and mild exponential single criterion utility functions a total of 11 out of 405 policy changes (n^*, c^*) occurred when switching from the multilinear to the additive bicriterion functional form. However, while the frequency of obtaining a nonoptimal policy by employing an incorrect bicriterion form is low, its potential impact on ALIC and AOQ is high. In all 11 cases, each policy switch was extreme, resulting in radically different sampling plans, ALIC's, and AOQ's. This phenomenon occurred most frequently when P_D was high, and may be a result of the utility independence assumption associated with the linear and multilinear bicriterion functions.

The strong exponential single criterion utility function showed 18 out of 135 changes in (n^*, c^*); but only 5 of the 18 policy switches were extreme. Thus, the changes in (n^*, c^*) for the additive and multilinear bicriterion utility functions for this single-criterion utility function were more frequent, but also more gentle. The difference between the results for the strong and mild exponential functions suggest that parameter selection may be a significant consideration in the construction of certain single-criterion utility functions. The frequency of (n^*, c^*) changes as a function of the convexity/concavity of the single-criterion utility functions is summarized in table 4. The impact of convexity/concavity is mediated by the scaling constants (k_1, k_2) (as shown by the MANOVA) and by the presence of a strong exponential single-criterion

Table 3

Effect	T^2	DF	F
Convexity of U(AOQ)	0.171	4/2786	59.71
Convexity of U(ALIC)	0.034	4/2786	11.87
Scaling constants	1.132	8/2786	197.10
Scaling constants by convexity of U(AOQ)	0.428	16/2786	37.26

Table 4
Frequency of changes in (n^*, c^*) with varying convexity/concavity.

Utility model	Scaling constants		Number of changes multilinear/additive
	k_1	k_2	out of a possible 45 [a]
Logarithmic	0.8	0.8	7/5
	0.1	0.9	24/24
	0.5	0.5	8/8
	0.4	0.3	5/5
	0.8	0.1	1/1
Quadratic	0.8	0.8	6/5
	0.1	0.9	19/19
	0.5	0.5	5/5
	0.4	0.3	5/5
	0.8	0.1	1/1
Mild exponential	0.8	0.8	7/5
	0.1	0.9	19/19
	0.5	0.5	5/5
	0.4	0.3	5/5
	0.8	0.1	1/1
Strong exponential	0.8	0.8	19/18
	0.1	0.9	20/20
	0.5	0.5	18/18
	0.4	0.3	14/15
	0.8	0.1	1/1

[a] All combinations of $P_D = 0.05, 0.10, 0.15, 0.20$ and 0.30; AOQ (and $ALIC$) being convex, linear and concave.

utility function, indicating that utility model and parameter selection must thus be considered conjointly.

5.3. Effect on local maxima

Using a $5 \times 3 \times 3$ factorial design, an ANOVA was performed with the frequency of local maxima as the dependent variable, and the scaling constants and the convexity/concavity of $U(AOQ)$ as the independent variables. The bicriterion and single-criterion forms of utility functions had no effect on the occurrence of local maxima.

The results showed that the frequency of local maxima were affected by the scaling constants ($F_{4,1395} = 43.1$, $p < 0.001$), convexity/concavity of $U(AOQ)$ ($F_{2,1395} = 31.9$, $p < 0.001$) and convexity/concavity of $U(ALIC)$ ($F_{2,1395} = 257.2$, $p < 0.001$). There were also the following two- and three-way interaction effects: (1) scaling constant by convexity/concavity of $U(AOQ)$

Table 5
Frequency of models with local maxima based on multilinear and additive bicriterion forms.

Model	Multilinear/additive out of a possible 180
Strong exponential	37/35
Logarithmic	22/20
Quadratic	25/25
Mild exponential	26/24

($F_{8,1395} = 2.6, p < 0.008$); (2) scaling constant by convexity/concavity of $U(ALIC)$ ($F_{8,1395} = 15.0, p < 0.001$); (3) convexity/concavity of $U(AOQ)$ by convexity/concavity of $U(ALIC)$ ($F_{4,1395} = 25.9, p < 0.001$); and (4) scaling constant by convexity/concavity of $U(AOQ)$ by convexity/concavity of $U(ALIC)$ ($F_{16,1395} = 3.1, p < 0.001$).

The frequency of local maxima and the optimal policy (n^*, c^*) are affected by the same factors, namely scaling constants and the convexity/concavity of the single criterion utility functions. The multilinear and additive forms yielded essentially the equivalent number of local optima (table 5). Table 6, however, shows that single-criterion utility functions that are convex yield bicriterion utility response surfaces that are not as well behaved. Only the use of the strong exponential also increased the likelihood of finding local maxima in the

Table 6
Frequency of local maxima based on convexity, concavity, and linearity of single-criterion utility functions (each item out of a possible 50 results).

		Concave $u_1(ALIC)$	Linear $u_1(ALIC)$	Convex $u_1(ALIC)$
Concave $u_2(AOQ)$	Logarithmic	0	0	2
	"Mild" exponential	0	0	2
	Quadratic	0	0	0
	"Strong" exponential	0	0	2
Linear $u_2(AOQ)$	Logarithmic	0	6	9
	"Mild" exponential	2	6	9
	Quadratic	2	6	13
	"Strong" exponential	0	6	15
Convex $u_2(AOQ)$	Logarithmic	0	9	16
	"Mild" exponential	6	9	17
	Quadratic	6	9	14
	"Strong" exponential	6	15	28

feasible region. The interaction effects are evidenced by the large increase in the frequency of local maxima in switching both single-criterion utility functions from concave to convex (table 6).

6. Conclusions and future research

In this paper a bicriterion model for acceptance sampling was presented that explicitly or implicitly incorporated a DM's preferences in selecting an optimal plan. Such a model allows explicit consideration of conflicting goals that are currently not considered in existing acceptance sampling schemes.

Simulation experiments were also conducted to determine the robustness of the sampling plan and response surface of the bicriterion model to various utility functions. The ultimate objective of these simulations was to reveal ways of simplifying the process of measuring an explicit utility function as well as optimizing the model.

Concerning future research, development of more efficient discrete optimization algorithms for solving the bicriterion utility function problem and the sequence of single-objective optimization problems in the interactive method are in progress. We are also investigating certain intelligent search methods for solving the discrete optimization problem. The COMPLEX search method of Box [2] is being modified to handle variables restricted to integers only.

Incorporation of a prior probability function for lot quality P_D is also being considered to give the bicriteron model a Bayesian flavor. Furthermore, we plan to modify some of our previous models, which were based on a criterion of expected cost, to accommodate multiple criteria and a DM's risk preferences [12–14].

Acknowledgements

The research was supported in part by a National Science Foundation grant no. 8007103. The authors would like to acknowledge the assistance provided by Sathiadev Makesh and B. Vinod on this project.

References

[1] R.H. Ailor, J.W. Schmidt and G.K. Bennett, "The Design of Economic Acceptance Sampling Plans for a Mixture of Variables and Attributes", *AIIE Transactions* 7,4 (1975) 370.
[2] M.J. Box, "A New Method of Constrained Optimization and a Comparison with Other Methods", *Computer Journal* 8 (1965) 42–52.
[3] K.E. Case, J.W. Schmidt and G.K. Bennett, "A Discrete Economic Multiattribute Acceptance Sampling Plan", *AIIE Transactions* 7,4 (1975) 363.

[4] R.D. Collins, K.E. Case and G.K. Bennett, "The Effects of Inspector Error on Single Sampling Inspection Plans", *International Journal of Production Research* 11 (1973) 289–298.
[5] A.J. Duncan, *Quality Control and Industrial Statistics* (Richard D. Irwin, Inc., Homewood, Illinois, 1959).
[6] A.M. Geoffrion, J.S. Dyer and A. Feinberg, "An Interactive Approach to Multi-Criterion Optimization with an Application to the Operation of an Academic Department", *Management Science* 19 (1976) 357–368.
[7] R.L. Keeney, "Multiplicative Utility Functions", *Operations Research* 22,1 (1974) 22–34.
[8] R.L. Keeney and H. Raiffa, *Decisions with Multiple Objectives: Preferences and Value Tradeoffs* (John Wiley and Sons, New York, 1976).
[9] B.A. Latimer, G.K. Bennett and J.W. Schmidt, "The Economic Design of a Dual Purpose Multicharacteristic Quality Control System," *AIIE Transactions* 5,3 (1973) 214.
[10] G. Minton, "Verification Error in Single Sampling Inspection Plans for Processing Survey Data, *Journal of the American Statistical Association* 67, 337 (1972) 46–54.
[11] C.W. Moreno, "A Performance Approach to Attribute Sampling and Multiple Action Decisions", *AIIE Transactions* 11,3 (1979) 183–197.
[12] H. Moskowitz and W.L. Berry, "A Bayesian Algorithm for Determining Optimal Single Sample Acceptance Plans for Product Attributes", *Management Science* 22 (1976) 1238–1250.
[13] H. Moskowitz and R.K. Fink, "A Bayesian Algorithm Incorporating Inspector Errors for Quality Control and Auditing, in: M.F. Neuts, ed., *Algorithmic Methods in Probability, TIMS Studies in the Managements Sciences*, vol. 7 (North-Holland Publishers, 1977) 7 pp 79–104.
[14] H. Moskowitz, A. Ravindran and J.M. Patton, "An Algorithm for Selecting an Optimal Acceptance Plan in Quality Control and Auditing", *International Journal of Production Research* 17 (1979) 581–594.
[15] S. Sadagopan and A. Ravindran, "Interactive Solution of Bicriteria Mathematical Programs", Research Memorandum 80-2, School of Industrial Engineering, Purdue University, West Lafayette, IN (1980). (Forthcoming in *Naval Res. Log. Qtrly.*)
[16] J.W. Schmidt and G.K. Bennett, "Economic Multiattribute Acceptance Sampling", *AIIE Transactions* 4,3 (1972).
[17] J.W. Schmidt, K.E. Case and G.K. Bennett, "The Choice of Variables Sampling Plans Using Cost Effective Criteria", *AIIE Transactions* 6,3 (1979).
[18] J.W. Schmidt and R.E. Taylor, "A Dual Purpose Cost Based Quality Control System", *Technometrics* 15 (1973) 166.
[19] J. Wallenius, "Comparative Evaluation of Some Interactive Approaches to Multicriteria Optimization", *Management Science* 21 (1975) 1387–1396.
[20] S. Zionts, ed., *Multiple Criteria Problem Solving* (Springer Verlag, New York, 1977).

NOTES ABOUT AUTHORS

Ronald D. Armstrong ("An Algorithm to Select the Best Subset for a Least Absolute Value Regression Problem") is an Associate Professor in the Department of Quantitative Business Analysis at the University of Georgia. His research interests include statistics, scheduling and menu planning with emphasis on computer implementations of mathematical programming algorithms. He has coauthored two books on data processing and is the author or coauthor of over 40 refereed publications.

Terry E. Dielman ("LAV (Least Absolute Value) Estimation in Regression: A Review") is an Assistant Professor of Decision Sciences at the M.J. Neeley School of Business at Texas Christian University, Fort Worth, Texas 76129. He received a B.A. in mathematics at Emporia State University, an M.S. in mathematics from the University of Cincinnati, and a Ph.D. in statistics and management science from the University of Michigan. His current research interests include optimal robust estimators in linear models and the analysis of pooled cross-sectional and time-series data. He is the author of the book *Pooled Data for Financial Markets* and has published in the *Journal of Financial and Quantitative Analysis*.

P.K. Eswaran ("A Bicriterion Model for Acceptance Sampling") is a Ph.D. student working in the area of multiple criteria decision making at Purdue University, in the School of Industrial Engineering. He has a bachelors degree in mechanical engineering and a masters degree in industrial management from the Indian Institute of Technology, Madras. Earlier he worked as a computer consultant in India.

Robert S. Garfinkel ("Error Localization for Erroneous Data: A Survey") is Professor and Chairman of the Management Science Program at the University of Tennessee, Knoxville, TN 37916. He holds a B.A. in mathematics from Brooklyn College and Ph.D. in operations research from Johns Hopkins. Professor Garfinkel has published in *Operations Research, Management Science, Naval Research Logistics Quarterly, Transportation Science, Mathematical Programming, JACM, Fibonacci Quarterly,* and *Networks*. He is coauthor of the book *Integer Programming* (with George Nemhauser) and does research in a number of areas of combinatorial optimization, including facility location, vehicle routing, and job scheduling, as well as continued work on the error localization problem.

Carl M. Harris ("Parameter Estimation under Progressive Censoring Conditions for a Finite Mixture of Weibull Distributions") is a Senior Research Fellow at the Center for Management and Policy Research in Washington, DC. Prior to joining CMPR, Dr. Harris was Chairman and Professor of Industrial Engineering and Operations Research at Syracuse University. He has a B.S. degree in mathematics from Queens College (N.Y.), and an M.S. and Ph.D. in mathematics from the Polytechnic Institute of New York, He is coauthor with Donald Gross of *Fundamentals of Queueing Theory* (Wiley, 1974), and is author or coauthor of approximately 40 articles in the professional literature. Dr. Harris's current research interests include statistics, applied probability (particularly in queueing), and criminal justice and energy modeling. He has been spending the 1981–82 academic year as a Visiting Professor of Engineering Science and Systems at the University of Virginia, Charlottesville.

Robert E. Hausman Jr. ("Constrained Multivariate Analysis") is a member of the Statistical Models, Planning and Characterization Group at Bell Laboratories, Holmdel, New Jersey. He earned his S.B. in electrical engineering at M.I.T., his M.S. in computer sciences from the University of Wisconsin, and his M.B.A. in management sciences and M.A. and Ph.D. in statistics from the University of Pennsylvania. His current research interests are in multivariate statistical methods. He is a member of the ASA, TIMS and Sigma Xi.

George Kimeldorf ("Concordant and Discordant Monotone Correlations and Their Evaluation by Nonlinear Optimization") received his A.B. from the University of Rochester and his M.A. and Ph.D. in mathematics from the University of Michigan. He has numerous published papers in statistics and mathematics journals. He is currently Professor of Mathematical Sciences at the University of Texas at Dallas.

Gary Klein ("A Bicriterion Model for Acceptance Sampling") is a doctoral student at the Krannert Graduate School of Management, Purdue University. He will soon assume the duties of Assistant Professor of Management with the MIS Group at the University of Arizona.

Darwin Klingman ("Generalized Network Approaches for Solving Least Absolute Value and Tchebycheff Regression Problems") is a Professor in the departments of Operations Research and Computer Sciences, and Director of Computer Science Research at the Center for Cybernetic Studies at the University of Texas at Austin. Professor Klingman is a former associate editor for *Management Science, Operations Research,* and *Naval Research Logistics Quarterly,* and is on the editorial board of *Discrete Applied Mathematics.* He is also the Vice-President-at-Large of The Institute of Management Sciences. Professor Klingman graduated Phi Beta Kappa and holds a B.A. and M.A. in mathematics from Washington State University and a Ph.D. in mathematics, computer sciences, and business administration from the University of Texas. He is the author of more than 80 published research papers on theory and computation of transportation and network algorithms, linked list structures, graphs, and integer programming. He is a coauthor of two books, coeditor of a special issue of *Mathematical Programming Study* on network applications, and coeditor of a special issue of *INFOR* on networks.

Research efforts have included the development and testing of codes for large-scale transportation and transshipment problems, generalized network problems, multi-commodity network problems, and fixed-charge problems. He has consulted widely for government and industry on the application of operations research and computer science to resource management, site location models, transportation planning, commodity distribution, network mathematical programming systems, telecommunications, and manpower problems.

Mabel T. Kung ("An Algorithm to Select the Best Subset for a Least Absolute Value Regression Problem") is an Assistant Professor in the Faculty of Business, McMaster University, Hamilton, Ontario. She received a Ph.D. from the University of Texas at Austin and her major research interest is mathematical programming applications. She has published articles in *Applied Statistics, Applied Mathematics and Computation*, and *Communications in Statistics.*

Anand S. Kunnathur ("Error Localization for Erroneous Data: A Survey") holds a Master's degree in phyiscs from the University of Delhi and a Master's degree in applied mathematics from York University. As a doctoral student in the Management Science Program at the University of Tennessee, he was engaged in error localization research aimed at data validation. He is currently at the University of Wisconsin at Milwaukee and is a consultant to the Oak Ridge National Laboratories and is a member of Phi Kappa Phi.

Naftali A. Langberg ("Extreme Points of the Class of Discrete Decreasing Failure Rate Average Life Distributions") is Professor of Statistics at the Haifa University in Haifa, Israel. He earned his

B.A. degree in mathematics and statistics at the Hebrew University in Jerusalem, Israel, and his M.Sc. and Ph.D. in mathematics form the Hebrew University. His current research interests are in reliability, epidemiology and applied statistics.

Ramón V. León ("Extreme Points of The Class of Discrete Decreasing Failure Rate Average Life Distributions") was born in Holguín, Cuba, on 29 September 1948. He received a B.S. degree in mathematics from Florida State University (FSU) in 1972, an M.S. degree in mathematics from Tulane University in 1975, an M.S. degree in statistics from FSU in 1976, and a Ph.D. degree in statistics from FSU in 1979. Dr. León received the Ralph A. Bradley Award of 1979 in recognition of outstanding achievement as a graduate student.

Dr. León is presently a member of the technical staff of Bell Laboratories. Holmdel, New Jersey. During 1978 to 1979, he was a visiting instructor in the Department of Statistics at FSU, and from 1979 to 1980, Assistant Professor in the Department of Statistics at Rutgers University. He has published papers on statistics, probability, and operations research.

Dr. León is a member of the Institute of Mathematical Statistics, the American Statistical Association, and the American Society for Quality Control.

Gunar E. Liepins ("Error Localization for Erroneous Data: A Survey") received his Ph.D. in mathematics from Dartmouth College in 1974 and a subsequent M.S. in engineering–economic systems from Stanford University in 1977. From 1974 to 1976 he served as a visiting lecturer in mathematics at Texas Tech. University, and has been with Oak Ridge National Laboratory since 1977. During the 1980–81 academic year he was on leave of absence at the Quantitative Business Analysis Department of the University of Georgia. His research interests include statistical error modelling, discrete optimization, and quantitative techniques to assess data quality.

James Lynch ("Extreme Points of the Class of Discrete Decreasing Failure Rate Average Life Distributions") received his B.S. from Marquette University, his M.S. and Ph.D. from Florida State University, is an Assistant Professor in the Department of Statistics at the Pennsylvania State University, and is presently a visiting Associate Professor in the Mathematics and Statistics Department at the University of South Carolina. His research interests are probability theory and reliability theory and he has published papers in the *Annals of Probability, Mathematics of Operations Research,* and *Sankhya*. He is a member of IMS and ASA.

Jay Mandelbaum ("Parameter Estimation under Progressive Censoring Conditions for a Finite Mixture of Weibull Distributions") is a senior mathematical statistician working for the U.S. Department of Transportation, Federal Railroad Administration. He is Deputy Chief of the Information and Statistics Division within the Policy Office. He received a B.A. in physics from Rutgers University and an M.Sc. in operations research from the George Washington University. This work is part of his ongoing D.Sc. dissertation research there. He is a member of TIMS, ORSA, TRF, and TRB.

Nancy R. Mann ("Optimal Outlier Tests for a Weibull Model – To Identify Process Changes or to Predict Failure Times") is Research Biomathematician (Research Professor) in the Biomathematics Department and Department of Psychiatry at UCLA while on leave of absence from the Science Center of Rockwell International, where she is Project Manager, Reliability and Statistics. She earned her B.A. and M.A. degrees in mathematics and her Ph.D. in biostatistics, all from UCLA. While her research interests have centered on methodology for analysis of reliability and survival data, she is at present developing this methodology as a tool for analysis and characterization of the recidivism process in drug addicts. She is the senior author of a book, *Methods for Statistical Analysis of Reliability and Life Data*, has contributed articles to books and encyclopedias and has published numerous articles in *Technometrics, Journal of the American Statistical Association, Naval Research Logistics Quarterly,* and *Communications in Statistics*.

Jerrold H. May ("Concordant and Discordant Monotone Correlations and Their Evaluation by Nonlinear Optimization") received his B.A. from Roosevelt University and his M.Phil. and Ph.D. in administrative sciences from Yale University. His recent research has concentrated on developing and implementing nonderivative nonlinear programming algorithms. He is currently an Associate Professor of Business Administration at the University of Pittsburgh.

Herbert Moskowitz ("A Bicriterion Model for Acceptance Sampling") Ph.D. University of California at Los Angeles, is a Professor of Management and Director, Professional Graduate Programs in Management, at the Krannert Graduate School of Management, Purdue University. He has authored three books in management science and statistics and has published over 40 papers in such journals as *Management Science, Operations Research, AIIE Transactions, Decision Sciences, Omega, Academy of Management, IEEE Transactions on Engineering Management, Policy Science* and *Organizational Behavior and Human Performance.* He has served as vice-president and member of the Executive Board of AIDs, ORSA liaison representative to AIDS, and is a member of TIMS.

John Mote ("Generalized Network Approaches for Solving Least Absolute Value and Tchebycheff Regression Problems") is the Executive Vice-President of Analysis, Research and Computation, Inc., where he serves as a consultant to both industry and government on the implementation of large-scale mathematical programming systems. His current activities involve the algorithmic design and implementation of network-based production, distribution and inventory planning systems. He received a Ph.D. in operations research from the University of Texas at Austin.

Clark A. Mount-Campbell ("Selection of Cost-Optimal Fractional Factorials, $2^{m-r} \, 3^{n-s}$ Series") is Assistant Professor of Industrial and Systems Engineering, The Ohio State University. His research interests are in the areas of optimization, optimal experimental design, optimization problems resulting from uncertainty in decision processes, and economic and capital budgeting modeling with optimization models. He holds a B.S. and M.S. from New Mexico State University and Ph.D. from the University of Oklahoma. Dr. Mount-Campbell is a member of AIIE, ORSA, ASEE, Sigma Xi and Alpha Pi Mu.

Subhash C. Narula ("Optimization Techniques in Linear Regression: A Review") is an Associate Professor in the School of Management at the Rensselaer Polytechnic Institute, Troy, New York 12181. He earned his B.E. in mechanical engineering from Delhi University, Delhi, India, and his M.S. and Ph.D. in industrial and management engineering from the University of Iowa, Iowa City, Iowa. His current research interests are in applied statistics and operations research. He has published in several statistics and operations research journals.

John B. Neuhardt ("Selection of Cost-Optimal Fractional Factorials, $2^{m-r} \, 3^{n-s}$ Series") is Professor, Industrial and Systems Engineering, the Ohio State University. Research activities include optimal engineering experimental design applications, including the area of highway safety. Dr. Neuhardt is coprincipal investigator on projects with FHWA and NCHRP in areas of highway safety, and was formerly director of system analysis and programming with AC Spark Plug, Division of General Motors. He received a B.A., M.S. in statistics and Ph.D. in industrial engineering from the University of Michigan.

Roger C. Pfaffenberger ("LAV (Least Absolute Value) Estimation in Linear Regression: A Review") is the Continental National Bank Professor of Management Science at the M.J. Neeley School of Business at Texas Christian University, Fort Worth, Texas 76129. He earned a B.S. in mathematics at California Polytechnic State University, and his M.S. and Ph.D. in statistics from Texas A&M University. His current research interests are in the area of optimal robust estimation in the linear model. He is coauthor of the books, *Mathematical Programming for Business and Economics* and *Statistical Methods for Business and Economics*, has contributed articles appearing in several books, and has published articles in *Biometrika, Biometrics, Journal of Financial and Quantitative Analysis*, and the *European Journal of Operations Research*.

Robert Plante ("Algorithmic Improvements for Obtaining the Upper Multinomial Bound") is an Assistant Professor of Management at the Krannert Graduate School of Management, Purdue University, West Lafayette, Indiana 47907. He has contributed to *The Accounting Review* and *Journal of the American Statistical Association*. His research interests are in auditing, quality control and marketing.

Frank Proschan ("Extreme Points of the Class of Discrete Decreasing Failure Rate Average Life Distributions") has been engaged in mathematical and statistical research and application for the past 40 years: from 1941 to 1952 for the government, from 1952 to 1960 for Sylvania Electric Products, Inc., from 1960 to 1970 for Boeing Scientific Research Laboratories, and presently with Florida State University as Professor of Statistics. For the past 25 years he has been actively engaged in research in the mathematical theory of system reliability. With Dr. Richard Barlow, he has written a monograph, *Mathematical Theory of Reliability* (Wiley, 1965), at the request of the Society for Industrial and Applied Mathematics (translated into Russian), and with Dr. Barlow, a text, *Statistical Theory of Reliability and Life Testing* (Holt, Rinehart and Winston, 1975), translated into Russian and German.

Dr. Proschan has written over 100 papers on statistics, statistical quality control, operations research, inventory theory, and reliability; his dissertation was selected as one of the award winners of the 1959 Ford Foundation Doctoral Dissertation Competition and published by Prentice-Hall. He is a Fellow of the Institute of Mathematical Statistics, a Fellow of the American Statistical Association, and a member of the International Statistical Institute. He has been an Associate Editor of the *Annals of Mathematical Statistics,* of *Technometrics,* and of *Mathematics of Operations Research*. He has also been a visiting lecturer for SIAM and for the statistical societies.

Dr. Proschan received a B.S. in mathematics from the City College of New York in 1941, an M.A. in statistics from George Washington University in 1948, and a Ph.D. in statistics from Stanford University in 1959.

A. Ravindran ("A Bicriterion Model for Acceptance Sampling") is a Professor at the School of Industrial Engineering at the University of Oklahoma, Norman, Oklahoma. He received his B.S. in electrical engineering from Birla Institute of Technology and Science, Pilani (India), and M.S. and Ph.D. degrees in industrial engineering and operations research from the University of California at Berkeley. His research interests include mathematical programming, multi-criteria optimization, goal programming, metal cutting, health planning, energy models and transportation analysis. His professional affiliations include AIIE, ORSA, TIMS, AIDS, Sigma Xi, and Alpha Pi Mu.

Jagdish S. Rustagi (coeditor) is Professor and Chairman, Department of Statistics, the Ohio State University, Columbus, Ohio. He received his B.A. and M.A. degrees in mathematics from the University of Delhi and Ph.D. in statistics from Stanford University. Prior to joining the Ohio State University, he had been on the faculties of Aligarh Muslim University, University of Cincinnati College of Medicine, Michigan State University and Carnegie Institute of Technology.

His current research interests include applied and theoretical statistics, teaching of statistics, operations research and environmental health. He is author of a book on *Variational Methods in Statistics,* editor of two volumes on *Optimizing Methods in Statistics* and coeditor of *Teaching of Statistics and Statistical Consulting* (with Douglas A. Wolfe), all published by Academic Press.

He is managing editor of *Annals of Statistics* and *Annals of Probability* and is a member of the editorial boards of *Communications in Statistics,* Series A and E. His publications have appeared in several journals including *Annals of Mathematical Statistics, Archives of Environmental Health, Communications in Statistics, Mathematical Biosciences* and *Operations Research*.

Allan R. Sampson ("Concordant and Discordant Monotone Correlations and Their Evaluation by Nonlinear Optimization") received his A.B. from UCLA and his M.A. and Ph.D. in statistics from Stanford University. He has a number of published papers in statistics and mathematics journals. Currently he is Associate Professor of Mathematics and Statistics, University of Pittsburgh.

Nozer D. Singpurwalla ("A Bayesian Scheme for Estimating Reliability Growth under Exponential Failure Times") is Professor of Operations Research and Research Professor of Statistics at The George Washington University, Washington, D.C. He is a Fellow of the American Statistical Association and an Elected Member of the International Statistical Institute. He has published over 40 papers and a book on topics dealing with probabilistic and statistical aspects of reliability.

Prabhakant Sinha ("Algorithmic Improvements for Obtaining the Upper Multinomial Bound") is an Associate Professor of Quantitative Studies at the Graduate School of Management, Rutgers University. He has contributed to *Management Science, Operations Research*, and *Mathematical Programming*. His research interests are in resource allocation in menu planning and marketing.

Larry E. Stanfel ("An Algorithm Using Lagrangian Relaxation and Column Generation for One-Dimensional Clustering Problems") is Professor of Management and Marketing at Clarkson College, Potsdam, New York 13676. He has an B.S. from the Illinois Institute of Technology and an M.S. and Ph.D. from Northwestern University. His research interests are in mathematical optimization and information systems. He is the coauthor of two books and his articles have appeared in such publications as the *European Journal of Operations Research, Information and Control, Information Systems, Optical Engineering*, and the *Journal* and *Communications of the A.C.M.*

Asher Tishler ("An Absolute Deviations Curve-Fitting Algorithm for Nonlinear Models") is presently a visiting Assistant Professor in the Economics Department at the University of Southern California. He is on leave from the Economics Department and the Faculty of Management, Tel-Aviv University, Israel. He received his Ph.D. in 1976 from the University of Pennsylvania. His current research is in econometrics and energy modeling. Professor Tishler has published articles in journals such as the *Journal of the American Statistical Association, Journal of Econometrics, Journal of International Economics* and *European Economic Review*.

Chiang Wang ("Derivation of a Maximum Rank Sum Statistic and Application to Discriminant Analysis") is an Assistant Professor at the Systems and Industrial Engineering Department, the University of Arizona. He received his B.S. degree from Tunghai University, Taiwan, his masters and Ph.D. from the University of Iowa. His current research interests are in mathematical programming, applied statistics, and decision analysis. He has published in *Management Science*. He is a member of ORSA, TIMS, an AIIE.

Chien-Fu Wu ("Some Algorithms for Concave Regression and Isotonic Regression") is Associate Professor of Statistics at the University of Wisconsin, Madison, Wisconsin. He holds a Ph.D. from the University of California, Berkeley. He is an Associate Editor of the *Annals of Statistics*. His research interests include statistical algorithms, experimental design, survey sampling and large sample theory. He has published articles in *Annals of Statistics, Biometrika, Journal of American Statistical Association, Technometrics* and other journals.

Notes about authors

Stelios H. Zanakis (coeditor) is Professor and Chairman of Decision Sciences at the College of Business Administration, Florida International University. Prior to joining FIU, he was Associate Professor and Director of the Industrial Engineering and Systems Analysis Program at the West Virginia College of Graduate Studies. He holds a Ph.D. in business administration (management science), an M.A. in statistics and an M.B.A., all from the Pennsylvania State University, and an M.A. in mechanical/electrical engineering from the Greek National Technical University.

His current research interests include the optimization/statistics interface, goal programming, capital budgeting, inventory management, health care administration, and real-world applications of management science. He is coauthor of a forthcoming AIIE monograph on *Mathematical Programming Applications in Production Planning and Scheduling* and has published articles in *Management Science, TIMS Studies in the Management Sciences, Journal of Operational Research Society, Operational Research Quarterly, Journal of Statistical Computation & Simulation, Interfaces, AIIE Transactions, International Journal for Production Research*, and elsewhere. He is listed in several *Who's Who* and has consulted for various industrial, hospital and governmental organizations. Dr. Zanakis is a member of TIMS, AIDS, AIIE and APICS.

Israel Zang ("An Absolute Deviations Curve-Fitting Algorithm for Nonlinear Models") is with the Faculty of Management of Tel Aviv University, Israel, where he has been chairman of the operations research program since 1978. Before joining Tel Aviv University in 1975, he spent a year at the Center for Operations Research and Econometrics (CORE), Louvain-la-Neuve, Belgium. He received his D.Sc. degree from the Technion, Haifa, Israel, in 1974. His current research is in mathematical programming and its interaction with economics. Professor Zang has published articles in journals such as *International Journal of Production Research, Journal of the American Statistical Association, Journal of Econometrics, Journal of Economic Theory, Mathematics of Operations Research, Mathematical Programming* and *Networks*.

AUTHORS' ADDRESSES

Ronald D. Armstrong
University of Georgia, Graduate School of Business, Dept. of Quantitative Analysis, Athens, GA 30602

Terry Dielman
Texas Christian University, M.J. Neeley School of Business, Fort Worth, TX 76129

P.K. Eswaran
Purdue University, School of Industrial Engineering, West Lafayette, IN 47907

Robert S. Garfinkel
University of Tennessee, College of Business Administration, Management Science Program, Knoxville, TN 37916

Carl M. Harris
Center for Management & Policy Research, Inc., 1625 Eye Street, N.W., Washington, DC 20006

Robert E. Hausman Jr.
WB IH-216, Bell Laboratories, Holmdel, NJ 07733

George Kimeldorf
University of Texas at Dallas, Dept. of Mathematical Sciences, Richardson, TX 75080

Gary Klein
Purdue University, Krannert Graduate School of Management, West Lafayette, IN 47907

Darwin Klingman
University of Texas at Austin, Department of General Business, Austin, TX 78712

Mabel T. Kung
McMaster University, Faculty of Business, Hamilton, Ontario, Canada L8S 4M4

A.S. Kunnathur
School of Business, University of Wisconsin, Milwaukee, WI 53201

Naftali A. Langberg
University of Haifa, Statistics Department, Haifa 31 999, Israel

Ramón V. León
Bell Laboratories, Holmdel, NJ 07733

Gunar E. Liepins
Oak Ridge National Laboratory, Oak Ridge, TN 37830

James Lynch
The Pennsylvania State University, Dept. of Statistics, University Park, PA 16802

Jay Mandelbaum
United States Department of Transportation, Washington, DC

Nancy R. Mann
UCLA, Department of Biomathematics, Los Angeles, CA 90024

Jerrold H. May
University of Pittsburgh, Graduate School of Business, Pittsburgh, PA 15260

Herbert Moskowitz
Purdue University, Krannert Graduate School of Management, West Lafayette, IN 47907

John Mote
Analysis, Research & Computation, Inc., P.O. Box 4067, Austin, TX 78765

Clark A. Mount-Campbell
Ohio State University, Dept. of Industrial & Systems Engineering, Columbus, OH 43210

Subhash C. Narula
Rensselaer Polytechnic Institute, School of Management, Troy, NY 12181

John B. Neuhardt
Ohio State University, Department of Industrial & Systems Engineering, Columbus, OH 43210

Roger C. Pfaffenberger
Texas Christian University, M.J. Neeley School of Business, Fort Worth, TX 76129

Robert Plante
Purdue University, Krannert Graduate School of Mangement, West Lafayette, IN 47907

Frank Proschan
Florida State University, Dept. of Statistics & Statistics Consulting Center, Tallahassee, FL 32306

A. Ravindran
School of Industrial Engineering, University of Oklahoma, Norman, OK 73019

Jagdish S. Rustagi
The Ohio State University, Dept. of Statistics, Columbus, OH 43210

Allan R. Sampson
University of Pittsburgh, Dept. of Mathematics & Statistics, Pittsburgh, PA 15260

Nozer D. Singpurwalla
George Washington University, Dept. of Operations Research, School of Engineering & Applied Science, Washington, DC 20052

Prabhakant Sinha
Rutgers University, Graduate School of Management, Newark, NJ

Larry E. Stanfel
Clarkson College, School of Management, Potsdam, NY 13676

Asher Tishler
Department of Economics-MC 0035, University of Southern California, University Park, Los Angeles, CA 90089

Chiang Wang
University of Arizona, Systems & Industrial Engineering Dept., Tucson, AZ 85721

Chien-Fu Wu
University of Wisconsin, Department of Statistics, Madison, WI 53706

Stelios H. Zanakis
Professor and Chairman, Decision Sciences, Florida International University, College of Business Administration, Management Department, Miami, FL 33199

Israel Zang
Faculty of Management, Tel Aviv University, Tel Aviv 69978, Israel

RAYMOND H. FOGLER LIBRARY
DATE DUE

BOOKS ARE SUBJECT TO